NAVIER–STOKES EQUATIONS IN PLANAR DOMAINS

NAVIER–STOKES EQUATIONS IN PLANAR DOMAINS

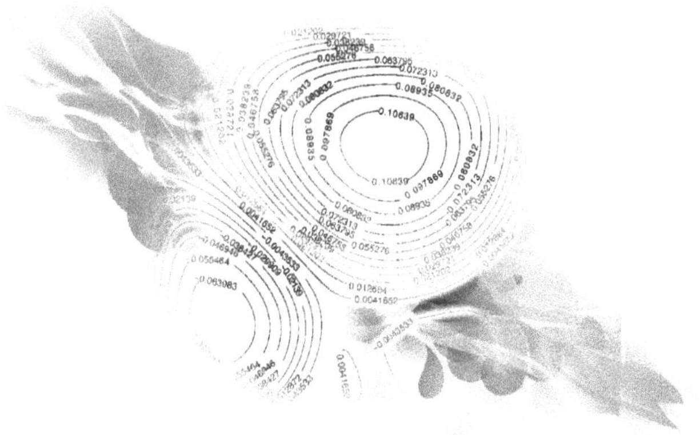

Matania Ben-Artzi
Hebrew University of Jerusalem, Israel

Jean-Pierre Croisille
Université de Lorraine, France

Dalia Fishelov
Afeka Tel-Aviv Academic College of Engineering, Israel

Imperial College Press

Published by

Imperial College Press
57 Shelton Street
Covent Garden
London WC2H 9HE

Distributed by

World Scientific Publishing Co. Pte. Ltd.
5 Toh Tuck Link, Singapore 596224
USA office: 27 Warren Street, Suite 401-402, Hackensack, NJ 07601
UK office: 57 Shelton Street, Covent Garden, London WC2H 9HE

British Library Cataloguing-in-Publication Data
A catalogue record for this book is available from the British Library.

NAVIER–STOKES EQUATIONS IN PLANAR DOMAINS

ISBN 978-1-84816-275-4

Printed in Singapore by Mainland Press Pte Ltd.

Preface

This monograph is devoted to the Navier–Stokes system in planar domains.

The Navier–Stokes system of equations, modeling incompressible, viscous flow in two or three space dimensions, is one of the most well-known classical systems of fluid dynamics and, in fact, of mathematical physics in general.

The pioneering work of Leray in the 1930s established the global (in time) well-posedness of the system in the two-dimensional case. It seemed therefore that this case was "closed." However, this was certainly not the case; the subsequent development of fluid dynamics in three directions, theoretical, numerical and experimental, led to a great interest in flow problems associated with "singular objects." These objects, including point vortices or vortex filaments, have been well-known since the early days of fluid dynamics in the eighteenth century. Their evolution in time serves as a central feature in the overall description of the flow.

A fact that is common to all these singular cases is that the L^2 norm of the associated velocity field (corresponding to the total energy of the fluid) is not finite. On the other hand, the basic premise of Leray's theory is the assumption that this norm is finite! Thus, there has been no rigorous mathematical theory applicable, say, to the motion of point vortices in two-dimensional flows, while the numerical treatment of such motion ("vortex methods") has increased in popularity since the 1970s.

The rigorous treatment of the two-dimensional Navier–Stokes system with such "rough initial data" was taken up only in the second half of the 1980s. The first part of this monograph gives a detailed exposition of these developments, based on a classical parabolic approach. It is based on the *vorticity formulation* of the system. A fundamental role is played by the integral operators associated with the heat kernel and the Biot–Savart kernel (relating the vorticity to the velocity).

The system is considered either in the whole plane $\mathbb{R}^2$ or in a square with periodic boundary conditions. Thus, physical boundary conditions (such as the "no-slip" condition) are not considered in this part. It is remarkable that the rigorous treatment of the motion of an initial vortex in a bounded planar domain remains an open problem!

On the numerical side, on the other hand, we cannot avoid the case of a bounded domain, subject to physically relevant boundary conditions. However, such boundary conditions are not readily translated into "vorticity boundary conditions." The progress made in the last twenty years in terms of "compact schemes" has produced very efficient algorithms for the approximation of the biharmonic operator. In turn, the use of the *streamfunction formulation* of the system has become a very attractive option; the system is reduced to a scalar equation and the boundary conditions are naturally implemented. The appearance of the biharmonic operator in this equation seems a reasonable price to pay. The second part of this monograph gives a detailed account of this approach, based primarily on the authors' work during the last decade.

We provide detailed introductions to the two parts, where background material is expounded. The first (theoretical) part is supplemented with an Appendix containing topics from functional analysis that are not readily found in basic books on partial differential equations. Thus, this part should be accessible not only to specialists in mathematical analysis, but also to graduate students and researchers in the physical sciences who are interested in the rigorous theory.

In the second (numerical) part, we have made an effort to make it wholly self-contained. The basic relevant facts concerning difference operators are expounded, making the passage to modern compact schemes accessible even to readers having no background in numerical analysis. The accuracy of the schemes as well as the convergence of discrete solutions to the continuous ones are discussed in detail.

This monograph grew out of talks, joint papers and short courses given by us at our respective institutes and elsewhere. Stimulating discussions with S. Abarbanel, C. Bardos, B. Bialecki, A. Chorin, A. Ditkowski, A. Ern, G. Fairweather, S. Friedlander, T. Gallay, J. Gibbon, R. Glowinski, J.-L. Guermond, G. Katriel, P. Minev, M. Schonbek, C.-W. Shu, R. Temam, S. Trachtenberg, E. Turkel, and D. Ye were very valuable to us.

One of us (D.F.) was the first graduate student of the late David Gottlieb. All three of us enjoyed his hospitality at Brown University during the summer of 2007. In spite of his failing health, he generously gave us his attention and his comments helped us shape Part II of the monograph.

The second author (J.-P.C.) thanks B. Courbet, D. Dutoya, J. Falcovitz, and F. Haider for illuminating discussions on the numerical treatment and the physical understanding of fluid flows.

Special thanks are due to I. Chorev. Section 11.2 is based on his M.Sc. thesis written under the supervision of M.B-A.

Our joint work demanded frequent trips between France and Israel. Support for these trips was provided through the French–Israeli scientific cooperation: we gratefully acknowledge the Arc-en-Ciel/Keshet program and the French-Israel High Council for Scientific and Technological Cooperation (French Ministry of Foreign Affairs, French Ministry of Research and Israeli Ministry of Science and Technology). We also acknowledge the support of the Hebrew University of Jerusalem, the University Paul Verlaine-Metz and the Afeka Tel Aviv Academic College of Engineering.

This work could not have been realized without the support of our families.

Jerusalem, Metz, Tel Aviv, August 2011

Contents

PART I
Basic Theory

Big whirls have little whirls that feed on their velocity, and little whirls have lesser whirls and so on to viscosity.

L. F. Richardson, *Weather Prediction by Numerical Process* (1922).

Chapter 1

Introduction

In this part of the book we establish the existence, uniqueness and regularity theory of solutions to the Navier–Stokes equations in two spatial dimensions. In addition, we also discuss the large-time asymptotic behavior of these solutions.

The main theme of the monograph (in both the theoretical and the numerical parts) is the **evolution of the vorticity** in the planar geometric setting. The recognition of the vorticity as a central object in the understanding of fluid flow dates back to the early days of this field. We refer the reader to the classic books [123, 124], where the physical significance, as well as numerous examples, are expounded.

More recently, a growing number of researchers, both in *Mathematical Fluid Dynamics* and in *Computational Fluid Dynamics*, have turned their attention to the vorticity in their studies. This increased interest is well reflected in the books [43, 133, 157].

The present monograph also highlights the fundamental role of *vorticity* (and its associated *streamfunction*) in the study of the Navier–Stokes system. However, the topics treated here are different from those addressed in the aforementioned books [43,133,157]. Indeed, Chorin's book is mostly devoted to the statistical physics aspects of vorticity (and turbulence), which we do not discuss here. The book by Majda and Bertozzi presents a unified treatment of the Navier–Stokes and the Euler equations, focusing primarily on the latter, whereas we deal exclusively with the Navier–Stokes equations in two dimensions. Our aim is to study the evolution of vorticity for rather general initial data, beyond the classical Leray theory. We shall return to this later in this chapter.

We refer to [44, 123, 124] for extensive discussion of the basic equations governing viscous (incompressible) fluid flow. We briefly recall these equations in the physical three-dimensional setting, and then restrict to the two-dimensional case, the subject matter of this monograph.

In order to distinguish between scalar and vector-valued functions we use boldface for vectors and vector functions in $\mathbb{R}^n$. Their components are labeled as $\mathbf{w} = (w^1, \ldots, w^n)$. In particular, for the Navier–Stokes equations this convention applies both to the planar case ($n = 2$) and the three-dimensional case. The scalar product is denoted by $\mathbf{a} \cdot \mathbf{b} = \sum_{i=1}^n a^i \cdot b^i$ and the Euclidean norm is $|\mathbf{w}|^2 = \sum_{i=1}^n (w^i)^2$.

Partial derivatives with respect to the time or spatial coordinates are denoted, respectively, by $\partial_t = \frac{\partial}{\partial t}$, $\partial_{x^i} = \frac{\partial}{\partial x^i}$. Using the gradient operator $\boldsymbol{\nabla} = \{\partial_{x^1}, ..., \partial_{x^n}\}$ we can represent the divergence and curl of a vector field $\mathbf{u}$ by, respectively, $\boldsymbol{\nabla} \cdot \mathbf{u}$ and $\boldsymbol{\nabla} \times \mathbf{u}$.

Occasionally (especially in integration) we will write $\boldsymbol{\nabla}_{\mathbf{x}}$ for clarity.

The Laplacian operator is $\Delta = \boldsymbol{\nabla} \cdot \boldsymbol{\nabla} = \sum_{i=1}^n \partial_{x^i}^2$.

If $\boldsymbol{\alpha} \in \mathbb{Z}_+^n$ is a multi-index, we let $\boldsymbol{\nabla}^{\boldsymbol{\alpha}} = \prod_{i=1}^n \partial_{x^i}^{\alpha^i}$ and $|\boldsymbol{\alpha}| = \sum_{i=1}^n \alpha^i$.

Denoting the velocity by $\mathbf{u}(\mathbf{x}, t)$, the pressure by $p(\mathbf{x}, t)$ and the (constant) coefficient of viscosity by ν ($\nu > 0$), the Navier–Stokes equations in a domain $\Omega \subseteq \mathbb{R}^n$, $n = 2, 3$, are

$$
\begin{aligned}
&\partial_t \mathbf{u} + (\mathbf{u} \cdot \boldsymbol{\nabla})\mathbf{u} = -\boldsymbol{\nabla} p + \nu \Delta \mathbf{u}, \\
&\nabla \cdot \mathbf{u} = 0.
\end{aligned}
\tag{1.1}
$$

The equations are supplemented by an initial condition

$$
\mathbf{u}(\mathbf{x}, 0) = \mathbf{u}_0(\mathbf{x}), \tag{1.2}
$$

and, if $\Omega \neq \mathbb{R}^n$, by boundary conditions (such as $\mathbf{u} = 0$, the "no-slip" condition) on the boundary $\partial\Omega$, for all $t \geq 0$. If $\Omega = \mathbb{R}^n$, a growth (or, rather, decay) condition must be imposed on $\mathbf{u}$ at infinity.

These equations should yield solutions $\mathbf{u}(x, t)$, $p(\mathbf{x}, t)$, for $\mathbf{x} \in \Omega$ and all positive time $t \in \mathbb{R}_+ = (0, \infty)$. The term "well-posedness" expresses the fact that, in a suitable functional framework, the solution should not only be unique but depend continuously on the initial and boundary data.

In the case that $\mathbf{u}_0 \in L^2(\Omega)$ (or $\mathbf{u}_0 \in H^1(\Omega)$) the existence of weak solutions to the problem (strong for $H^1(\Omega)$) has been known since the pioneering work of Leray [125], (see also [128] for the case of the full plane). Strong well-posedness is only local in time if $n = 3$, and is global in time if $n = 2$. We refer to [47, 54, 121, 171] where the Galerkin approach (originally used by Leray) is expounded.

Regarding whether the system (1.1)–(1.2) is well-posed beyond the L^2 framework, we refer to [75, 116] and references therein, as well as to earlier works by Kato and Ponce using commutator estimates in various Sobolev spaces [107, 110–112, 151].

We next recall the vorticity and streamfunction formulations of the equations. The reader is referred to the books [44, 133] for extensive discussions concerning the role of these functions in fluid dynamics. Taking the curl of the first equation in Equation (1.1), and denoting by $\boldsymbol{\omega} = \boldsymbol{\nabla} \times \mathbf{u}$ the vorticity, we get

$$\partial_t \boldsymbol{\omega} + (\mathbf{u} \cdot \boldsymbol{\nabla})\boldsymbol{\omega} - (\boldsymbol{\omega} \cdot \boldsymbol{\nabla})\mathbf{u} = \nu \Delta \boldsymbol{\omega}, \tag{1.3}$$

$$\boldsymbol{\nabla} \times \mathbf{u} = \boldsymbol{\omega}. \tag{1.4}$$

Since, in view of the second equation in (1.1), the vector function $\mathbf{u}$ is divergence-free, we have (assuming Ω is simply-connected) a vector function $\boldsymbol{\psi}$ such that

$$\mathbf{u} = -\boldsymbol{\nabla} \times \boldsymbol{\psi}. \tag{1.5}$$

The vector function $\boldsymbol{\psi}$ is the **streamfunction** (introduced by Lagrange in 1781 [123, Chapter IV]).

Combining Equations (1.4) and (1.5) and assuming that $\boldsymbol{\nabla} \cdot \boldsymbol{\psi} = 0$, we obtain

$$\boldsymbol{\omega} = \boldsymbol{\nabla} \times \mathbf{u} = \Delta \boldsymbol{\psi}. \tag{1.6}$$

Note that the fact (implied clearly by (1.4)) that $\boldsymbol{\nabla} \cdot \boldsymbol{\omega} = 0$ for all $t \geq 0$ is compatible with the "gauge" requirement $\boldsymbol{\nabla} \cdot \boldsymbol{\psi} = 0$.

In the case $\Omega = \mathbb{R}^3$ we can take, under mild growth assumptions,

$$\boldsymbol{\psi} = \Gamma * \boldsymbol{\omega}, \tag{1.7}$$

where Γ is the fundamental solution (Green's kernel) of Δ.

Notice that when $\mathbf{u}$ is given by (1.5), then automatically $\boldsymbol{\nabla} \cdot \mathbf{u} = 0$, so that Equations (1.3)–(1.4) are equivalent to (1.1), at least for sufficiently regular solutions. The system is supplemented by the initial condition

$$\boldsymbol{\omega}(\mathbf{x}, 0) = \boldsymbol{\omega}_0(\mathbf{x}), \quad \mathbf{x} \in \mathbb{R}^3. \tag{1.8}$$

From the point of view of hydrodynamical phenomena, an interesting case is that of the evolution of the vorticity (and its associated velocity field) when it is initially given by isolated vortices, vortex filaments or sheets. Since, in the "zero viscosity limit" (i.e., $\nu = 0$, leading to the Euler equations) the circulation is preserved (Kelvin's theorem), the use of vorticity in numerical methods has become very popular. In particular, in "vortex methods" [50, 133], even smooth initial data are replaced by a distribution of singular "vortical objects." Mathematically speaking, we need to study the system (1.3)–(1.4), and (1.8), when $\boldsymbol{\omega}_0(\mathbf{x})$ is a measure.

We now focus on the vorticity formulation in the two-dimensional (planar) case. The flow is invariant with respect to translations in one direction, say the x^3 axis. In other words, the velocity field is parallel to the (x^1, x^2) plane and can be expressed as $\mathbf{u}(\mathbf{x}, t) = (u^1(x^1, x^2, t), u^2(x^1, x^2, t), 0)$. The vorticity is given by $\boldsymbol{\omega}(\mathbf{x}, t) = \omega(\mathbf{x}, t)\mathbf{k}$, where $\mathbf{k}$ is the unit vector in the x^3 direction, and

$$\omega(\mathbf{x}, t) = \partial_{x^1} u^2 - \partial_{x^2} u^1. \tag{1.9}$$

We henceforth refer to ω as the **scalar vorticity**.

The term $(\boldsymbol{\omega} \cdot \boldsymbol{\nabla})\mathbf{u}$ vanishes identically, so that Equation (1.3) reduces to a (nonlinear) convection–diffusion equation for ω,

$$\partial_t \omega + (\mathbf{u} \cdot \boldsymbol{\nabla})\omega = \nu \Delta \omega, \quad \omega(\mathbf{x}, 0) = \omega_0(\mathbf{x}), \quad \mathbf{x} \in \Omega. \tag{1.10}$$

For simplicity, we have avoided adding a source term (external force) in (1.1) or (1.10). In fact, for issues considered here such as existence, uniqueness and regularity, the results can be extended to the non-homogeneous case in a rather standard way.

In the presence of finite boundaries (namely, $\Omega \neq \mathbb{R}^2$), boundary conditions must be provided for $\omega(\mathbf{x}, t)$, $\mathbf{x} \in \partial\Omega, t \geq 0$. This is obviously the case when we seek to simulate physical flow problems, where the domain is either bounded or "external" (such as flow outside an obstacle). The most common physically plausible boundary conditions are stated in terms of $\mathbf{u}$ (such as the "no-slip" condition $\mathbf{u} = 0$). Casting these conditions in terms of ω is quite involved, and in fact has hardly been treated in theoretical studies. In the numerical literature, on the other hand, the methods used for the implementation of such vorticity boundary conditions (or, in the hydrodynamical language, "generation of vorticity") are quite diverse. We refer the reader to the book [50] for more details.

The scalar vorticity leads to a *scalar streamfunction* related to it by (1.6). If we replace ω with $\Delta\psi$ in (1.10) it becomes a fourth-order equation; $\nu\Delta\omega$ is replaced by $\nu\Delta^2\psi$. Significant properties of the convection–diffusion equation, such as the maximum principle, are lost in this transition. In numerical simulations this loss is offset by the fact that the boundary conditions are very naturally expressed in terms of the streamfunction. In fact, in the second part of this monograph the "pure streamfunction" formulation will serve as the starting point in the search for discrete approximations displaying a high order of accuracy up to the boundary.

In order to avoid the boundary problem in the theoretical part of the monograph, we consider here only the full plane, $\Omega = \mathbb{R}^2$. Our treatment applies also to the periodic case, in which periodic boundary conditions are imposed on the flow in a square.

The assumption that our domain is the full plane enables us to replace expressions (1.5)–(1.7) with the more explicit formula

$$\mathbf{u}(\mathbf{x}, t) = (\mathbf{K} * \omega)(\mathbf{x}, t) = \int_{\mathbb{R}^2} \mathbf{K}(\mathbf{x} - \mathbf{y})\omega(\mathbf{y}, t)d\mathbf{y}, \tag{1.11}$$

where the "Biot–Savart" kernel $\mathbf{K}$ is given by [133, Chapter 2]

$$\mathbf{K}(\mathbf{x}) = \frac{1}{2\pi}|\mathbf{x}|^{-2}(-x^2, x^1). \tag{1.12}$$

We note that all vectors are now two dimensional.

The fact that $\nabla \cdot \mathbf{K} = 0$ (which can be verified directly), yields (by (1.11)) the incompressibility condition $\nabla \cdot \mathbf{u} = 0$. In Remark 2.12 we state this in a more rigorous fashion.

The main focus of this theoretical part of the monograph is the study of global solutions to Equation (1.10), where $\mathbf{u}$ is related to ω by (1.11). By "global solutions" we mean solutions that can be shown to exist for all $t \in [0, \infty)$, and that evolve continuously in time, in a suitable function space. The choice of this space reflects the generality of our admissible initial data ω_0. The classical Leray theory, already mentioned above, requires "finite energy" initial data, namely, $\mathbf{u}_0 \in L^2(\mathbb{R}^2)$. On the other hand, our interest here lies in the evolution of vorticity that is initially highly concentrated in "small sets," such as "point vortices" or "vortex filaments." Mathematically speaking, we need to study solutions of (1.10) subject to initial data that are measures. Such initial data are, in general, of *infinite energy* and thus outside the scope of Leray's theory. In fact, we make the following claim [133, Proposition 3.3].

CLAIM. *Suppose that $\omega_0(\mathbf{x})$ is compactly supported and bounded. Then the associated velocity field $\mathbf{u}_0(\mathbf{x})$ has finite energy, $\int\limits_{\mathbb{R}^2} |\mathbf{u}_0(\mathbf{x})|^2 d\mathbf{x} < \infty$, if and only if $\int\limits_{\mathbb{R}^2} \omega_0(\mathbf{x})d\mathbf{x} = 0$.*

In order to deal with the infinite energy case, Majda and Bertozzi [133, Chapter 3] observe that for a wide class of problems of interest the velocity field is *locally* square-integrable. In such cases they make a further hypothesis that the (smooth) velocity field has a *radial-energy decomposition*. Roughly speaking, this means the initial vorticity can be expressed as a sum $\omega_0(\mathbf{x}) = \tilde{\omega}(|\mathbf{x}|) + \omega_1(\mathbf{x})$, where $\tilde{\omega}(|\mathbf{x}|)$ is smooth and radial, while the velocity field $\mathbf{u}_1(\mathbf{x}, t)$ associated with $\omega_1(\mathbf{x})$ has finite energy. The radial part of the velocity can be explicitly expressed in terms of $\tilde{\omega}$, giving rise to an equation for $\mathbf{u}_1(\mathbf{x}, t)$. The existence and uniqueness of the solutions to this equation are proved by smooth approximations (mollification). We note that this treatment applies also to the Euler equation (zero viscosity), as well as the (short-time) three-dimensional case.

The velocity field due to a point vortex, however, does not belong in this category. Indeed, assuming that the vortex is located at the origin, the corresponding velocity behaves like $|\mathbf{x}|^{-1}$, according to (1.11), hence is not locally square-integrable.

In this monograph we present a theory pertaining to the general case of a measure-valued vorticity. In doing so, we emphasize the *parabolic character* of the vorticity equation (1.10), that is, the fact that it is a *convection–diffusion* equation. Two of the most powerful tools for dealing with such equations are the *maximum principle* and *duality*, meaning that the dual equation is in the same class. These tools play a decisive role in our treatment.

We study the well-posedness of (1.10) in various functional spaces X. This means (at the least) that given the initial vorticity $\omega_0 \in X$ the solution evolves along a continuous trajectory in X. We move gradually from a very restrictive space of smooth functions (Chapter 2) to the general space of measures. The treatment in Chapter 2 is entirely based on the parabolic nature of the (scalar) vorticity equation. For the sake of completeness we have included a fairly detailed treatment of the linear convection–diffusion equation with "rough" data (both convection coefficient and initial data). In order to carry out the passage to more general initial data, one needs precise space-time estimates for smooth solutions, expressed in terms of L^p norms. Such estimates are established in Chapter 3. They are the main tools used in the extension of the solution operator to initial vorticities in $L^1(\mathbb{R}^2)$, as in Chapter 4.

Chapter 5 deals with the further extension to measure-valued initial data. It is here that we establish the existence (for all time) of solutions to the vorticity equation, when the initial data are objects such as a collection of point vortices or vortex filaments (curves).

It should be pointed out that Chapters 2–5 constitute a self-contained treatment of the existence and uniqueness theory of Navier–Stokes equations in the plane, ranging from very smooth initial data to very rough ones, including, in particular, point vortices and vortex lines.

In Chapter 6 we address the asymptotic behavior of the vorticity for large time. This behavior has two aspects; the decay in norm and the detailed convergence to some (scaled) shape. We outline the theorem of Gallay and Wayne concerning the global decay to a multiple of the Oseen vortex. Thus the vorticity equation is globally stable. This is of course far from the case of a bounded (non-periodic) domain, where rigorous results are rather scarce. These observations connect this theoretical part to the final chapter of Part II, where the stability of the steady state for a driven cavity is numerically studied.

For the material in this part, we assume that the reader is familiar with basic functional analysis, partial differential equations and (Lebesgue) measure theory. Practically all the background material can be found in the books [58,102]. Additional relevant facts and theorems are collected in Appendix A, with appropriate references.

1.1 Functional notation

Here we introduce the basic function spaces and norms that will be extensively used in Part I of this monograph.

- $C_0^\infty(\mathbb{R}^n)$ are functions that are smooth (namely, continuously differentiable of any order) and compactly supported.

- $C^k(\mathbb{R}^n)$ are functions f such that $\nabla^{\boldsymbol{\alpha}} f$ is continuous for $|\boldsymbol{\alpha}| \le k$. A sequence $\{f_j\}_{j=1}^\infty \subseteq C^k(\mathbb{R}^n)$ is convergent if $\{\nabla^{\boldsymbol{\alpha}} f_j\}_{j=1}^\infty$ converges uniformly in every compact subset of $\mathbb{R}^n$, for $|\boldsymbol{\alpha}| \le k$.

- $C_b^k(\mathbb{R}^n) \subseteq C^k(\mathbb{R}^n)$ are functions f such that $\nabla^{\boldsymbol{\alpha}} f$ is continuous and bounded for $|\boldsymbol{\alpha}| \le k$.
  Its norm is defined by
  $$\|\psi\|_{C_b^k} = \sum_{j=0}^{k} \sum_{|\boldsymbol{\alpha}|=j} \sup_{\mathbf{x}\in\mathbb{R}^n} |\nabla^{\boldsymbol{\alpha}} \psi(\mathbf{x})|.$$

- $C^\lambda(\mathbb{R}^n)$ are uniformly Hölder continuous functions satisfying
  $$|f(\mathbf{x}) - f(\mathbf{y})| \le M_f |\mathbf{x} - \mathbf{y}|^\lambda, \quad \mathbf{x}, \mathbf{y} \in \mathbb{R}^n, \quad \lambda \in (0,1)$$

- $C^{k,\lambda}(\mathbb{R}^n)$ are functions $f \in C^k(\mathbb{R}^n)$, such that

$$\nabla^{\boldsymbol{\alpha}} f \in C^\lambda(\mathbb{R}^n), \quad |\boldsymbol{\alpha}| \le k.$$

- $D^0(\mathbb{R}^n) \subseteq C(\mathbb{R}^n)$ is the subspace of continuous functions f such that $\sup\limits_{|\mathbf{x}|>R} |f(\mathbf{x})| \xrightarrow{R\to\infty} 0.$

  It is normed by

$$\|f\|_\infty = \sup_{\mathbf{x}\in\mathbb{R}^n} |f(\mathbf{x})|.$$

- The norm in $L^p(\mathbb{R}^n)$, $1 \le p < \infty$, is denoted by

$$\|\psi\|_p = \left[\int_{\mathbb{R}^n} |\psi(\mathbf{x})|^p d\mathbf{x}\right]^{1/p}$$

  with the usual (ess-sup) modification for $p = +\infty$.

  If ψ is defined on a domain $\Omega \ne \mathbb{R}^n$ we denote for clarity

$$\|\psi\|_{L^p(\Omega)} = \left[\int_{\Omega} |\psi(\mathbf{x})|^p d\mathbf{x}\right]^{1/p}$$

  with the usual (ess-sup) modification for $p = +\infty$.

- The space $W^{s,p}(\mathbb{R}^n)$ (where s is a positive integer) is the L^p Sobolev space, normed by

$$\|\psi\|_{W^{s,p}} = \sum_{k=0}^{s} \sum_{|\boldsymbol{\alpha}|=k} \|\nabla^{\boldsymbol{\alpha}}\psi\|_p.$$

- If X is a Banach space, normed by $\|\cdot\|_X$, and $I \subseteq \overline{\mathbb{R}}_+$ is a finite or infinite interval, we define the following spaces of X-valued functions $f : I \to X$.

$C(I;X)$	Continuous functions (not necessarily bounded), topologized by uniform convergence over compact subintervals of I.
$L^p(I;X)$	Strongly measurable functions, normed by $(\int_I \|f(t)\|_X^p dt)^{1/p}$, $1 \le p < \infty$, with the usual modification for $p = \infty$.
$L^p_{loc}(I;X)$	Strongly measurable functions such that $\phi f \in L^p(I, X)$ for all $\phi \in C_0^\infty(I)$.

- If X_1, X_2 are Banach spaces, then $X = X_1 \cap X_2$ is normed by $\|\cdot\|_X = \|\cdot\|_{X_1} + \|\cdot\|_{X_2}$.

- $Y_p = L^1(\mathbb{R}^2) \cap L^p(\mathbb{R}^2), \quad 1 < p < \infty,$
  equipped with the sum of norms.

- $Y_\infty = L^1(\mathbb{R}^2) \cap D^0(\mathbb{R}^2).$

Chapter 2

Existence and Uniqueness of Smooth Solutions

In this chapter we study the vorticity equation in the full plane, subject to smooth and compactly supported initial data. Thus, the system (1.10)–(1.11) is now written as

$$\partial_t\omega + (\mathbf{u}\cdot\nabla)\omega = \nu\Delta\omega, \quad \omega(\mathbf{x},0) = \omega_0(\mathbf{x}), \quad \mathbf{x}\in\mathbb{R}^2, \tag{2.1}$$

$$\mathbf{u}(\mathbf{x},t) = (\mathbf{K}*\omega)(\mathbf{x},t) = \int_{\mathbb{R}^2}\mathbf{K}(\mathbf{x}-\mathbf{y})\omega(\mathbf{y},t)d\mathbf{y}, \tag{2.2}$$

where the Biot–Savart kernel $\mathbf{K}$ is given in (1.12).

We denote

$$\mathbb{R}_+ = (0,\infty), \quad \overline{\mathbb{R}_+} = [0,\infty).$$

We claim that solutions exist for nonnegative time $t\in\overline{\mathbb{R}_+}$ and prove the following theorem.

Theorem 2.1. *Assume $\omega_0(\mathbf{x})\in C_0^\infty(\mathbb{R}^2)$. Then the system (2.1)–(2.2) has a unique bounded solution $\omega(\boldsymbol{x},t)\in C^\infty(\mathbb{R}^2\times\overline{\mathbb{R}_+})$.*

This solution satisfies, for all $t\in\overline{\mathbb{R}_+}$,

$$||\omega(\cdot,t)||_p \le ||\omega_0||_p, \qquad 1\le p\le\infty. \tag{2.3}$$

Furthermore, the solution is smooth and for every integer k and double-index $\boldsymbol{\alpha}$ the functions $\partial_t^k\nabla^{\boldsymbol{\alpha}}\omega$ and $\partial_t^k\nabla^{\boldsymbol{\alpha}}\mathbf{u}$ are decaying in the sense that (with g denoting any one of them), for any fixed $T>0$,

$$\sup_{0\le t\le T,|\mathbf{x}|>R}|g(\mathbf{x},t)|\to 0 \quad \textit{as} \quad R\to\infty. \tag{2.4}$$

We prove the theorem in Section 2.2 below by an iterative procedure, constructing a sequence of functions that converges to the solution. Let $\omega^{(-1)}\equiv\mathbf{u}^{(-1)}\equiv 0$ and then, for $k=0,1,2,\ldots$, solve

$$\begin{aligned}\partial_t\omega^{(k)}(\mathbf{x},t) - \nu\Delta_{\mathbf{x}}\omega^{(k)}(\mathbf{x},t) &= -(\mathbf{u}^{(k-1)}(\mathbf{x},t)\cdot\nabla_{\mathbf{x}})\omega^{(k)}(\mathbf{x},t),\\ \mathbf{u}^{(k)}(\mathbf{x},t) &= \mathbf{K}*\omega^{(k)}(\mathbf{x},t),\end{aligned} \tag{2.5}$$

subject to the initial condition

$$\omega^{(k)}(\mathbf{x}, 0) = \omega_0(\mathbf{x}), \quad \mathbf{x} \in \mathbb{R}^2, \quad k = 0, 1, 2, \ldots \tag{2.6}$$

We seek solutions to (2.5)–(2.6) that decay, along with all their derivatives, in every strip $\Omega_T = \mathbb{R}^n \times [0, T]$, in the sense of (2.4).

Equation (2.5) is a linear equation (for $\omega^{(k)}$) combining a "diffusion process" (the heat operator $\partial_t \omega^{(k)} - \nu \Delta \omega^{(k)}$) and a "convection process" $(-(\mathbf{u}^{(k-1)} \cdot \boldsymbol{\nabla})\omega^{(k)})$. We shall therefore need some basic facts concerning such "convection–diffusion" equations. These facts are discussed in the following section.

2.1 The linear convection–diffusion equation

We will establish an existence and uniqueness result that is more general than is actually required for the specific treatment of smooth solutions of the Navier–Stokes vorticity equation. In particular, the equation is posed in $\mathbb{R}^n$, not necessarily $\mathbb{R}^2$. We do so because the method of proof will be useful later on, when dealing with irregular initial data. In addition, the more general proof serves to highlight the significance of the heat kernel and the fundamental role it plays in the treatment of this problem.

Thus we consider the scalar linear equation

$$\phi_t(\mathbf{x}, t) - \Delta_{\mathbf{x}} \phi(\mathbf{x}, t) = \mathbf{a}(\mathbf{x}, t) \cdot \boldsymbol{\nabla}_{\mathbf{x}} \phi(\mathbf{x}, t), \qquad (\mathbf{x}, t) \in \mathbb{R}^n \times \overline{\mathbb{R}_+}, \tag{2.7}$$

subject to the initial condition

$$\phi(\mathbf{x}, 0) = \phi_0(\mathbf{x}), \qquad \mathbf{x} \in \mathbb{R}^n. \tag{2.8}$$

For notational simplicity we set the heat coefficient to be equal to unity in Equation (2.7).

Note that $\mathbf{a}(\mathbf{x}, t) = (a^1(\mathbf{x}, t), \ldots, a^n(\mathbf{x}, t)) \in \mathbb{R}^n$.

Theorem 2.2. *Assume that*

$$\mathbf{a}(\mathbf{x}, t) \in L^\infty(\mathbb{R}^n \times \mathbb{R}_+), \tag{2.9}$$

$$\phi_0(\mathbf{x}) \in L^1(\mathbb{R}^n). \tag{2.10}$$

Then Equation (2.7), subject to the initial condition (2.8), has a unique global solution

$$\phi(\cdot, t) \in C(\overline{\mathbb{R}_+}; L^1(\mathbb{R}^n)) \cap C(\mathbb{R}_+; W^{1,1}(\mathbb{R}^n)). \tag{2.11}$$

We refer to [58, Chapter 5] for the definition and basic properties of Sobolev spaces.

Remark 2.3. If all the derivatives appearing in Equation (2.7) are continuous, we say that the solution is *classical.* However, the requirement that the convection coefficient $\mathbf{a}$ be bounded (and not necessarily continuous) means that in general there are non-classical solutions to the equation. The exact meaning of the solution is discussed as a first step of the proof.

Proof. Let

$$G(\mathbf{x}, t) = (4\pi t)^{-\frac{n}{2}} \exp\Big(- \frac{|\mathbf{x}|^2}{4t} \Big) \tag{2.12}$$

be the heat kernel, so that the solution to the Cauchy problem for the heat equation [$\mathbf{a} \equiv 0$ in (2.7)] is given by [58, Section 2.3]

$$H(t)\phi_0 = \int\limits_{\mathbb{R}^n} G(\mathbf{x} - \mathbf{y}, t)\phi_0(\mathbf{y})d\mathbf{y}. \tag{2.13}$$

Now using the Duhamel principle [58, Section 2.3], [102, Chapter 5], we can *formally* express the solution to (2.7) as the solution to the integral equation

$$\begin{aligned} \phi(\mathbf{x}, t) &= H(t)\phi_0 + \int\limits_0^t \int\limits_{\mathbb{R}^n} G(\mathbf{x} - \mathbf{y}, t - s)\mathbf{a}(\mathbf{y}, s) \cdot \nabla_{\mathbf{y}}\phi(\mathbf{y}, s)d\mathbf{y}ds, \\ \nabla_{\mathbf{x}}\phi(\mathbf{x}, t) &= \nabla_{\mathbf{x}}(H(t)\phi_0) \\ &\quad + \int\limits_0^t \int\limits_{\mathbb{R}^n} \nabla_{\mathbf{x}}G(\mathbf{x} - \mathbf{y}, t - s)(\mathbf{a}(\mathbf{y}, s) \cdot \nabla_{\mathbf{y}}\phi(\mathbf{y}, s))d\mathbf{y}ds. \end{aligned} \tag{2.14}$$

We now **define** a solution to (2.7) as a function $\phi(\cdot, t)$ such that

$$\phi(\cdot, t) \in C([0, \infty); L^1(\mathbb{R}^n)) \cap C((0, \infty); W^{1,1}(\mathbb{R}^n))$$

satisfying (2.14) in the sense of $L^1(\mathbb{R}^n \times \mathbb{R}_+)$, as follows.

The integrals on the right-hand sides converge for all $t > 0$, and equality holds as elements of $L^1(\mathbb{R}^n)$, for all $t > 0$.

The heat kernel is a *positive summability kernel*, so that for all $t > 0$,

$$\int\limits_{\mathbb{R}^n} G(\mathbf{x}, t)d\mathbf{x} = 1. \tag{2.15}$$

Furthermore, using a scaling argument, we obtain the following

$$\int_{\mathbb{R}^n} |\nabla_{\mathbf{x}} G(\mathbf{x},t)| d\mathbf{x} = \beta t^{-\frac{1}{2}}, \quad t > 0, \tag{2.16}$$

where $\beta = \int_{\mathbb{R}^n} |\nabla_{\mathbf{x}} G(\mathbf{x},1)| d\mathbf{x}$.

Note that this equality ensures that the integrals in (2.14) are well defined.

We start the proof by obtaining a solution in a strip $\Omega_T = \mathbb{R}^n \times [0,T]$, where $T > 0$ is to be selected later.

Let $X(T) \subseteq C([0,T]; L^1(\mathbb{R}^n)) \cap C((0,T]; W^{1,1}(\mathbb{R}^n))$ be a space of functions $r(\mathbf{x},t)$ normed by

$$\|r\|_{X(T)} = \max(\sup_{0<t\leq T} \|r(\cdot,t)\|_1, \sup_{0<t\leq T} t^{\frac{1}{2}} \|\nabla r(\cdot,t)\|_1) < \infty. \tag{2.17}$$

The equalities (2.15) and (2.16) imply that the solution $H(t)\phi_0$ to the heat equation satisfies, by the Young inequality (see Appendix A, Section A.2),

$$\|H(t)\phi_0\|_{X(T)} \leq c_0 \|\phi_0\|_1, \tag{2.18}$$

where $c_0 = \max(1,\beta) > 0$ is a constant independent of T.

Given $\psi \in X(T)$ we define, with $\phi_0 \in L^1(\mathbb{R}^n)$,

$$\Lambda_T(\psi)(\mathbf{x},t) = H(t)\phi_0 + \int_0^t \int_{\mathbb{R}^n} G(\mathbf{x}-\mathbf{y}, t-s)\mathbf{a}(\mathbf{y},s) \cdot \nabla_{\mathbf{y}} \psi(\mathbf{y},s) d\mathbf{y} ds. \tag{2.19}$$

It is easy to see that

$$\phi = \Lambda_T(\psi) \subseteq C([0,T]; L^1(\mathbb{R}^n)) \cap C((0,T]; W^{1,1}(\mathbb{R}^n)).$$

Indeed, an application of (2.15) and (2.16) yields,

$$\begin{aligned} &\|\phi\|_{X(T)} \\ &\leq c_0\|\phi_0\|_1 + c_0 A \|\psi\|_{X(T)} \Big\{ \int_0^T s^{-\frac{1}{2}} ds + \sup_{0<t<T} t^{\frac{1}{2}} \int_0^t s^{-\frac{1}{2}} (t-s)^{-\frac{1}{2}} ds \Big\} \\ &= c_0(\|\phi_0\|_1 + (2+\pi) A \|\psi\|_{X(T)} T^{\frac{1}{2}}), \end{aligned} \tag{2.20}$$

where $A = \|\mathbf{a}\|_{L^\infty(\Omega_T)}$.

In particular, $\phi \in X(T)$, and in the sense defined above, it is a solution (in Ω_T) to the equation

$$\phi_t(\mathbf{x},t) - \Delta_{\mathbf{x}} \phi(\mathbf{x},t) = \mathbf{a}(\mathbf{x},t) \cdot \nabla_{\mathbf{x}} \psi(\mathbf{x},t), \qquad (\mathbf{x},t) \in \Omega_T, \tag{2.21}$$

subject to the initial condition (2.8).

Let $\gamma = 2c_0\|\phi_0\|_1$ and let $X_\gamma(T)$ be the ball of radius γ in $X(T)$, centered at 0. It follows from (2.20) that if $T_1 < \left(\frac{1}{2Ac_0(2+\pi)}\right)^2$ then Λ_{T_1} maps $X_\gamma(T_1)$ into itself. Moreover, if also $\widetilde{\psi} \in X(T_1)$ then

$$\|\Lambda_{T_1}(\widetilde{\psi}) - \Lambda_{T_1}(\psi)\|_{X(T_1)} \leq (2+\pi)c_0 A T_1^{\frac{1}{2}} \|\widetilde{\psi} - \psi\|_{X(T_1)} \leq \frac{1}{2}\|\widetilde{\psi} - \psi\|_{X(T_1)}.$$

It follows that Λ_{T_1} is a contraction. Applying Banach's fixed point theorem [58, Section 9.2], we conclude that Λ_{T_1} has a unique fixed point $\phi \in X_\gamma(T_1)$, which is a solution to (2.7)–(2.8) in the strip Ω_{T_1}. Furthermore, it follows from $\|\phi\|_{X(T_1)} < \gamma$ that

$$\limsup_{t \uparrow T_1} \|\phi(\cdot, t)\|_1 < \infty.$$

Since T_1 was independent of $\|\phi_0\|_1$ it follows that the solution can be continued to the time interval $[T_1, 2T_1]$ and so on, thus proving that the solution is global in time.

To prove the uniqueness of this global solution ϕ, we suppose that

$$\widetilde{\phi} \in C([0,\infty); L^1(\mathbb{R}^n)) \cap C((0,\infty); W^{1,1}(\mathbb{R}^n))$$

is also a solution of (2.7)–(2.8) in the sense of (2.14). If $\widetilde{\phi} \in X(T)$ for some small $T > 0$, then it coincides with ϕ due to the uniqueness of the fixed point. We could then proceed by time intervals, obtaining $\widetilde{\phi} = \phi$ for all $t > 0$. However, we do not know *a priori* that $\widetilde{\phi} \in X(T)$, and we need to establish this.

Take $T_1 > 0$ as in the existence proof above and let $\tau \in (0, \frac{T_1}{2})$. Set $\widetilde{\phi}_\tau(\mathbf{x}, t) = \widetilde{\phi}(\mathbf{x}, t+\tau)$. By the assumptions on $\widetilde{\phi}$ it is clear that $\widetilde{\phi}_\tau \in X(T)$ for $0 < T < \frac{T_1}{2}$. Let $\phi_\tau \in X(T)$ be the solution constructed above (as a fixed point), subject to initial data $\phi_\tau(\mathbf{x}, 0) = \widetilde{\phi}(\mathbf{x}, \tau)$. By the uniqueness of the solution in $X(T)$ we conclude that $\widetilde{\phi}_\tau(\mathbf{x}, t) = \phi_\tau(\mathbf{x}, t)$, $t \in [0, T]$. We now let $\tau \to 0$. By the assumed continuity of $\widetilde{\phi}$ we get, for $t > 0$,

$$\widetilde{\phi}_\tau(\cdot, t) \xrightarrow[\tau \to 0]{} \widetilde{\phi}(\cdot, t) \quad \text{in } L^1(\mathbb{R}^n).$$

On the other hand, we note that the fixed point solution is continuous with respect to the initial data (indeed, the map Λ_T above depends continuously on ϕ_0). Thus

$$\phi_\tau(\cdot, t) \xrightarrow[\tau \to 0]{} \phi(\cdot, t) \quad \text{in } L^1(\mathbb{R}^n).$$

We conclude that $\widetilde{\phi} = \phi \in X(T)$, and the uniqueness is established. ■

The method of proof actually yields a better result, as expressed in the following corollary.

Corollary 2.4. *Under the conditions of Theorem 2.2 the solution also satisfies*

$$\phi(\cdot, t) \in C(\mathbb{R}_+; W^{1,\infty}(\mathbb{R}^n)). \tag{2.22}$$

Proof. Let $\varepsilon > 0$. Using the result already proved, we know that $\phi(\cdot, \varepsilon) \in W^{1,1}(\mathbb{R}^n)$. The Sobolev embedding theorem [58, Section 5.6] implies that $\phi(\cdot, \varepsilon) \in L^q(\mathbb{R}^n)$, where $\frac{1}{q} = 1 - \frac{1}{n}$ or $q = \infty$ if $n = 1$.

Let us now reconsider Equation (2.7) on the time interval $[\varepsilon, \infty)$. Shifting $t \to t - \varepsilon$, we are back to the initial value problem (2.7)–(2.8), now with $\phi_0 \in L^q(\mathbb{R}^n)$.

We can repeat verbatim the proof of the theorem, with the space $X(T)$ now defined as

$$X(T) \subseteq C([0, T]; L^q(\mathbb{R}^n)) \cap C((0, T]; W^{1,q}(\mathbb{R}^n)),$$

normed by

$$\|r\|_{X(T)} = \max\Big(\sup_{0<t\leq T} \|r(\cdot, t)\|_q, \sup_{0<t\leq T} t^{\frac{1}{2}} \|\nabla r(\cdot, t)\|_q \Big) < \infty. \tag{2.23}$$

We conclude that the solution $\phi(\cdot, t)$ satisfies

$$\phi(\cdot, t) \in C([\varepsilon, \infty); L^q(\mathbb{R}^n)) \cap C((\varepsilon, \infty); W^{1,q}(\mathbb{R}^n)). \tag{2.24}$$

This procedure can obviously be repeated, raising the exponent q, until the value $q > n$ is attained. Then, by Morrey's inequality [58, Section 5.6.2], $\phi(\cdot, \varepsilon) \in C^{0,\lambda}(\mathbb{R}^n)$, namely, a bounded and uniformly Hölder continuous function, with exponent $\lambda \in (0, 1)$, for every $\varepsilon > 0$. Once again, the proof of the theorem can be repeated, defining

$$X(T) \subseteq C([0, T]; C^{0,\lambda}(\mathbb{R}^n)) \cap C((0, T]; W^{1,\infty}(\mathbb{R}^n)),$$

normed as in (2.23) with $q = \infty$, thus establishing (2.22). ■

Remark 2.5. Theorem 2.2 was formulated so as to give a global (in time) solution. However, it is clear from the proof that if assumption (2.9) holds in the strip $\Omega_T = \mathbb{R}^n \times [0, T]$ then the (unique) solution exists in $X(T)$.

We now study the solution of the convection–diffusion equation when the convection coefficient is in $C_b^k(\mathbb{R}^n \times [0,T])$, for every k and every T, namely, continuously differentiable of any order, with all derivatives bounded in the strip. We show that the solution obtained above is smooth "off the initial line" $t = 0$, with all derivatives decaying at infinity.

If in addition the initial data is smooth and compactly supported, we show that the solution, now smooth "up to" $t = 0$, is exponentially decaying. This situation is the one we shall encounter when dealing with the recursion equation (2.5). To obtain the precise analog with this equation we reinstall the diffusion coefficient $\nu > 0$, and consider the heat equation (in $\mathbb{R}^n$) $z_t - \nu \Delta z = 0$, $\nu > 0$. Its kernel is obtained from the kernel (2.12), replacing t by νt. Thus

$$G_\nu(\mathbf{x}, t) = (4\pi\nu t)^{-\frac{n}{2}} e^{-\frac{|\mathbf{x}|^2}{4\nu t}}. \tag{2.25}$$

Notice that, by scaling,

$$\int_{\mathbb{R}^n} |\nabla_{\mathbf{x}} G_\nu(\mathbf{x}, t)| d\mathbf{x} = \beta_\nu t^{-\frac{1}{2}}, \quad t > 0, \tag{2.26}$$

where $\beta_\nu = \beta \nu^{-\frac{1}{2}}$ and β is as in (2.16).

Lemma 2.6. *Let $\mathbf{a}(\mathbf{x}, t)$ be a C^∞ (vector) function in $\mathbb{R}^n \times \overline{\mathbb{R}_+}$. Assume that in every strip $\Omega_T = \mathbb{R}^n \times [0,T]$ and for every integer k and double-index $\boldsymbol{\alpha}$,*

$$\sup_{(\mathbf{x},t) \in \Omega_T} |\nabla_{\mathbf{x}}^{\boldsymbol{\alpha}} \partial_t^k \mathbf{a}(\mathbf{x}, t)| < \infty. \tag{2.27}$$

Let $\phi(\cdot, t)$ be the unique solution (obtained in Theorem 2.2) to the equation

$$\partial_t(\mathbf{x}, t) - \nu \Delta_{\mathbf{x}} \phi(\mathbf{x}, t) = \mathbf{a}(\mathbf{x}, t) \cdot \nabla_{\mathbf{x}} \phi(\mathbf{x}, t) \quad in \quad \mathbb{R}^2 \times \mathbb{R}_+,$$
$$\phi(\boldsymbol{x}, 0) = \phi_0(\mathbf{x}) \in L^1(\mathbb{R}^n).$$

Then $\phi(\mathbf{x}, t) \in C^\infty(\mathbb{R}^n \times \mathbb{R}_+)$. Furthermore, it is decaying, along with all its derivatives, in every strip $\mathbb{R}^n \times [\varepsilon, T]$, $\varepsilon > 0$, in the sense of (2.4).

Assume in addition that $\phi_0 \in C_0^\infty(\mathbb{R}^n)$ and it is supported in the ball $|\boldsymbol{x}| \le R$. Let $M = \|\phi_0\|_\infty$ and $A_T = \sup_{\Omega_T} |\mathbf{a}(\mathbf{x}, t)|$. Then

$$|\phi(\mathbf{x}, t)| \le \min(M, Me^{((A_T+1)t+R-|\boldsymbol{x}|)/\nu}), \quad (\mathbf{x}, t) \in \Omega_T. \tag{2.28}$$

Proof. The smoothness property of ϕ follows from well-known results in the theory of parabolic equations [69].

Nevertheless, we indicate briefly how this property can be proved using the methodology of the proof of Theorem 2.2. As was the case in the proof of Corollary 2.4, the basic idea is to reconsider Equation (2.14), beginning at the initial time $t = \varepsilon > 0$. We already know that $\phi(\cdot, \varepsilon) \in W^{1,1}(\mathbb{R}^n) \cap W^{1,\infty}(\mathbb{R}^n)$, and by the smoothness hypothesis on $\mathbf{a}(\mathbf{x}, t)$ this is also true for the product $\phi\mathbf{a}$.

Take $1 \le j \le n$ and differentiate in (2.14) (shifted by ε) with respect to x^j. After integration by parts (in the sense of distributions) in the integrals, noting that $\partial_{x^j} G_\nu(\mathbf{x}-\mathbf{y}, t-s) = -\partial_{y^j} G_\nu(\mathbf{x}-\mathbf{y}, t-s)$, we obtain

$$\begin{aligned}
\partial_{x^j}\phi(\mathbf{x},t) &= \int_{\mathbb{R}^n} G_\nu(\mathbf{x}-\mathbf{y}, t-\varepsilon)\partial_{y^j}\phi(\mathbf{y},\varepsilon)d\mathbf{y} \\
&\quad + \int_\varepsilon^t \int_{\mathbb{R}^n} G_\nu(\mathbf{x}-\mathbf{y}, t-s)\partial_{y^j}(\mathbf{a}(\mathbf{y},s)\cdot \nabla_{\mathbf{y}}\phi(\mathbf{y},s))d\mathbf{y}ds, \qquad (2.29)\\
\nabla_{\mathbf{x}}\partial_{x^j}\phi(\mathbf{x},t) &= \int_{\mathbb{R}^n} \nabla_{\mathbf{x}} G_\nu(\mathbf{x}-\mathbf{y}, t-\varepsilon)\partial_{y^j}\phi(\mathbf{y},\varepsilon)d\mathbf{y} \\
&\quad + \int_\varepsilon^t \int_{\mathbb{R}^n} \nabla_{\mathbf{x}} G_\nu(\mathbf{x}-\mathbf{y}, t-s)\partial_{y^j}(\mathbf{a}(\mathbf{y},s)\cdot \nabla_{\mathbf{y}}\phi(\mathbf{y},s))d\mathbf{y}ds.
\end{aligned}$$

As in the proof of Corollary 2.4 we shift back by $t \to t - \varepsilon$ and assume $\phi_0 \in W^{1,1}(\mathbb{R}^n) \cap W^{1,\infty}(\mathbb{R}^n)$. We obtain a solution $\partial_{x^j}\phi(\mathbf{x}, t)$ to (2.29), satisfying

$$\begin{aligned}
\partial_{x^j}\phi(\cdot,t) \in X(T) \subseteq C([0,T]; L^1(\mathbb{R}^n)\cap L^\infty(\mathbb{R}^n)) \\
\cap C((0,T]; W^{1,1}(\mathbb{R}^n)\cap W^{1,\infty}(\mathbb{R}^n)),
\end{aligned}$$

equipped with norm

$$\begin{aligned}
\|r\|_{X(T)} = \max\Big[&\sup_{0<t\le T} (\|r(\cdot,t)\|_1 + \|r(\cdot,t)\|_\infty), \\
&\sup_{0<t\le T} t^{\frac12}(\|\nabla r(\cdot,t)\|_1 + \|\nabla r(\cdot,t)\|_\infty)\Big] < \infty,
\end{aligned}$$

and thus infer that

$$\phi(\cdot,t) \in C\big((0,T]; W^{2,1}(\mathbb{R}^n)\cap W^{2,\infty}(\mathbb{R}^n)\big).$$

It is now clear how to treat all the spatial derivatives. Observe that the t-derivatives are obtained by resorting to the original equation (2.7).

The decay property stated in the lemma relies on the smoothness of the solution. We show it for ϕ as follows.

Recall that $\phi(\cdot,t) \in C\big([\varepsilon,\infty); W^{1,1}(\mathbb{R}^n) \cap W^{1,\infty}(\mathbb{R}^n)\big)$ (see Corollary 2.4). We use once again the "shifting idea," expressing the solution as in the first equation of (2.14), but with the initial time $t=0$ shifted to $t=\varepsilon$.

The integral $\int_{\mathbb{R}^n} G_\nu(\mathbf{x}-\mathbf{y},t-\varepsilon)\phi(\mathbf{y},\varepsilon)d\mathbf{y}$ is easily seen to vanish as $|\mathbf{x}| \to \infty$, uniformly for $t \in [2\varepsilon, T]$. The details are left to the reader (note that only $\phi(\cdot,\varepsilon) \in L^1(\mathbb{R}^n)$ is needed here; just split the integration inside and outside a large ball).

Consider next the double integral in the first equation of (2.14). Let

$$Q = \sup_{(\mathbf{y},s)\in\mathbb{R}^n\times[\varepsilon,T]} |\mathbf{a}(\mathbf{y},s)\cdot\nabla_{\mathbf{y}}\phi(\mathbf{y},s)|.$$

We can estimate, for $t \in [2\varepsilon, T]$ and small $0<\eta<t-\varepsilon$,

$$\begin{aligned}
&\Big|\int_{\varepsilon}^{t}\int_{\mathbb{R}^n} G_\nu(\mathbf{x}-\mathbf{y},t-s)(\mathbf{a}(\mathbf{y},s)\cdot\nabla_{\mathbf{y}}\phi(\mathbf{y},s))d\mathbf{y}ds\Big| \\
&\le \Big|\int_{t-\eta}^{t}\int_{\mathbb{R}^n} G_\nu(\mathbf{x}-\mathbf{y},t-s)(\mathbf{a}(\mathbf{y},s)\cdot\nabla_{\mathbf{y}}\phi(\mathbf{y},s))d\mathbf{y}ds\Big| \\
&\quad + \Big|\int_{\varepsilon}^{t-\eta}\Big\{\int_{|\mathbf{y}|\le\lambda} + \int_{|\mathbf{y}|\ge\lambda} G_\nu(\mathbf{x}-\mathbf{y},t-s)(\mathbf{a}(\mathbf{y},s)\cdot\nabla_{\mathbf{y}}\phi(\mathbf{y},s))d\mathbf{y}\Big\}ds\Big| \\
&\le Q\eta + (4\pi\nu\eta)^{-\frac{n}{2}}\int_{\varepsilon}^{t-\eta}\int_{|\mathbf{y}|\ge\lambda} |\mathbf{a}(\mathbf{y},s)\cdot\nabla_{\mathbf{y}}\phi(\mathbf{y},s)|d\mathbf{y}ds \\
&\quad + Q\int_{\varepsilon}^{t-\eta}\int_{|\mathbf{y}|\le\lambda} G_\nu(\mathbf{x}-\mathbf{y},t-s)d\mathbf{y}ds = I_1+I_2+I_3.
\end{aligned}$$

All three terms in the last sum can be made small: in fact, let $\delta>0$. First, taking $\eta<\frac{\delta}{Q}$ yields $|I_1|<\delta$. Then taking λ sufficiently large (note that $|\mathbf{a}(\mathbf{y},s)\cdot\nabla_{\mathbf{y}}\phi(\mathbf{y},s)|$ is integrable) we ensure $|I_2|<\delta$ and, finally, taking $|\mathbf{x}|$ sufficiently large, we have $|I_3|<\delta$ by the exponential decay of G_ν.

In the same way, the decay of the derivatives of ϕ of all orders is established, using repeated differentiation as in (2.29).

To prove the estimate (2.28) we let $\theta(\mathbf{x})$ be a smooth concave function, so that $\Delta\theta \le 0$. Assume further that $|\nabla\theta(\mathbf{x})| \le 1$, namely, that $\theta(\mathbf{x})$ is uniformly Lipschitz with constant ≤ 1. The explicit construction of such a function is deferred to the end of the proof.

It is readily verified that

$$\Delta\left(\exp\left(\frac{\theta}{\nu}\right)\right) = \exp\left(\frac{\theta}{\nu}\right)\left(\frac{1}{\nu^2}|\nabla\theta|^2 + \frac{1}{\nu}\Delta\theta\right) \le \frac{1}{\nu^2}\exp\left(\frac{\theta}{\nu}\right)|\nabla\theta|^2.$$

Let $\psi(\mathbf{x},t) = \exp\left(\frac{(A_T+1)t+\theta(\mathbf{x})}{\nu}\right)$. Then

$$\Delta_{\mathbf{x}}\psi(\mathbf{x},t) \le \frac{1}{\nu^2}\psi(\mathbf{x},t)|\nabla\theta|^2.$$

It follows that

$$\begin{aligned}&\partial_t\psi(\mathbf{x},t) - \nu\Delta_{\mathbf{x}}\psi(\mathbf{x},t) - \mathbf{a}(\mathbf{x},t)\cdot\nabla_{\mathbf{x}}\psi(\mathbf{x},t)\\ &\qquad\qquad \ge \frac{1}{\nu}\{A_T + 1 - |\nabla\theta|^2 - A_T|\nabla\theta|\}\psi(\mathbf{x},t) \ge 0,\end{aligned}$$

where we have used the assumption $|\nabla\theta(\mathbf{x})| \le 1$. Assume now that $\theta(\mathbf{x}) \ge 0$ in the ball $|\mathbf{x}| \le R$. Then clearly $M\psi(\mathbf{x},0) \ge |\phi_0(\mathbf{x})|$, $\mathbf{x} \in \mathbb{R}^n$, and the maximum principle for linear parabolic equations [69], [102, Section 7.2], yields

$$|\phi(\mathbf{x},t)| \le M\psi(\mathbf{x},t), \quad (\mathbf{x},t) \in \Omega_T. \tag{2.30}$$

We now construct a special function $\theta(\mathbf{x})$ that will enable us to conclude the proof.

Take $\widetilde{\theta}(\mathbf{x}) = R - |\mathbf{x}|$ and let $0 \le \eta(\mathbf{x}) \in C_0^\infty(\mathbb{R}^n)$ be a mollifier [58, Appendix C], namely, a smooth function with support in the unit ball $\{|\mathbf{x}| \le 1\}$ and such that $\int_{\mathbb{R}^n}\eta(\mathbf{x})d\mathbf{x} = 1$. We set

$$\theta(\mathbf{x}) = \int_{\mathbb{R}^n}\eta(\mathbf{y})\widetilde{\theta}(\mathbf{x}-\mathbf{y})d\mathbf{y} = \int_{\mathbb{R}^n}\eta(\mathbf{x}-\mathbf{y})\widetilde{\theta}(\mathbf{y})d\mathbf{y},$$

and note that it satisfies the following two properties:

(i) $|\theta(\mathbf{x}_1) - \theta(\mathbf{x}_2)| \le |\mathbf{x}_1 - \mathbf{x}_2|$, $\mathbf{x}_1, \mathbf{x}_2 \in \mathbb{R}^n$,
which follows from the (obvious) same inequality for $\widetilde{\theta}(\mathbf{x})$.

(ii) $\Delta\theta(\mathbf{x}) \le 0$.

Indeed, by a well-known approximation formula for the Laplacian (which is expounded in detail in Section 9.2 of this monograph), if e_j is the unit vector in the x^j direction, then

$$\begin{aligned}\partial^2_{x^j}\theta(\mathbf{x}) = \lim_{h\to 0}\frac{1}{h^2}\int_{\mathbb{R}^n}\eta(\mathbf{y})\Big[&\widetilde{\theta}(\mathbf{x}+he_j-\mathbf{y}) + \widetilde{\theta}(\mathbf{x}-he_j-\mathbf{y})\\ &- 2\widetilde{\theta}(\mathbf{x}-\mathbf{y})\Big]d\mathbf{y} \le 0,\end{aligned}$$

where the last inequality follows from the concavity of $\widetilde{\theta}(\mathbf{x})$ and the non-negativity of η.

The same facts hold when $\eta(\mathbf{y})$ is replaced by $\eta_\varepsilon(\mathbf{y}) = \varepsilon^{-n}\eta\left(\frac{\mathbf{y}}{\varepsilon}\right)$ and the corresponding

$$\theta_\varepsilon(\mathbf{x}) = \int_{\mathbb{R}^n} \eta_\varepsilon(\mathbf{y})\widetilde{\theta}(\mathbf{x}-\mathbf{y})d\mathbf{y},$$

so that we can take in (2.30), $\psi(\mathbf{x},t) = \psi_\varepsilon(\mathbf{x},t) = \exp\left(\frac{(A_T+1)t+\theta_\varepsilon(\mathbf{x})}{\nu}\right)$.

Now observe [58, Appendix C] that $\theta_\varepsilon(\mathbf{x}) \xrightarrow{\varepsilon\to 0} \widetilde{\theta}(\mathbf{x})$, so (2.28) follows from (2.30) as $\varepsilon \to 0$, and the fact that the maximum principle also implies $|\phi(\mathbf{x},t)| \le M$. ■

We also state the following result, supplementing Lemma 2.6 for the case that $\mathbf{a}(\mathbf{x},t)$ is a solenoidal (divergence-free) field.

Lemma 2.7. *In addition to the assumptions of Lemma 2.6 assume that,*

$$\nabla_{\mathbf{x}} \cdot \mathbf{a}(\mathbf{x},t) = \sum_{i=1}^{n} \frac{\partial}{\partial x^i} a^i = 0 \quad \text{in} \quad \Omega_T. \tag{2.31}$$

Then for any $0 \le t \le T$, *the solution to* (2.7) *satisfies*

$$\|\phi(\cdot,t)\|_p \le \|\phi_0\|_p, \quad 1 \le p \le \infty. \tag{2.32}$$

Proof. In view of (2.31) we have $(\mathbf{a}\cdot\nabla)\phi = \nabla\cdot(\mathbf{a}\phi)$. Let $\tau \in [0,T]$ and consider the *dual equation*

$$\psi_t + \nu\Delta\psi + (\mathbf{a}\cdot\nabla)\psi = 0,$$

subject to the "terminal" condition

$$\psi(\mathbf{x},\tau) = \psi_0(\mathbf{x}) \in C_0^\infty(\mathbb{R}^n).$$

This equation is solved backwards in time, from $t = \tau$ down to $t = 0$. Lemma 2.6 determines the existence and decay properties of ψ in Ω_τ. Integrating by parts and noting the spatial decay of ϕ and ψ (ensuring that all boundary terms vanish) we obtain

$$\begin{aligned} 0 &= \int_0^\tau \int_{\mathbb{R}^n} \big(\phi_t(\mathbf{x},t) - \nu\Delta_{\mathbf{x}}\phi(\mathbf{x},t) - (\mathbf{a}\cdot\nabla_{\mathbf{x}})\phi(\mathbf{x},t)\big)\psi(\mathbf{x},t)d\mathbf{x}dt \\ &= \int_{\mathbb{R}^n} \big(\phi(\mathbf{x},\tau)\psi_0(\mathbf{x}) - \phi_0(\mathbf{x})\psi(\mathbf{x},0)\big)d\mathbf{x}. \end{aligned}$$

The maximum principle yields $\|\psi(\cdot,0)\|_\infty \le \|\psi_0\|_\infty$. Thus

$$\Big|\int_{\mathbb{R}^n} \big(\phi(\mathbf{x},\tau)\psi_0(\mathbf{x})\big)d\mathbf{x}\Big| \le \|\phi_0\|_1\|\psi_0\|_\infty.$$

It follows that $\|\phi(\cdot,\tau)\|_1 \leq \|\phi_0\|_1$, proving (2.32) for $p = 1$. The case $p = \infty$ is of course the maximum principle . Let $S(\tau)$ be the solution (linear) operator $S(\tau)(\phi_0) = \phi(\mathbf{x},\tau)$. The arguments above show that it is a contraction in $L^p(\mathbb{R}^n)$ when $p = 1$ and $p = \infty$. By the Riesz–Thorin interpolation theorem (see Appendix A, Section A.3) it is indeed a contraction for all $p \in [0,\infty]$, which proves (2.32). ■

Remark 2.8. The method of proof employed here is called *a duality argument.* It is closely related to Holmgren's method for uniqueness [102, Section 3.5].

2.1.1 *Unbounded coefficient at* $t = 0$

In this subsection we address again the convection–diffusion equations (2.7)-(2.8) and consider it in the two-dimensional case $n = 2$.

The need for the results discussed here will arise only in Section 5.1, so the reader can skip this subsection on first reading.

The initial data used here are finite Borel measures in the space $\mathcal{M}$. The definition of this space, as well as the notion of weak* convergence, are given in Appendix A, Section A.4.

Consider the equation

$$(2.33)\quad \partial_t\phi(\mathbf{x},t) - \nu\Delta_{\mathbf{x}}\phi(\mathbf{x},t) = \mathbf{a}(\mathbf{x},t)\cdot\nabla_{\mathbf{x}}\phi(\mathbf{x},t), \qquad (\mathbf{x},t) \in \mathbb{R}^2\times(0,T).$$

As an initial condition for this equation we allow any $\phi_0 \in \mathcal{M}$, and require that

$$(2.34)\qquad \phi(\cdot,t) \xrightarrow[t\to 0]{weak^*} \phi_0.$$

The convection coefficient is smooth, $\mathbf{a}(\mathbf{x},t) \in C^\infty(\mathbb{R}^2\times(0,T))$, and satisfies in every strip $\mathbb{R}^2 \times (\tau, T)$, $\tau > 0$, the conditions of both Lemmas 2.6 and 2.7. However, we allow it to be unbounded as $t \to 0$. Its growth rate at $t = 0$, while not the most general, is the one we shall encounter for the velocity field.

We assume that with some constant $N > 0$,

$$(2.35)\qquad |\mathbf{a}(\mathbf{x},t)| \leq \frac{N}{\sqrt{t}}, \quad (\mathbf{x},t) \in \mathbb{R}^2 \times (0,T).$$

As in the case of (2.14), we consider the integral version of (2.33)

$$\phi(\mathbf{x},t) = H_\nu(t)\phi_0 + \int_0^t \int_{\mathbb{R}^2} G_\nu(\mathbf{x}-\mathbf{y},t-s)\mathbf{a}(\mathbf{y},s)\cdot\nabla_{\mathbf{y}}\phi(\mathbf{y},s)d\mathbf{y}ds,$$

where

$$H_\nu(t)\phi_0 = \int_{\mathbb{R}^2} G_\nu(\mathbf{x}-\mathbf{y},t)\phi_0(\mathbf{y})d\mathbf{y}. \tag{2.36}$$

The heat kernel G_ν is defined in (2.25), with $n = 2$.

Using the fact that $\mathbf{a}(\mathbf{x},t)$ is divergence-free and integrating (formally) by parts, we **define** a (mild) solution to (2.33) as a function $\phi(\cdot,t) \in C((0,T);L^1(\mathbb{R}^2))$ such that, for any $t_0 \in (0,T)$,

(2.37)

$$\phi(\mathbf{x},t) = H_\nu(t-t_0)\phi(\cdot,t_0) + \int_{t_0}^{t}\int_{\mathbb{R}^2} \nabla_{\mathbf{x}} G_\nu(\mathbf{x}-\mathbf{y},t-s)\cdot\mathbf{a}(\mathbf{y},s)\phi(\mathbf{y},s)d\mathbf{y}ds,$$

where the equality is understood in the sense of $C([t_0,T);L^1(\mathbb{R}^2))$, namely, the two sides are equal as $L^1(\mathbb{R}^2)$ functions for every $t \in [t_0,T)$.

Lemma 2.9. *Let $\mathbf{a}(\mathbf{x},t)$ satisfy the hypotheses of both Lemmas 2.6 and 2.7 in every strip $\mathbb{R}^2\times(\varepsilon,T)$, $\varepsilon > 0$, and assume that it satisfies also* (2.35). *Given any $\phi_0 \in \mathcal{M}$, there exists a unique mild solution $\phi(\cdot,t) \in C((0,T);L^1(\mathbb{R}^2))\cap C^\infty(\mathbb{R}^2\times(0,T))$ to* (2.37), *satisfying the initial condition* (2.34).

Furthermore, this solution decays in every strip $\mathbb{R}^2\times[\varepsilon,T)$, $\varepsilon > 0$, in the sense of (2.4).

Proof. Let $\phi_{0,k}$ be the restriction of ϕ_0 to the disk $\{|\mathbf{x}| < k\}$. Using Claim A.8 in Appendix A (and its proof), we construct a sequence $\{\phi_0^{(k)} = \eta_k * \phi_{0,k}\}_{k=1}^\infty \subseteq C_0^\infty(\mathbb{R}^2)$, such that

$$\phi_0^{(k)} \xrightarrow[k\to\infty]{weak^*} \phi_0, \tag{2.38}$$

and $\|\phi_0^{(k)}\|_1 \le \|\phi_0\|_{\mathcal{M}}$.

Let $0 \le \theta(t) \in C^\infty(\mathbb{R})$ be a monotone nondecreasing function such that $\theta(t) = 0$ for $t \le 0$ and $\theta(t) = 1$ for $t \ge 1$.

Let $\{t_k\}_{k=1}^\infty$ be a sequence decreasing to zero. Define

$$\mathbf{a}_k(\mathbf{x},t) = \theta\Big(\frac{t-t_k}{t_k}\Big)\mathbf{a}(\mathbf{x},t), \quad k = 1,2,\ldots$$

Clearly $\mathbf{a}_k(\mathbf{x},t)$ satisfies the conditions of Lemmas 2.6 and 2.7. Considering Equation (2.37), with $\mathbf{a}$ replaced by $\mathbf{a}_k$, subject to the initial condition $\phi_0^{(k)}$. It has a unique solution $\phi_k(\cdot,t) \in C([0,T);L^1(\mathbb{R}^2))\cap C^\infty(\mathbb{R}^2\times(0,T))$. The norm $\|\phi_k(\cdot,t)\|_1$ is a nonincreasing function of t.

In every strip $\mathbb{R}^2 \times (\varepsilon, T)$ the functions $\{\mathbf{a}_k\}_{k=1}^{\infty}$, as well as their derivatives of any order, are uniformly bounded. It follows from Lemma 2.6 that for any pair of indices $\boldsymbol{\alpha}$, j,

$$\sup_{1\le k<\infty} \sup_{(\mathbf{x},t)\in\mathbb{R}^2\times(\varepsilon,T)} |\nabla_{\mathbf{x}}^{\boldsymbol{\alpha}} \partial_t^j \phi_k(\mathbf{x},t)| < \infty.$$

Indeed, as in the proof of Theorem 2.2, the bounds for derivatives of the solution depend only on bounds for the derivatives of the convection coefficient in the strip (and on $\|\phi_0\|_{\mathcal{M}}$).

Invoking the Arzelà–Ascoli theorem [58, Appendix C.7], together with an obvious diagonal process, we get a function $\phi(\mathbf{x},t) \in C^\infty(\mathbb{R}^2 \times (0,T))$ and a subsequence (relabeled as ϕ_k), such that, for all indices $\boldsymbol{\alpha}$, j,

$$\nabla_{\mathbf{x}}^{\boldsymbol{\alpha}} \partial_t^j \phi_k(\mathbf{x},t) \to \nabla_{\mathbf{x}}^{\boldsymbol{\alpha}} \partial_t^j \phi(\mathbf{x},t), \quad (\mathbf{x},t) \in \mathbb{R}^2 \times (0,T), \tag{2.39}$$

uniformly in any compact subset.

In addition, Fatou's lemma [58, Appendix E.3] implies that

$$\|\phi(\cdot,t)\|_1 \le \liminf_{k\to\infty} \|\phi_k(\cdot,t)\|_1 \le \|\phi_0\|_{\mathcal{M}}.$$

The solution ϕ_k satisfies, for any $\varepsilon \in (0,T)$,

(2.40)
$$\phi_k(\mathbf{x},t) = H_\nu(t-\varepsilon)\phi_0^{(k)}(\cdot,\varepsilon)$$
$$+ \int_\varepsilon^t \int_{\mathbb{R}^2} \nabla_{\mathbf{x}} G_\nu(\mathbf{x}-\mathbf{y}, t-s) \cdot \mathbf{a}_k(\mathbf{y},s)\phi_k(\mathbf{y},s) d\mathbf{y} ds, \quad (\mathbf{x},t) \in \mathbb{R}^2 \times [\varepsilon, T).$$

For any $s \in [\varepsilon, T)$ let

$$M_s = \sup_{1\le k<\infty} \|\phi_k(\cdot,s)\|_\infty < \infty.$$

Let

$$f_k(\mathbf{x},s;t) = \int_{\mathbb{R}^2} \nabla_{\mathbf{x}} G_\nu(\mathbf{x}-\mathbf{y}, t-s) \cdot \mathbf{a}_k(\mathbf{y},s)\phi_k(\mathbf{y},s) d\mathbf{y}.$$

We have, for any $\varepsilon \le s < t$, in view of (2.26) and the assumption (2.35),

$$\|f_k(\cdot,s;t)\|_\infty \le N M_s \beta_\nu (t-s)^{-\frac{1}{2}} s^{-\frac{1}{2}}. \tag{2.41}$$

Using the uniform (in every compact subset) convergence (2.39) and the exponential decay of the heat kernel we get, for fixed $(\mathbf{x},t) \in \mathbb{R}^2 \times [\varepsilon, T)$, the limit

$$f(\mathbf{x},s;t) = \lim_{k\to\infty} f_k(\mathbf{x},s;t) = \int_{\mathbb{R}^2} \nabla_{\mathbf{x}} G_\nu(\mathbf{x}-\mathbf{y}, t-s) \cdot \mathbf{a}(\mathbf{y},s)\phi(\mathbf{y},s) d\mathbf{y}. \tag{2.42}$$

Since the right-hand side of (2.41) is integrable in $s \in [\varepsilon, t)$, the dominated convergence theorem [58, Appendix E.3] can be invoked in order to pass to the limit in (2.40), yielding, for any $\varepsilon > 0$,

(2.43)
$$\phi(\mathbf{x},t) = H_\nu(t-\varepsilon)\phi(\cdot,\varepsilon) + \int_\varepsilon^t \int_{\mathbb{R}^2} \nabla_{\mathbf{x}} G_\nu(\mathbf{x}-\mathbf{y}, t-s)\cdot \mathbf{a}(\mathbf{y},s)\phi(\mathbf{y},s)d\mathbf{y}ds, \quad (\mathbf{x},t)\in \mathbb{R}^2\times[\varepsilon,T).$$

Note that by (2.26), (2.35) and Young's inequality

$$\|f(\cdot,s;t)\|_1 \le N\|\phi_0\|_{\mathcal{M}}\beta_\nu(t-s)^{-\frac{1}{2}}s^{-\frac{1}{2}}, \tag{2.44}$$

so that, for every $t\in[\varepsilon,T)$, the equality (2.43) holds in the sense of $L^1(\mathbb{R}^2)$.

In particular, $\phi(\cdot,t)\in C((0,T);L^1(\mathbb{R}^2))$.

We now address the convergence to the initial data as expressed in (2.34). The notation $< , >$ designates the $(\mathcal{M}, D^0(\mathbb{R}^2))$ pairing (see Section A.4 in Appendix A).

We need to prove that, given any $\psi \in C_0^\infty(\mathbb{R}^2)$, the function

$$g(t) = \begin{cases} \int_{\mathbb{R}^2}\phi(\mathbf{x},t)\psi(\mathbf{x})d\mathbf{x}, & t>0,\\ <\phi_0,\psi>, & t=0,\end{cases}$$

is continuous at $t=0$. This will follow if we prove that the sequence of continuous functions

$$g_k(t) = \int_{\mathbb{R}^2}\phi_k(\mathbf{x},t)\psi(\mathbf{x})d\mathbf{x}, \quad k=1,2,\ldots, t\ge 0,$$

converges to $g(t)$ uniformly in some interval $[0,\delta]$, $\delta < T$.

Since $g_k(t)\to g(t)$ as $k\to\infty$, for any fixed $t\ge 0$, the uniform convergence will follow by invoking the Arzelà–Ascoli theorem [58, Appendix C.7].

Thus it suffices to show that the family $\{g_k(t)\}_{k=1}^\infty$ is equicontinuous in $[0,\delta]$.

Take the scalar product of the two sides of the equality

$$\partial_t\phi_k(\mathbf{x},t) - \nu\Delta_{\mathbf{x}}\phi_k(\mathbf{x},t) = \mathbf{a}_k(\mathbf{x},t)\cdot\nabla_{\mathbf{x}}\phi_k(\mathbf{x},t)$$

with ψ, to get

$$\frac{d}{dt}g_k(t) = \nu\int_{\mathbb{R}^2}\phi_k(\mathbf{x},t)\Delta\psi(\mathbf{x})d\mathbf{x} - \int_{\mathbb{R}^2}\phi_k(\mathbf{x},t)\mathbf{a}_k(\mathbf{x},t)\cdot\nabla\psi(\mathbf{x})d\mathbf{x}. \tag{2.45}$$

The estimate

$$\|\phi_k(\cdot,t)\mathbf{a}_k(\cdot,t)\|_1 \le N\|\phi_0\|_{\mathcal{M}} t^{-1/2}, \quad t>0,$$

implies the uniform integrability of the derivatives in $[0,\delta]$, hence the equicontinuity of the family $\{g_k(t)\}_{k=1}^{\infty}$, as claimed.

We use the weak* continuity at $t=0$ to show that (2.43) is satisfied also with $\varepsilon = 0$, namely, in the sense of $L^1(\mathbb{R}^2)$,

(2.46)

$$\phi(\mathbf{x},t) = G_\nu(\cdot,t) * \phi_0 + \int_0^t \int_{\mathbb{R}^2} \nabla_{\mathbf{x}} G_\nu(\mathbf{x}-\mathbf{y},t-s)\cdot \mathbf{a}(\mathbf{y},s)\phi(\mathbf{y},s)d\mathbf{y}ds, \quad (\mathbf{x},t)\in\mathbb{R}^2\times(0,T).$$

In fact, combining this continuity with $G_\nu(\cdot,t)\in D^0(\mathbb{R}^2)$, $t>0$, we get, in the sense of L^1,

$$\lim_{\varepsilon\to 0} H_\nu(t-\varepsilon)\phi(\cdot,\varepsilon) = G_\nu(\cdot,t)*\phi_0.$$

On the other hand, if $0<\varepsilon_1<\varepsilon_2$, it follows from (2.44) that,

$$\int_{\varepsilon_1}^{\varepsilon_2} \|f(\cdot,s;t)\|_1 ds \le N\|\phi_0\|_{\mathcal{M}}\beta_\nu \int_{\varepsilon_1}^{\varepsilon_2} (t-s)^{-\frac12} s^{-\frac12} ds,$$

which tends to zero as $\varepsilon_2\to 0$.

To conclude the proof of the lemma, it remains to prove the uniqueness of the solution.

Assume that $\widetilde{\phi}$ is another solution, satisfying the same initial condition. The equality (2.46) is satisfied when ϕ is replaced by $\widetilde{\phi}$. Subtracting the two expressions we obtain

$$\phi(\mathbf{x},t)-\widetilde{\phi}(\mathbf{x},t) = \int_0^t\int_{\mathbb{R}^2} \nabla_{\mathbf{x}} G_\nu(\mathbf{x}-\mathbf{y},t-s)\cdot\mathbf{a}(\mathbf{y},s)[\phi(\mathbf{y},s)-\widetilde{\phi}(\mathbf{y},s)]d\mathbf{y}ds.$$

With $L>0$ yet to be determined we define

$$\Psi(\tau) = \sup_{0<s\le\tau} e^{-Ls}\|\phi(\cdot,s)-\widetilde{\phi}(\cdot,s)\|_1,$$

so that the last equality yields

$$\Psi(t) \le N\beta_\nu \int_0^t e^{-L(t-s)}(t-s)^{-\frac12}s^{-\frac12}\Psi(s)ds$$
$$\le N\beta_\nu \int_0^1 e^{-Lt(1-s)}(1-s)^{-\frac12}s^{-\frac12}ds\cdot\Psi(t).$$

Fixing any $t=t_0\in(0,T)$, we take L sufficiently large so that,

$$N\beta_\nu \int_0^1 e^{-Lt_0(1-s)}(1-s)^{-\frac{1}{2}}s^{-\frac{1}{2}}ds < \frac{1}{2},$$

implying that $\Psi(t_0) = 0$. Hence $\phi(\mathbf{x}, t) = \widetilde{\phi}(\mathbf{x}, t)$, $t \in (0, T)$. ■

The properties of the solution of (2.33), as stated in Lemma 2.9, can also be deduced from properties of the fundamental solution of this equation [79]. This fundamental solution has Gaussian behavior, analogous to that of the heat kernel. In particular, assuming $T = \infty$, and taking $\phi_0 \in L^1(\mathbb{R}^2)$, it can be shown that

Theorem 2.10. *Let $\phi(\mathbf{x}, t)$ be the solution given by Lemma 2.9. Then for any $\kappa \in (0, 1)$ there exists $C_\kappa > 0$, depending on N and $\|\phi_0\|_1$, such that*

$$(2.47)\qquad |\phi(\mathbf{x}, t)| \le \frac{C_\kappa}{t} \int_{\mathbb{R}^2} \exp\Big(-\kappa\frac{|\mathbf{x}-\mathbf{y}|^2}{4\nu t}\Big)|\phi_0(\mathbf{y})|d\mathbf{y}, \quad \mathbf{x} \in \mathbb{R}^2,\ t > 0.$$

For a proof, see [35, Theorem 3].

2.2 Proof of Theorem 2.1

We will first establish some properties of the solutions to our iterative system (2.5)–(2.6), which basically follow by a straightforward application of Lemmas 2.6 and 2.7. However, we shall also require uniform estimates for the sequence of solutions thus obtained.

Lemma 2.11. *Suppose that $\omega_0(\mathbf{x}) \in C_0^\infty(\mathbb{R}^2)$ and let $T > 0$, $\Omega_T = \mathbb{R}^2 \times [0, T]$. Then the system (2.5)–(2.6) defines successively, in Ω_T, a sequence of C^∞ solutions, $\{\omega^{(k)},\ \mathbf{u}^{(k)}\}_{k=0}^\infty$.*

These solutions are decaying in Ω_T, along with all their derivatives in the sense of (2.4), and satisfy for every $1 \le p \le \infty$,

$$(2.48)\qquad ||\omega^{(k)}(\cdot, t)||_p \le ||\omega_0||_p, \qquad 0 \le t \le T, \quad k = 0, 1, 2, \ldots$$

Furthermore, for every integer l and every double-index $\boldsymbol{\alpha}$,

$$(2.49)\qquad \sup_{\substack{k=0,1,2,\ldots\\ (\mathbf{x},t)\in\Omega_T}} \Big[|\partial_t^l \nabla_{\mathbf{x}}{}^{\boldsymbol{\alpha}} \omega^{(k)}(\mathbf{x}, t)| + |\partial_t^l \nabla_x{}^{\boldsymbol{\alpha}} \mathbf{u}^{(k)}(\mathbf{x}, t)|Big] < \infty.$$

Proof of the Lemma. We use induction on k. The claim is clear for $\omega^{(0)}$, from the explicit solution to the heat equation (2.13).

Suppose that $\{\omega^{(j)}, \mathbf{u}^{(j-1)}\}$, $j = 0, \ldots, k$, have been shown to be C^∞ solutions in Ω_T, decaying with all their derivatives, and satisfying for $p \in [0, \infty]$,

$$(2.50) \qquad ||\omega^{(j)}(\cdot, t)||_p \le ||\omega_0||_p, \qquad 0 \le t \le T, \quad 0 \le j \le k,$$

$$(2.51) \qquad \sup_{\substack{0\le j\le k \\ (\mathbf{x},t)\in\Omega_T}} \left[|\partial_t^l \nabla_{\mathbf{x}}{}^{\boldsymbol{\alpha}} \omega^{(j)}(\mathbf{x}, t)| + |\partial_t^l \nabla_{\mathbf{x}}{}^{\boldsymbol{\alpha}} \mathbf{u}^{(j-1)}(\mathbf{x}, t)| \right] < \infty.$$

Let $\chi(\mathbf{x}) \in C_0^\infty(\mathbb{R}^2)$ be a cutoff function such that $0 \le \chi(\mathbf{x}) \le 1$, $\chi(\mathbf{x}) = 1$ for $|\mathbf{x}| \le 1$. Let $\psi(\mathbf{x}) = 1 - \chi(\mathbf{x})$. In view of (2.2),

$$(2.52) \qquad \begin{aligned} \mathbf{u}^{(k)}(\mathbf{x}, t) &= \int_{\mathbb{R}^2} \psi(\mathbf{x} - \mathbf{y}) \mathbf{K}(\mathbf{x} - \mathbf{y}) \omega^{(k)}(\mathbf{y}, t) d\mathbf{y} \\ &+ \int_{\mathbb{R}^2} \chi(\mathbf{y}) \mathbf{K}(\mathbf{y}) \omega^{(k)}(\mathbf{x} - \mathbf{y}, t) d\mathbf{y} = \mathbf{I}_1(\mathbf{x}, t) + \mathbf{I}_2(\mathbf{x}, t). \end{aligned}$$

The integration in $\mathbf{I}_1(\mathbf{x}, t)$ extends over $|\mathbf{x} - \mathbf{y}| \ge 1$. The induction hypothesis on $\omega^{(k)}$ and the explicit expression (1.12) for $\mathbf{K}$ imply that $\mathbf{I}_1(\mathbf{x}, t)$ is C^∞ and decays, along with all derivatives, as $|\mathbf{x}| \to \infty$.

The same conclusion is valid for $\mathbf{I}_2(\mathbf{x}, t)$, again using the induction hypothesis and the finite domain of integration.

Now Lemma 2.6, applied to Equation (2.5) with k replaced by $k + 1$, yields the existence, regularity and decay properties of $\omega^{(k+1)}$, as expressed in (2.51).

We next claim that the velocity field $\mathbf{u}^{(k)}(\mathbf{x}, t)$ is divergence-free:

$$(2.53) \qquad \nabla_{\mathbf{x}} \cdot \mathbf{u}^{(k)}(\mathbf{x}, t) = 0.$$

A straightforward calculation yields $\nabla_{\mathbf{x}} \cdot \mathbf{K}(\mathbf{x}) = 0$ (for $\mathbf{x} \ne 0$), a fact which is at the core of proving this equation.

First, we let $\theta \in C_0^\infty(\overline{\mathbb{R}_+})$ be a real-valued function such that $\theta(r) = 1$ for $r \in [0, 1]$ and $\theta(r) = 0$ for $r > 2$. Let the above cutoff function be radial, $\chi(\mathbf{x}) = \theta(|\mathbf{x}|)$. Next, in the decomposition (2.52) replace the function $\chi(\mathbf{y})$ with $\chi_\varepsilon(\mathbf{y}) = \chi(\frac{\mathbf{y}}{\varepsilon})$ and $\psi_\varepsilon = 1 - \chi_\varepsilon$. Denote the corresponding terms by $\mathbf{I}_1^\varepsilon(\mathbf{x}, t)$ and $\mathbf{I}_2^\varepsilon(\mathbf{x}, t)$.

Thus, since the integral defining $\mathbf{I}_1^\varepsilon(\mathbf{x}, t)$ extends over $|\mathbf{x} - \mathbf{y}| \ge \varepsilon$,

$$\nabla_{\mathbf{x}} \cdot \mathbf{I}_1^\varepsilon(\mathbf{x}, t) = \int_{\mathbb{R}^2} \nabla_{\mathbf{x}} \psi_\varepsilon(\mathbf{x} - \mathbf{y}) \cdot \mathbf{K}(\mathbf{x} - \mathbf{y}) \omega^{(k)}(\mathbf{y}, t) d\mathbf{y}.$$

However, since $\psi_\varepsilon(\mathbf{x} - \mathbf{y})$ is radial (with respect to the center $\mathbf{y}$) while $\mathbf{K}(\mathbf{x} - \mathbf{y})$ is tangential [see Equation (1.12)], it follows that $\nabla_{\mathbf{x}} \psi_\varepsilon(\mathbf{x} - \mathbf{y}) \cdot \mathbf{K}(\mathbf{x} - \mathbf{y}) = 0$, hence

$$\nabla_{\mathbf{x}} \cdot \mathbf{I}_1^\varepsilon(\mathbf{x}, t) = 0.$$

As for $\mathbf{I}_2^\varepsilon(\mathbf{x},t)$, we have, since $\mathbf{K}(\mathbf{y})$ is integrable near $\mathbf{y}=0$,

$$\nabla_{\mathbf{x}} \cdot \mathbf{I}_2^\varepsilon(\mathbf{x},t) = \int_{\mathbb{R}^2} \chi_\varepsilon(\mathbf{y})\mathbf{K}(\mathbf{y}) \cdot \nabla_{\mathbf{x}}\omega^{(k)}(\mathbf{x}-\mathbf{y},t)d\mathbf{y} \xrightarrow{\varepsilon\to 0} 0.$$

This establishes (2.53), since $\mathbf{u}^{(k)}(\mathbf{x},t) = \mathbf{I}_1^\varepsilon(\mathbf{x},t) + \mathbf{I}_2^\varepsilon(\mathbf{x},t)$ for all $\varepsilon > 0$.

The estimate (2.50) (for $j = k+1$) now follows from Lemma 2.7.

Finally, we establish the uniform estimates (2.49). The reader is advised to compare this part of the proof with the proof of the regularity claim in Lemma 2.6. Note that in that proof we needed to demonstrate the existence of derivatives, while here we know that $\omega^{(k)}(\mathbf{x},t)$, $\mathbf{u}^{(k)}(\mathbf{x},t) \in C^\infty(\mathbb{R}^2 \times \overline{\mathbb{R}_+})$, so only the uniform estimates need to be proved.

In the decomposition (2.52) we estimate the two terms using (2.48) for $p=1$ and $p=\infty$, and also noting $|\mathbf{K}(\mathbf{y})| \le (2\pi)^{-1}|\mathbf{y}|^{-1}$, so that

$$|\mathbf{u}^{(k)}(\mathbf{x},t)| \le \|\omega_0\|_\infty + \frac{1}{2\pi}\|\omega_0\|_1. \tag{2.54}$$

Recall that the heat kernel (2.25) for our case ($n=2$) is

$$G_\nu(\mathbf{x},t) = (4\pi\nu t)^{-1}e^{-\frac{|\mathbf{x}|^2}{4\nu t}}.$$

It follows from (2.5) and the Duhamel principle [see Equation (2.14) where it was assumed that $\nu = 1$] that,

$$\begin{aligned}\omega^{(k)}(\mathbf{x},t) &= \omega^{(0)}(\mathbf{x},t)\\ &- \int_0^t \int_{\mathbb{R}^2} G_\nu(\mathbf{x}-\mathbf{y},t-s)(\mathbf{u}^{(k-1)}(\mathbf{y},s)\cdot\nabla_{\mathbf{y}})\omega^{(k)}(\mathbf{y},s)d\mathbf{y}ds,\end{aligned} \tag{2.55}$$

where $\omega^{(0)}(\mathbf{x},t) = G_\nu(\cdot,t) * \omega_0$. Differentiating with respect to $\mathbf{x}$,

(2.56)

$$\begin{aligned}\nabla_{\mathbf{x}}\omega^{(k)}(\mathbf{x},t) &= \nabla\omega^{(0)}(\mathbf{x},t)\\ &- \int_0^t \int_{\mathbb{R}^2} \nabla_{\mathbf{x}}G_\nu(\mathbf{x}-\mathbf{y},\ t-s)(\mathbf{u}^{(k-1)}(\mathbf{y},\ s)\cdot\nabla_{\mathbf{y}})\omega^{(k)}(\mathbf{y},\ s)d\mathbf{y}ds.\end{aligned}$$

Recall that from (2.20)

$$\int_{\mathbb{R}^2} |\nabla_{\mathbf{x}}G_\nu(\mathbf{x},t)|d\mathbf{x} = \beta_\nu t^{-\frac{1}{2}}, \quad t>0.$$

Letting $N_k(t) = \sup\{|\nabla_{\mathbf{y}}\omega^{(k)}(\mathbf{y},s)|,\ (\mathbf{y},s)\in\Omega_t\}$, and using (2.26) and (2.54) in (2.56), we obtain

$$N_k(t) \le C\Big[1+\int_0^t (t-s)^{-\frac{1}{2}}N_k(s)ds\Big], \quad 0 \le t \le T. \tag{2.57}$$

Here and in the sequel we use "C" to denote a generic constant, independent of k, $\mathbf{x}$ and t. Of course, C may still depend on general data of the problem such as ν, T, R or ω_0. When we need to indicate certain dependencies, we write $C = C(R, T, \dots)$.

If $L > 0$ is such that $C\int_0^T e^{-Ls}s^{-\frac{1}{2}}ds < \frac{1}{2}$ then (2.57) implies

$$N_k(t) \le 2Ce^{Lt}, \quad 0 \le t \le T. \tag{2.58}$$

We have therefore established the uniform boundedness of

$$\{|\nabla_{\mathbf{x}}\omega^{(k)}(\mathbf{x}, t)|,\ k = 0, 1, 2, \dots\}, \quad \text{in } \Omega_T.$$

From (2.52) we get, differentiating with respect to x^j,

$$\begin{aligned} \partial_{x^j}\mathbf{u}^{(k)}(\mathbf{x}, t) &= \int_{\mathbb{R}^2} \partial_{x^j}[\psi(\mathbf{x}-\mathbf{y})\mathbf{K}(\mathbf{x}-\mathbf{y})]\omega^{(k)}(\mathbf{y}, t)d\mathbf{y} \\ &\quad + \int_{\mathbb{R}^2} \chi(\mathbf{y})\mathbf{K}(\mathbf{y})\partial_{x^j}\omega^{(k)}(\mathbf{x}-\mathbf{y}, t)d\mathbf{y}, \end{aligned} \tag{2.59}$$

which clearly yields the uniform boundedness of

$$\{|\nabla_{\mathbf{x}}\mathbf{u}^{(k)}(\mathbf{x}, t)|,\ k = 0, 1, 2, \dots\}, \quad \text{in } \Omega_T$$

(note Equation (2.48) for $p = 1$).

Turning back to (2.56) and differentiating with respect to x^j we get

$$\begin{aligned} \nabla_{\mathbf{x}}\partial_{x^j}\omega^{(k)}(\mathbf{x}, t) &= \nabla_{\mathbf{x}}\partial_{x^j}\omega^{(0)}(\mathbf{x}, t) \\ &\quad - \int_0^t \int_{\mathbb{R}^2} \partial_{x^j}G_\nu(\mathbf{x}-\mathbf{y},\ t-s)\nabla_{\mathbf{y}}[(\mathbf{u}^{(k-1)}(\mathbf{y}, s)\cdot\nabla_{\mathbf{y}})\omega^{(k)}(\mathbf{y}, s)]d\mathbf{y}ds, \end{aligned}$$

where we have used

$$\nabla_{\mathbf{x}}G_\nu(\mathbf{x}-\mathbf{y},\ t-s) = -\nabla_{\mathbf{y}}G_\nu(\mathbf{x}-\mathbf{y},\ t-s),$$

and integrated by parts (with respect to $\mathbf{y}$).

Using the (already established) uniform boundedness of

$$\{|\nabla_{\mathbf{x}}\omega^{(k)}|,\ |\nabla_{\mathbf{x}}\mathbf{u}^{(k)}|,\ k = 0, 1, 2, \dots\} \text{ in } \Omega_T,$$

an argument similar to that used in (2.57)–(2.58) yields, for $|\boldsymbol{\alpha}| = 2$, the uniform boundedness of

$$\{|\nabla_{\mathbf{x}}{}^{\boldsymbol{\alpha}}\omega^{(k)}(\mathbf{x}, t)|,\ k = 0, 1, 2, \dots\} \text{ in } \Omega_T.$$

The uniform boundedness of

$$\{|\nabla_{\mathbf{x}}{}^{\boldsymbol{\alpha}}\mathbf{u}^{(k)}(\mathbf{x}, t)|,\ k = 0, 1, 2, \dots\}\ (|\boldsymbol{\alpha}| = 2) \text{ in } \Omega_T$$

now follows as in (2.59). From (2.5) we now obtain the uniform boundedness of

$$\{|\partial_t \omega^{(k)}|,\ |\partial_t \mathbf{u}^{(k)}|,\ \ k = 0, 1, 2, \ldots\} \text{ in } \Omega_T.$$

It is now clear how to establish (2.49) for higher values of l and $\boldsymbol{\alpha}$. ∎

Remark 2.12. Observe that the proof of (2.53) actually shows that

$$\nabla \cdot \mathbf{K}(\mathbf{x}) = 0, \tag{2.60}$$

in the sense of distributions (see [133, Section 2.4.3]).

We can now complete the proof of Theorem 2.1.

Proof of Theorem 2.1. Fix $T > 0$ and let $\{\omega^{(k)}, \mathbf{u}^{(k)}\}_{k=0}^{\infty}$ be the sequence of solutions to (2.5)–(2.6), as given by Lemma 2.11.

Let $Y_\infty = L^1(\mathbb{R}^2) \cap D^0(\mathbb{R}^2)$, where $D^0(\mathbb{R}^2)$ is the space of continuous functions decaying as $|\mathbf{x}| \to \infty$. It is normed by

$$\|f\|_{Y_\infty} = \|f\|_1 + \|f\|_\infty. \tag{2.61}$$

(We choose the notation $D^0(\mathbb{R}^2)$ instead of the more common $C_0(\mathbb{R}^2)$ because the latter is often used to denote the space of compactly supported functions.)

For $0 \le t \le T$ and $k = 0, 1, 2, \ldots$ we set

$$N_k(t) = \sup_{0 \le s \le t} ||\omega^{(k+1)}(\cdot, s) - \omega^{(k)}(\cdot, s)||_{Y_\infty}. \tag{2.62}$$

Recall that $||\omega^{(k)}(\cdot, t)||_\infty \le M = ||\omega_0||_\infty$ and the estimate (2.54) yields

$$A = \sup\{||\mathbf{u}^{(k)}(\cdot, s)||_\infty,\ 0 \le s \le t,\ k = 0, 1, 2, \ldots\} \le M + \frac{1}{2\pi}||\omega_0||_1. \tag{2.63}$$

Since by (2.53) we have $\nabla_{\mathbf{x}} \cdot \mathbf{u}^{(k)} = 0$ we can rewrite Equation (2.55), after integration by parts, as

$$\begin{aligned}\omega^{(k)}(\mathbf{x}, t) &= \omega^{(0)}(\mathbf{x}, t)\\ &\quad + \int_0^t \int_{\mathbb{R}^2} \nabla_{\mathbf{y}} G_\nu(\mathbf{x} - \mathbf{y}, t - s) \cdot (\mathbf{u}^{(k-1)}(\mathbf{y}, s)\omega^{(k)}(\mathbf{y}, s)) d\mathbf{y} ds.\end{aligned}$$

It follows that

(2.64)

$$\omega^{(k+1)}(\mathbf{x},t)-\omega^{(k)}(\mathbf{x},t)$$

$$=\int_0^t\int_{\mathbb{R}^2}\nabla_{\mathbf{y}}G_\nu(\mathbf{x}-\mathbf{y},t-s)\cdot(\omega^{(k+1)}(\mathbf{y},s)-\omega^{(k)}(\mathbf{y},s))\mathbf{u}^{(k)}(\mathbf{y},s)d\mathbf{y}ds$$

$$+\int_0^t\int_{\mathbb{R}^2}\nabla_{\mathbf{y}}G_\nu(\mathbf{x}-\mathbf{y},t-s)\cdot\omega^{(k)}(\mathbf{y},s)(\mathbf{u}^{(k)}(\mathbf{y},s)-\mathbf{u}^{(k-1)}(\mathbf{y},s))d\mathbf{y}ds.$$

In view of (2.48) (for $p=1$), (2.26) and (2.63) we obtain from (2.64),

$$\begin{aligned}&\|\omega^{(k+1)}(\cdot,t)-\omega^{(k)}(\cdot,t)\|_\infty\\ &\le A\beta_\nu\int_0^t(t-s)^{-\frac12}\|\omega^{(k+1)}(\cdot,s)-\omega^{(k)}(\cdot,s)\|_\infty ds\\ &\quad+M\beta_\nu\int_0^t(t-s)^{-\frac12}\|\mathbf{u}^{(k)}(\cdot,s)-\mathbf{u}^{(k-1)}(\cdot,s)\|_\infty ds,\end{aligned}\tag{2.65}$$

and

$$\begin{aligned}&\|\omega^{(k+1)}(\cdot,t)-\omega^{(k)}(\cdot,t)\|_1\\ &\le A\beta_\nu\int_0^t(t-s)^{-\frac12}\|\omega^{(k+1)}(\cdot,s)-\omega^{(k)}(\cdot,s)\|_1 ds\\ &\quad+\beta_\nu\|\omega_0\|_1\int_0^t(t-s)^{-\frac12}\|\mathbf{u}^{(k)}(\cdot,s)-\mathbf{u}^{(k-1)}(\cdot,s)\|_\infty ds.\end{aligned}\tag{2.66}$$

However, as in the derivation of (2.54),

$$\|\mathbf{u}^{(k)}(\cdot,s)-\mathbf{u}^{(k-1)}(\cdot,s)\|_\infty\le\|\omega^{(k)}(\cdot,s)-\omega^{(k-1)}(\cdot,s)\|_{Y_\infty},\tag{2.67}$$

so that, adding (2.65) and (2.66), and using the definition (2.62), we obtain

(2.68)

$$N_k(t)\le A\beta_\nu\int_0^t(t-s)^{-\frac12}N_k(s)ds+\beta_\nu(M+\|\omega_0\|_1)\int_0^t(t-s)^{-\frac12}N_{k-1}(s)ds.$$

We now apply an approach similar to that used in handling (2.57). First choose $L>0$ so that $A\beta_\nu\int_0^T s^{-\frac12}e^{-Ls}ds<\frac12$. It follows that, with $Q_k(t)=e^{-Lt}N_k(t)$,

$$Q_k(t)\le C\beta_\nu(M+\|\omega_0\|_1)\int_0^t(t-s)^{-\frac12}Q_{k-1}(s)ds.\tag{2.69}$$

(See the paragraph following (2.57) for the use of the "generic constant" C.)

An inductive argument will now be used to prove the following estimate, for $k = 0, 1, 2, \ldots$,

$$(2.70) \qquad Q_k(t) \leq N_0(T) \cdot \left[C\beta_\nu(M + ||\omega_0||_1)\Gamma(\tfrac{1}{2})\right]^k \Gamma\left(\frac{k+2}{2}\right)^{-1} t^{k/2},$$

where Γ is the Gamma function. Indeed, (2.70) is valid for $k = 0$ and to pass from k to $k+1$ we use the well-known identity

$$(2.71) \qquad \int_0^t (t-s)^{-\frac{1}{2}} s^{k/2} ds = B\left(\frac{1}{2}, \frac{k+2}{2}\right) t^{\frac{k+1}{2}} = \frac{\Gamma(\frac{1}{2})\Gamma\left(\frac{k+2}{2}\right)}{\Gamma\left(\frac{k+3}{2}\right)} t^{\frac{k+1}{2}},$$

where B is the Beta function.

We conclude that $\sum_{k=0}^{\infty} N_k(t)$ converges uniformly for $0 \leq t \leq T$. Hence

$$(2.72) \qquad \omega(\cdot, t) = \lim_{k\to\infty} \omega^{(k)}(\cdot, t)$$

exists in $C([0,T]; Y_\infty)$.

From Equation (2.49) we know that the sequences

$$\left\{\partial_t^l \nabla_{\mathbf{x}}{}^{\boldsymbol{\alpha}} \omega^{(k)}(\mathbf{x}, t)\right\}_{k=1}^{\infty}, \quad \text{and} \quad \left\{\partial_t^l \nabla_{\mathbf{x}}{}^{\boldsymbol{\alpha}} \mathbf{u}^{(k)}(\mathbf{x}, t)\right\}_{k=1}^{\infty}$$

are uniformly bounded and equicontinuous in Ω_T, for every fixed l and $\boldsymbol{\alpha}$. We can therefore invoke the Arzelà–Ascoli theorem [58, Appendix C.7], combined with a diagonal process over expanding compact subsets of Ω_T, in order to obtain the following strengthening of (2.72).

The limits,

$$(2.73) \qquad \omega(\boldsymbol{x}, t) = \lim_{k\to\infty} \omega^{(k)}(\boldsymbol{x}, t), \qquad \boldsymbol{u}(\boldsymbol{x}, t) = \lim_{k\to\infty} \boldsymbol{u}^{(k)}(\boldsymbol{x}, t),$$

exist in $C^\infty(\Omega_T)$, namely, uniform convergence, along with derivatives of all orders, in every compact subset of Ω_T. Furthermore, all derivatives of $\omega(\mathbf{x}, t)$ and $\mathbf{u}(\mathbf{x}, t)$ are in $L^\infty(\Omega_T)$ and Equations (2.1)–(2.2) *are satisfied in Ω_T.*

By Lemma 2.6, the uniform boundedness of $\mathbf{u}(\mathbf{x}, t)$ and its derivatives yields the decay of $\omega(\mathbf{x}, t)$ and its derivatives. Now, as in (2.52), we have

$$(2.74) \qquad \begin{aligned} \partial_t^\ell \nabla_{\mathbf{x}}^{\boldsymbol{\alpha}} \mathbf{u}(\mathbf{x}, t) &= \int_{\mathbb{R}^2} \nabla_{\mathbf{x}}^{\boldsymbol{\alpha}} [\psi(\mathbf{x} - \mathbf{y}) \mathbf{K}(\mathbf{x} - \mathbf{y})] \partial_t^\ell \omega(\mathbf{y}, t) d\mathbf{y} \\ &+ \int_{\mathbb{R}^2} \chi(\mathbf{y}) \mathbf{K}(\mathbf{y}) \partial_t^\ell \nabla_{\mathbf{x}}^{\boldsymbol{\alpha}} \omega(\mathbf{x} - \mathbf{y}, t) d\mathbf{y}, \end{aligned}$$

which proves the decay of $\partial_t^\ell \nabla_{\mathbf{x}}^{\boldsymbol{\alpha}} \mathbf{u}(\mathbf{x}, t)$.

This concludes the proof of the existence of smooth decaying solutions of (2.1)–(2.2) in every strip Ω_T, where the estimate (2.3) follows from (2.48).

Observe in particular that $\nabla_{\mathbf{x}} \cdot \mathbf{u}(\mathbf{x}, t) = 0$, by Remark 2.12.

We have established the existence of a solution in Ω_T, for every $T > 0$. To complete the proof of Theorem 2.1 it suffices to prove the uniqueness of such solutions in Ω_T. Indeed, if this uniqueness is established, we can consider the solutions in Ω_T and Ω_{2T}, satisfying the same initial condition. By uniqueness they coincide in Ω_T, which means that the solution in Ω_{2T} is an extension of the solution in Ω_T. Continuing in this fashion we obtain the existence of a unique global solution in $\mathbb{R}^2 \times \overline{\mathbb{R}_+}$, as claimed.

Thus, suppose that $\theta(\mathbf{x}, t)$ and $\mathbf{v}(\mathbf{x}, t)$ solve

$$
\begin{aligned}
&\partial_t \theta + (\mathbf{v} \cdot \nabla)\theta = \nu \Delta \theta \quad \text{in } \Omega_T, \\
&\mathbf{v} = \mathbf{K} * \theta,
\end{aligned} \tag{2.75}
$$

$$
\theta(\mathbf{x}, 0) = \omega_0(\mathbf{x}), \qquad \mathbf{x} \in \mathbb{R}^2. \tag{2.76}
$$

We assume that $\theta(\mathbf{x}, t)$ and $\mathbf{v}(\mathbf{x}, t)$ are C^∞ and uniformly bounded in Ω_T, along with their derivatives of any order. Both $\mathbf{u}$ and $\mathbf{v}$ are divergence-free, by Remark 2.12. Subtracting (2.75) from (2.1) and using $\nabla \cdot \mathbf{u} = \nabla \cdot \mathbf{v} = 0$, we get

$$
\partial_t(\omega - \theta) - \nu\Delta(\omega - \theta) = -\nabla \cdot (\omega - \theta)\mathbf{u} + \nabla \cdot [\theta(\mathbf{u} - \mathbf{v})],
$$

so that, integrating by parts,

$$
\begin{aligned}
&\omega(\mathbf{x}, t) - \theta(\mathbf{x}, t) \\
&= \int_0^t \int_{\mathbb{R}^2} \nabla_{\mathbf{y}} G_\nu(\mathbf{x} - \mathbf{y}, t - s) \cdot (\omega(\mathbf{y}, s) - \theta(\mathbf{y}, s))\mathbf{u}(\mathbf{y}, s) d\mathbf{y} ds \\
&\quad + \int_0^t \int_{\mathbb{R}^2} \nabla_{\mathbf{y}} G_\nu(\mathbf{x} - \mathbf{y}, t - s) \cdot (\mathbf{u}(\mathbf{y}, s) - \mathbf{v}(\mathbf{y}, s))\theta(\mathbf{y}, s) d\mathbf{y} ds.
\end{aligned} \tag{2.77}
$$

Set

$$
N(t) = \sup_{0 \le s \le t} ||\omega(\cdot, s) - \theta(\cdot, s)||_{Y_\infty}.
$$

Note that, as in (2.54),

$$
||\mathbf{u}(\cdot, s) - \mathbf{v}(\cdot, s)||_\infty \le ||\omega(\cdot, s) - \theta(\cdot, s)||_{Y_\infty},
$$

so that (2.77) yields, using the same procedure leading to (2.68),

$$
N(t) \le C \int_0^t (t - s)^{-\frac{1}{2}} N(s) ds \le 2CN(t)t^{\frac{1}{2}}. \tag{2.78}
$$

In particular $N(t) = 0$ for $0 \le t \le \frac{1}{4C^2}$. Proceeding with finitely many time steps of this size we obtain $N(t) = 0$ for $t \in [0, T]$.

The proof of Theorem 2.1 is complete. ∎

2.3 Existence and uniqueness in Hölder spaces

In Theorem 2.1 we established the global existence and uniqueness of a solution to (2.1)–(2.2), assuming that the initial data $\omega_0(\mathbf{x}) \in C_0^\infty(\mathbb{R}^2)$. We next mention a theorem of McGrath [135], that deals with the global existence and uniqueness of the solution in Hölder spaces. We note that while the proof of Theorem 2.1 relied heavily on the parabolic character of the equation (thus forcing $\nu > 0$), McGrath's theorem applies also to the Euler system ($\nu = 0$). Since we shall not need this theorem in later chapters, we will only give a sketch of the proof.

Theorem 2.13. *Assume that for some* $0 < \lambda < 1$, $\omega_0(\mathbf{x}) \in L^1(\mathbb{R}^2) \cap C^{2,\lambda}(\mathbb{R}^2)$. *Then there exists a solution to* (2.1)–(2.2) *such that*

(a) The solution is classical; all derivatives in (2.1) *are continuous in* $\mathbb{R}^2 \times (0, \infty)$.

(b) $\omega(\mathbf{x}, t)$ *and* $\mathbf{u}(\mathbf{x}, t)$ *are continuous and uniformly bounded in* $\mathbb{R}^2 \times [0, \infty)$.

(c) $\omega(\mathbf{x}, \cdot) \in L^\infty([0, \infty); L^1(\mathbb{R}^2))$.

(d) For every $T > 0$,

$$\sup_{0 \le t \le T, |\mathbf{x}| > R} |\mathbf{u}(\mathbf{x}, t)| \to 0 \quad \textit{as} \quad R \to \infty.$$

Under conditions (a)–(d) the solution is unique.

Proof (outline). Fix $T > 0$ and let,

$$\begin{aligned} &\Omega_T = \mathbb{R}^2 \times [0, T], \\ &X(T) = C([0, T]; C_b(\mathbb{R}^2)) \cap C([0, T]; L^1(\mathbb{R}^2)), \end{aligned} \tag{2.79}$$

where $\|\omega(\mathbf{x}, t)\|_{X(T)} = \|\omega\|_{L^\infty(\Omega_T)} + \sup_{0 \le t \le T} \|\omega(\cdot, t)\|_1$.

Let ω_0 satisfy the assumption of the theorem and let $B_0 \subseteq X(T)$ be the ball

$$B_0 = \{\omega \in X(T), \ \ \|\omega\|_{X(T)} \le \|\omega_0\|_1 + \|\omega_0\|_\infty\}.$$

For $\xi \in B_0$, define the map $\mathbf{A}_1 \xi = \mathbf{v}$ by means of (2.2), that is, $\mathbf{v} = \mathbf{K} * \xi$, $0 \le t \le T$. In particular it is easily seen [see the decomposition (2.52)] that $\mathbf{v} \in C(\Omega_T) \cap L^\infty(\Omega_T)$.

Invoking Theorem 2.2 we see that the linear parabolic equation

$$\partial_t \theta + (\mathbf{v} \cdot \mathbf{\nabla})\theta = \nu \Delta \theta, \quad \theta(\mathbf{x}, 0) = \omega_0(\mathbf{x}), \tag{2.80}$$

has a unique solution in Ω_T. This solution there was shown to be in $C([0,T];L^1(\mathbb{R}^2)) \cap C((0,T];W^{1,1}(\mathbb{R}^2))$. By the Duhamel principle (see (2.14)) we have

$$\theta(\mathbf{x},t) = G_\nu * \omega_0 - \int_0^t \int_{\mathbb{R}^2} G_\nu(\mathbf{x}-\mathbf{y},t-s)\mathbf{v}(\mathbf{y},s)\cdot\nabla_{\mathbf{y}}\theta(\mathbf{y},s)d\mathbf{y}ds. \tag{2.81}$$

This equation, combined with the assumed uniform (Hölder) continuity of ω_0, implies that $\theta \in C([0,T];C_b(\mathbb{R}^2))$, so that $\theta \in X(T)$. We therefore obtain a map $A : \xi \in B_0 \to \theta \in X(T)$. Note also that the equation implies that $\theta(\mathbf{x},t)$ decays in the sense of (2.4).

We next take a more restricted space

$$Y(T) = C([0,T];C_b^2(\mathbb{R}^2)) \cap C([0,T];L^1(\mathbb{R}^2)).$$

Observe that if $\xi \in Y(T)$ then $\mathbf{v} = \mathbf{A}_1\xi \in C([0,T];C_b^2(\mathbb{R}^2))$ [use again the decomposition (2.52)]. Following the proof of Lemma 2.6, for up to second-order spatial derivatives, we conclude that if $\xi \in B_0 \cap Y(T)$ then $\theta = A\xi \in Y(T)$. In particular, $\theta(\mathbf{x},t)$ is a classical solution of (2.80).

Note that $\nabla\cdot\mathbf{A}_1\xi = 0$ by Remark 2.12, so that Lemma 2.7 is applicable. In particular it follows that A maps $B_0 \cap Y(T)$ into B_0 and the density of $B_0 \cap Y(T)$ in B_0 (in the $X(T)$ topology) yields $A : B_0 \hookrightarrow B_0$.

Taking into account the regularity hypothesis on ω_0, we have also

$$\nabla_{\mathbf{x}}\theta(\mathbf{x},t) = G_\nu * \nabla\omega_0 - \int_0^t \int_{\mathbb{R}^2} \nabla_{\mathbf{x}}G_\nu(\mathbf{x}-\mathbf{y},t-s)\mathbf{v}(\mathbf{y},s)\cdot\nabla_{\mathbf{y}}\theta(\mathbf{y},s)d\mathbf{y}ds,$$

from which we obtain $\nabla_{\mathbf{x}}\theta(\mathbf{x},t) \in C([0,T];C_b(\mathbb{R}^2))$ [see (2.16) for the estimate of $\nabla_{\mathbf{x}}G_\nu(\mathbf{x},t)$].

The decay of $\theta(\mathbf{x},t)$ as $|\mathbf{x}| \to \infty$ implies that

$$\{\theta(\cdot,t),\ \theta = A\xi,\ \xi \in B_0\}$$

is compactly embedded in $C_b(\mathbb{R}^2)$, for every $t \in [0,T]$.

In addition, it is obvious from (2.81) that the family of functions

$$t \in [0,T] \to \theta(\cdot,t) \in B_0, \quad \xi \in B_0,$$

is equicontinuous. We therefore infer from the Arzelà–Ascoli theorem that $A(B_0)$, the image of B_0 by A, is compactly embedded in B_0.

Since A is continuous, the Schauder fixed point theorem [58, Section 9.2.2] yields an $\omega \in B_0$ such that $\omega = A\omega$. This ω is a solution to (2.1) with $\mathbf{u} = \mathbf{A}_1\omega$. The uniqueness is shown by an argument similar to that employed in the uniqueness part of the proof of Theorem 2.1. ■

2.4 Notes for Chapter 2

- A solution defined by means of the "Duhamel integral," as in (2.14), is known as a "mild solution." This notion is commonly used in the case of nonlinear evolution equations, see [18] and references therein. In the case of a linear equation, it is difficult to find a proof of Theorem 2.2 with this generality in the literature. The reader is referred to [102, Section 7.2] for a proof using finite differences. The coefficients are assumed (at least) to be continuously differentiable.
- The estimate (2.28) is a somewhat simplified version of a decay estimate for quasi-linear parabolic equations given in [98, Lemma 3.2.1].
- The proof of Theorem 2.1 is taken from [9].

Chapter 3

Estimates for Smooth Solutions

In Theorem 2.1 we established the existence and uniqueness, as well as the regularity, of the solution $\omega(\mathbf{x}, t)$ to the vorticity equation (2.1) for initial data $\omega_0 \in C_0^\infty(\mathbb{R}^2)$. These properties extended also to the associated velocity field $\mathbf{u}(\mathbf{x}, t)$ given by (2.2). The solutions are global for all nonnegative time $t \in \overline{\mathbb{R}_+} = [0, \infty)$.

In this chapter we establish some basic estimates of this solution. Such estimates will be crucial in the process of extending the admissible space of initial data. In fact, the procedure is quite common in the study of partial differential equations; solutions are established under rather restrictive conditions and are then estimated using "weaker" norms, namely, norms in more general spaces. Consequently, the class of solutions can be extended to those spaces. Typically, the extended solutions lose some of the "classical properties," as was the case with the linear convection–diffusion equation considered in Section 2.1. Considering $C_0^\infty(\mathbb{R}^2)$ as the space of initial data, we designate by

$$S : C_0^\infty(\mathbb{R}^2) \to C^\infty(\mathbb{R}^2 \times \overline{\mathbb{R}}_+) \tag{3.1}$$

the solution operator to (2.1)–(2.2), which we write as $(S\omega_0)(t) = \omega(\cdot, t)$. The corresponding velocity field is given by (2.2), and we denote it as

$$\begin{aligned} &\mathbf{U} : C_0^\infty(\mathbb{R}^2) \to \mathbf{C}^\infty(\mathbb{R}^2 \times \overline{\mathbb{R}}_+), \\ &(\mathbf{U}\omega_0)(t) = \mathbf{K} * (S\omega_0)(t) = \mathbf{u}(\cdot, t). \end{aligned} \tag{3.2}$$

When there is no risk of confusion we shall write simply $\omega(t)$ instead of $\omega(\cdot, t)$.

The decay property of the solution, as stated in Theorem 2.1, justifies the vanishing of all boundary terms in the integrations by parts performed below.

We will split the basic estimates over three sections. In Section 3.1 we deal with estimates involving only $\|\omega_0\|_1$. In analogy with the linear heat equation, such estimates involve a "blow-up" factor as $t \to 0$. In Section 3.2 we use also L^p norms of the initial vorticity, which entails better regularity of the vorticity and velocity fields near the initial time. In Section 3.3 we give space-time estimates for the gradients of the velocity and vorticity fields.

Remark 3.1. Given $\omega_0 \in C_0^\infty(\mathbb{R}^2)$, the solution $\omega(\cdot,t) = (S\omega_0)(t)$ exists for all $t > 0$. Furthermore, if $\tau > 0$ is fixed, then it follows from (the much more general) Theorem 2.2 that $(S\omega_0)(t+\tau), (U\omega_0)(t+\tau)$ is the unique solution to Equations (2.1)–(2.2), subject to the initial condition $\omega(\cdot,\tau)$ (at $t = 0$). Since this solution and all its derivatives decay exponentially as $|\mathbf{x}| \to \infty$, all the estimates obtained below are valid with $\omega(\cdot,\tau)$ as initial data. In particular, if we estimate some norm of $\omega(\cdot,t)$ with respect to a (possibly different) norm of ω_0, the same estimate holds for the norm of $\omega(\cdot,t+\tau)$ with respect to the suitable norm of $\omega(\cdot,\tau)$.

3.1 Estimates involving $\|\omega_0\|_{L^1(\mathbb{R}^2)}$

Multiplying (2.1) by ω and integrating over $\mathbb{R}^2$ we obtain

$$\partial_t \|\omega(\cdot,t)\|_2^2 = -2\nu \|\nabla\omega(\cdot,t)\|_2^2. \tag{3.3}$$

Observe that the nonlinear (convective) term in (2.1) vanished in the integration

$$\int_{\mathbb{R}^2} \omega(\mathbf{x},t)(\mathbf{u}(\mathbf{x},t)\cdot\nabla_{\mathbf{x}})\omega(\mathbf{x},t)d\mathbf{x} = 0, \tag{3.4}$$

because $\nabla_x \cdot \mathbf{u}(\mathbf{x},t) = 0$. Indeed, $(\mathbf{u}(\mathbf{x},t)\cdot\nabla_{\mathbf{x}})\omega(\mathbf{x},t) = \nabla_{\mathbf{x}}\cdot(\omega(\mathbf{x},t)\mathbf{u}(\mathbf{x},t))$ and integration by parts yields (at any fixed t)

$$\int_{\mathbb{R}^2} \omega(\mathbf{x},t)(\mathbf{u}(\mathbf{x},t)\cdot\nabla_{\mathbf{x}})\omega(\mathbf{x},t)d\mathbf{x} = -\int_{\mathbb{R}^2} \omega(\mathbf{x},t)(\mathbf{u}(\mathbf{x},t)\cdot\nabla_{\mathbf{x}})\omega(\mathbf{x},t)d\mathbf{x} = 0.$$

We will now recall the Nash inequality [34, 61]; if ϕ is a smooth decaying function in $\mathbb{R}^2$, then, for some $\eta > 0$,

$$\|\phi\|_2^2 \le \eta^{-1}\|\phi\|_1\|\nabla\phi\|_2. \tag{3.5}$$

Using this inequality in (3.3) (with $\phi = \omega(\cdot,t)$) and noting (2.3) with $p = 1$ we get

$$\partial_t\left(\|\omega(\cdot,t)\|_2^2\right) \le -2\nu\eta\|\omega_0\|_1^{-2}\|\omega(\cdot,t)\|_2^4, \tag{3.6}$$

which can be rewritten as

$$\partial_t\big(\|\omega(\cdot,t)\|_2^{-2}\big) \geq 2\nu\eta\|\omega_0\|_1^{-2}.$$

Integration of this inequality yields

$$\|\omega(\cdot,t)\|_2^{-2} \geq 2\nu\eta\|\omega_0\|_1^{-2}t + \|\omega_0\|_2^{-2},$$

hence

$$\|\omega(\cdot,t)\|_2 \leq (2\nu\eta t)^{-1/2}\|\omega_0\|_1. \tag{3.7}$$

We now use a duality argument, similar to that used in the proof of Lemma 2.7 (see Remark 2.8).

Fix $\tau > 0$ and let $\xi_\tau \in \mathbf{C}_0^\infty(\mathbb{R}^2)$. Using Lemma 2.6 we solve *backwards* (in time) the equation

$$\partial_t\xi(\mathbf{x},t) + (\mathbf{u}(\mathbf{x},t)\cdot\nabla_{\mathbf{x}})\xi(\mathbf{x},t) = -\nu\Delta_{\mathbf{x}}\xi(\mathbf{x},t), \quad \xi(\mathbf{x},\tau) = \xi_\tau(\mathbf{x}), \quad \mathbf{x}\in\mathbb{R}^2.$$

Multiplying Equation (2.1) by ξ and integrating over the strip $\Omega_\tau = \mathbb{R}^2 \times [0,\tau]$, we obtain after integration by parts (and using again $\nabla_{\mathbf{x}}\cdot\mathbf{u}(\mathbf{x},t) = 0$)

$$\int_{\mathbb{R}^2}\omega(\mathbf{x},\tau)\xi_\tau(\mathbf{x})d\mathbf{x} = \int_{\mathbb{R}^2}\omega_0(\mathbf{x})\xi(\mathbf{x},0)d\mathbf{x}.$$

The solution $\xi(\mathbf{x},t)$ satisfies (backwards in time) the estimate (3.7) so that by the Cauchy–Schwarz inequality

$$\Big|\int_{\mathbb{R}^2}\omega(\mathbf{x},\tau)\xi_\tau(\mathbf{x})d\mathbf{x}\Big| \leq (2\nu\eta\tau)^{-1/2}\|\xi_\tau\|_1\|\omega_0\|_2.$$

Since $L^\infty(\mathbb{R}^2)$ is the dual of $L^1(\mathbb{R}^2)$ we infer that,

$$\|\omega(\cdot,t)\|_\infty \leq (2\nu\eta t)^{-1/2}\|\omega_0\|_2,$$

so that

$$\|\omega(\cdot,t)\|_\infty \leq (\nu\eta t)^{-1/2}\|\omega(\cdot,t/2)\|_2 \leq (\nu\eta t)^{-1}\|\omega_0\|_1, \quad t > 0. \tag{3.8}$$

This estimate can be combined with the case $p = 1$ (see Equation (2.3)), by using the interpolation inequality

$$\|g\|_{L^p(\mathbb{R}^n)} \leq \|g\|_{L^\infty(\mathbb{R}^n)}^{1-\frac{1}{p}}\|g\|_{L^1(\mathbb{R}^n)}^{\frac{1}{p}}.$$

We thus obtain

$$\|\omega(\cdot,t)\|_p \leq (\nu\eta t)^{-1+\frac{1}{p}}\|\omega_0\|_1, \quad t > 0,\ p\in[1,\infty]. \tag{3.9}$$

To estimate $\|\mathbf{u}(\cdot,t)\|_\infty$ in terms of $\|\omega_0\|_1$, note that $|\mathbf{K}(\mathbf{y})| \leq (2\pi)^{-1}|\mathbf{y}|^{-1}$, so that

$$
\begin{aligned}
|\mathbf{u}(\mathbf{x},t)| &\leq \int_{|\mathbf{y}|\leq(\frac{\nu\eta t}{2\pi})^{1/2}} |\mathbf{K}(\mathbf{y})\omega(\mathbf{x}-\mathbf{y},t)|d\mathbf{y} \\
&\quad + \int_{|\mathbf{y}|\geq(\frac{\nu\eta t}{2\pi})^{1/2}} |\mathbf{K}(\mathbf{y})\omega(\mathbf{x}-\mathbf{y},t)|d\mathbf{y} \\
&\leq \left(\frac{\nu\eta t}{2\pi}\right)^{1/2} \|\omega(\cdot,t)\|_\infty + (2\pi)^{-1/2}(\nu\eta t)^{-1/2}\|\omega(\cdot,t)\|_1 \\
&\leq \left(\frac{2}{\pi}\right)^{1/2} (\nu\eta t)^{-1/2}\|\omega_0\|_1.
\end{aligned}
\tag{3.10}
$$

The bound $|\mathbf{K}(\mathbf{y})| \leq (2\pi)^{-1}|\mathbf{y}|^{-1}$ allows us to use the Hardy–Littlewood–Sobolev inequality (see Appendix A, Section A.2), obtaining

$$
\|\mathbf{u}(\cdot,t)\|_q \leq \xi_p\|\omega(\cdot,t)\|_p, \quad 1<p<2, \quad \frac{1}{q}=\frac{1}{p}-\frac{1}{2}. \tag{3.11}
$$

Note that this estimate could also be inferred from the fact that $\nabla\mathbf{K}$ is a Calderón–Zygmund kernel (see Appendix A, Section A.1).

In some cases, including the case $p=\frac{4}{3}$ (which will be important later on in our treatment of the vorticity equation, see Corollary 5.6), there exists an "optimal" explicit value for the constant ξ_p [127].

Combining (3.11) with (3.9), we obtain the following estimate for $\|\mathbf{u}(\cdot,t)\|_q$, $q\in(2,\infty]$, in terms of $\|\omega_0\|_1$,

$$
\begin{aligned}
&\|\mathbf{u}(\cdot,t)\|_q \leq \xi_p\,\|\omega(\cdot,t)\|_p \leq \xi_p(\nu\eta t)^{-1+\frac{1}{p}}\|\omega_0\|_1, \quad t>0, \\
&\qquad 1<p\leq 2, \quad \frac{1}{q}=\frac{1}{p}-\frac{1}{2}.
\end{aligned}
\tag{3.12}
$$

The bound for $p=2, q=\infty$ follows from (3.10).

Remark 3.2. The endpoint case $p=1$, $q=2$ in (3.11) is false, even if $\omega_0\in C_0^\infty$. Indeed, according to the Claim in the Introduction, a necessary condition for $\|\mathbf{u}(\cdot,t)\|_2<\infty$ is $\int\limits_{\mathbb{R}^2}\omega_0(\mathbf{x})d\mathbf{x}=0$.

We collect the estimates obtained so far in the following frame (retaining their equation numbers above), where we employ the solution operators in (3.1),(3.2):

(3.9) $\|(S\omega_0)(t)\|_p \leq (\nu\eta t)^{-1+\frac{1}{p}}\|\omega_0\|_1, \quad t>0,\ p\in[1,\infty]$,
(3.12) $\|(\mathbf{U}\omega_0)(t)\|_q \leq \xi_p(\nu\eta t)^{-\frac{1}{2}+\frac{1}{q}}\|\omega_0\|_1, \quad t>0,\ q\in(2,\infty]$.

3.1.1 *Refinement for short time*

The estimate (3.9) says that $t^{1-\frac{1}{p}}\|\omega(\cdot,t)\|_p$ is bounded by $C\|\omega_0\|_1$. However, we shall now establish the fact that $t^{1-\frac{1}{p}}\|\omega(\cdot,t)\|_p$ *vanishes* as $t \to 0$, for $1 < p \leq \infty$. This refined estimate will be crucial in the next chapter, when discussing the well-posedness of the vorticity equation (2.1) in the $L^1(\mathbb{R}^2)$ framework. The required refinement is based on a closer look at the action of the heat kernel $G_\nu(\mathbf{x},t)$ [see (2.25)] in $L^p(\mathbb{R}^2)$ spaces, as will be seen in Equation (3.16) below.

We first introduce the following notational convention.

The constant $C > 0$ stands for a generic positive constant and $\delta(t)$ stands for a monotone nondecreasing, uniformly bounded, generic function defined for $t \geq 0$, such that $\lim_{t\to 0}\delta(t) = 0$. Both C and $\delta(t)$ may depend on various parameters $(p,\nu,\ldots)$ but not on the solution functions. However, they may depend on certain subsets of initial data. We sometimes indicate specific dependencies by adding parameters, for example, $C(p)$ or $\delta(t;\Lambda)$.

Using the Duhamel principle (2.14) and $\nabla_{\mathbf{x}} \cdot \mathbf{u}(\mathbf{x},t) = 0$, the solution $\omega(\mathbf{x},t)$ to the vorticity equation (2.1) satisfies,

$$\begin{aligned}\omega(\mathbf{x},t) &= \textstyle\int_{\mathbb{R}^2} G_\nu(\mathbf{x}-\mathbf{y},t)\omega_0(\mathbf{y})d\mathbf{y} \\ &\quad + \textstyle\int_0^t \int_{\mathbb{R}^2} \nabla_{\mathbf{y}} G_\nu(\mathbf{x}-\mathbf{y},t-s)\cdot\mathbf{u}(\mathbf{y},s)\omega(\mathbf{y},s)d\mathbf{y}ds.\end{aligned} \tag{3.13}$$

Our aim is to derive uniform estimates not only for solutions having a *specific* initial vorticity but also for solutions having initial data $\omega_0 \in \Lambda \subseteq C_0^\infty(\mathbb{R}^2)$, where Λ is precompact in the $L^1(\mathbb{R}^2)$ topology.

We first apply the interpolation estimate $\|g\|_p \leq \|g\|_\infty^{1-\frac{1}{p}}\|g\|_1^{\frac{1}{p}}$ to the heat kernel $G_\nu(\mathbf{x},t)$ [see (2.25)]. Since $\|G_\nu(\cdot,t)\|_1 = 1$ and $\|G_\nu(\cdot,t)\|_\infty = (4\pi\nu t)^{-\frac{1}{2}}$ we have

$$t^{1-\frac{1}{p}}\|G_\nu(\cdot,t)\|_p \leq (4\pi\nu)^{-1+\frac{1}{p}}, \quad t>0,\ p\in[1,\infty]. \tag{3.14}$$

For any fixed $p \in [1,\infty]$ we consider the family of convolution operators (parametrized by $t > 0$),

$$\omega_0 \hookrightarrow t^{1-1/p}G_\nu(\cdot,t)*\omega_0 = \int_{\mathbb{R}^2} t^{1-1/p}G_\nu(\mathbf{x}-\mathbf{y},t)\omega_0(\mathbf{y})d\mathbf{y}, \quad t>0.$$

Due to Young's inequality (see Appendix A, Section A.2) it is uniformly bounded from $L^1(\mathbb{R}^2)$ into $L^p(\mathbb{R}^2)$.

On the other hand, recall the basic property of the kernel $G_\nu(\cdot,t)$ (which is a "positive summability kernel"):

For $1 \leq p < \infty$, we have

$$G_\nu(\cdot, t) * \omega_0 \xrightarrow{t \to 0} \omega_0$$

in $L^p(\mathbb{R}^2)$ if $\omega_0 \in L^p(\mathbb{R}^2)$.

It follows that in this case

$$t^{1-1/p} G_\nu(\cdot, t) * \omega_0 \xrightarrow{t \to 0} 0 \tag{3.15}$$

in $L^p(\mathbb{R}^2)$, $1 < p \leq \infty$. In particular this holds for every smooth ω_0.

This convergence may now be extended to certain *sets of initial data* as follows.

Lemma 3.3. *Let $\Lambda \subseteq C_0^\infty(\mathbb{R}^2)$ be precompact in $L^1(\mathbb{R}^2)$. Then, for $1 < p \leq \infty$,*

$$t^{1-1/p} \|G_\nu(\cdot, t) * \omega_0\|_p \leq \delta(t; \Lambda), \qquad \omega_0 \in \Lambda \tag{3.16}$$

(where $\delta(t; \Lambda)$ depends on p, ν).

Proof. Given $\varepsilon > 0$ we find finitely many $\left\{\omega_0^1, ..., \omega_0^j\right\} \subseteq \Lambda$ so that the balls of radius ε (in the L^1-norm) centered at these points cover all of Λ.

From Equation (3.15) we have

$$t^{1-1/p} \|G_\nu(\cdot, t) * \omega_0^l\|_p \xrightarrow{t \to 0} 0, \quad 1 \leq l \leq j,$$

and on the other hand, for any index l and any $\omega_0 \in \Lambda$,

$$t^{1-1/p} \|G_\nu(\cdot, t) * (\omega_0 - \omega_0^l)\|_p \leq C \|\omega_0 - \omega_0^l\|_1.$$

In particular, if ω_0 is in the ball of radius ε centered at ω_0^l,

$$\begin{aligned} t^{1-1/p} &\|G_\nu(\cdot, t) * \omega_0\|_p \\ &\leq t^{1-1/p} \|G_\nu(\cdot, t) * \omega_0^l\|_p + t^{1-1/p} \|G_\nu(\cdot, t) * (\omega_0 - \omega_0^l)\|_p \\ &\leq C\varepsilon + t^{1-1/p} \|G_\nu(\cdot, t) * \omega_0^l\|_p \leq 2C\varepsilon, \end{aligned}$$

for sufficiently small t, uniformly in $\omega_0 \in \Lambda$.

This proves the claim of the lemma, in view of the definition of $\delta(t; \Lambda)$. ■

We now claim that the solution to the vorticity equation shares with the solution to the heat equation the same property as expressed in Lemma 3.3.

Lemma 3.4. *Let $\Lambda \subseteq C_0^\infty(\mathbb{R}^2)$ be precompact in $L^1(\mathbb{R}^2)$. Then, for $1 < p \leq \infty$,*

$$\|\omega(\cdot, t)\|_p \leq \delta(t; \Lambda) t^{-1+\frac{1}{p}}, \quad \omega_0 \in \Lambda, \quad 1 < p \leq \infty \tag{3.17}$$

(where $\delta(t; \Lambda)$ depends on p, ν).

Proof. We use Equation (3.13).

The first term on the right-hand side of (3.13) is already estimated by (3.16).

Note that by interpolating (2.26) and $\|\nabla G_\nu(\cdot,t)\|_\infty = Ct^{-\frac{3}{2}}$ we get

$$\|\nabla G_\nu(\cdot,t)\|_r = C(r,\nu)t^{-\frac{3}{2}+\frac{1}{r}}, \quad 1 \le r \le \infty, \tag{3.18}$$

where $C(r,\nu) > 0$ can be explicitly evaluated.

For the second term on the right-hand side of (3.13) we have from (3.18) and Young's inequality (see Appendix A, Section A.2),

$$\begin{aligned}\Big\| \int_{\mathbb{R}^2} \nabla_{\mathbf{y}} G_\nu(\mathbf{x}-\mathbf{y},t-s)\cdot \mathbf{u}(\mathbf{y},s)\omega(\mathbf{y},s)d\mathbf{y}\Big\|_p \\ \le C(r,\nu)(t-s)^{-\frac{3}{2}+\frac{1}{r}}\|\omega(\cdot,s)\mathbf{u}(\cdot,s)\|_q,\end{aligned} \tag{3.19}$$

where $\frac{1}{q}+\frac{1}{r}=\frac{1}{p}+1$.

Invoking (3.11) and Hölder's inequality we have

$$\|\omega(\cdot,s)\mathbf{u}(\cdot,s)\|_q \le \xi_p \|\omega(\cdot,s)\|_p^2,$$

where $\frac{1}{q}=\frac{2}{p}-\frac{1}{2}$, and $1<p<2$.

Solving for $\frac{1}{r}$ we get

$$\frac{1}{r}=\frac{1}{p}+1-\frac{1}{q}=\frac{3}{2}-\frac{1}{p}, \quad 1<p<2.$$

Collecting these estimates we obtain from (3.13), uniformly in $\omega_0\in\Lambda$, with $C=C(r,\nu)\xi_p$,

$$\begin{aligned}\|\omega(\cdot,t)\|_p &\le \delta(t;\Lambda)t^{-1+\frac{1}{p}} + C\int_0^t (t-s)^{-\frac{3}{2}+\frac{1}{r}}\|\omega(\cdot,s)\mathbf{u}(\cdot,s)\|_q ds \\ &\le \delta(t;\Lambda)t^{-1+\frac{1}{p}} + C\int_0^t (t-s)^{-\frac{1}{p}}\|\omega(\cdot,s)\|_p^2 ds,\end{aligned} \tag{3.20}$$

where $1<p<2$ and $\frac{1}{r}=\frac{3}{2}-\frac{1}{p}$. Note that the restriction $p\in(1,2)$ ensures the existence of suitable r,q and the convergence of the integral.

Let $M_p(t;\omega_0) = \sup_{0\le\tau\le t}\tau^{1-\frac{1}{p}}\|\omega(\cdot,\tau)\|_p$ for $\omega_0\in\Lambda$. Noting that $M_p(t;\omega_0)$ is continuous (recall Theorem 2.1) and $M_p(0;\omega_0)=0$, we infer from (3.20),

$$\begin{aligned}M_p(t;\omega_0) &\le \delta(t;\Lambda) + Ct^{1-\frac{1}{p}}\int_0^t (t-s)^{-\frac{1}{p}} s^{-2+\frac{2}{p}} s^{2-\frac{2}{p}}\|\omega(\cdot,s)\|_p^2 ds \\ &\le \delta(t;\Lambda) + CM_p(t;\omega_0)^2.\end{aligned}$$

Let $\bar{t}>0$ be sufficiently small, so that $4C\delta(t;\Lambda)<1$ for $t\in[0,\bar{t}\,]$.

Define

$$t_0 = \sup\left\{\tau \in [0,\bar{t}], \quad M_p(\tau;\omega_0) \le 2\delta(\tau;\Lambda)\right\}.$$

Then

$$M_p(t_0;\omega_0) \le \delta(t_0;\Lambda) + CM_p(t_0;\omega_0)^2 \le \delta(t_0;\Lambda) + 4C\delta(t_0;\Lambda)^2 < 2\delta(t_0;\Lambda).$$

It follows that $t_0 = \bar{t}$, since otherwise there exists $t_0 < \tau \le \bar{t}$ for which still $M_p(\tau;\omega_0) \le 2\delta(\tau;\Lambda)$, thus contradicting the definition of t_0.

Thus $M_p(t;\omega_0) \le \delta(t;\Lambda)$ $(1 < p < 2)$, namely,

$$\|\omega(\cdot,t)\|_p \le \delta(t;\Lambda)t^{-1+\frac{1}{p}}, \quad 0 < t \le \bar{t}, \quad \omega_0 \in \Lambda, \quad 1 < p < 2, \tag{3.21}$$

which is (3.17) for $1 < p < 2$.

We now consider the vorticity equation (2.1) for a given initial function $\omega_0(\mathbf{x})$. The unique solution is given in Theorem 2.1. We now **"freeze"** the resulting velocity field $\mathbf{u}(\mathbf{x},t)$ so that the equation can be viewed as a **linear convection–diffusion equation**, like Equation (2.7). For our given initial function its solution is clearly the vorticity $\omega(\mathbf{x},t)$. Of course for other initial data the solution of the linear equation will differ from that of the (nonlinear) equation. The linear solution operator satisfies the L^1-L^∞ estimate given by (3.8) (which is independent of $\mathbf{u}$) and an L^∞-L^∞ estimate (namely, the maximum principle). We may therefore apply the Riesz–Thorin theorem (see Appendix A, Section A.3) to get

$$\|\omega(\cdot,t)\|_\infty \le (\nu\eta t)^{-\frac{1}{q}}\|\omega_0\|_q, \quad t > 0, \ q \in [1,\infty]. \tag{3.22}$$

Remark that this estimate could be obtained as the dual estimate to (3.9), (using the duality procedure leading to (3.8).) Restricting to $q \in (1,2)$ and combining the estimates (3.21) and (3.22) we get

$$\|\omega(\cdot,t)\|_\infty \le \left(\frac{1}{2}\nu\eta t\right)^{-\frac{1}{q}}\|\omega(\cdot,t/2)\|_q \le \delta(t;\Lambda)t^{-1}, \quad t > 0, \ \omega_0 \in \Lambda.$$

Finally, using $\|g\|_p \le \|g\|_\infty^{1-\frac{1}{p}}\|g\|_1^{\frac{1}{p}}$ we obtain (3.17) for $p \in [2,\infty]$. ■

The estimate (3.10) can be strengthened to yield

$$\|\mathbf{u}(\cdot,t)\|_\infty \le \delta(t;\Lambda)t^{-\frac{1}{2}}, \quad \omega_0 \in \Lambda \tag{3.23}$$

($\Lambda \subseteq C_0^\infty(\mathbb{R}^2)$, precompact in the $L^1(\mathbb{R}^2)$ topology). Indeed, this follows from (3.10) by replacing the term $(\frac{\nu\eta t}{2\pi})^{1/2}$ with $(\frac{\nu\eta t}{2\pi\delta(t;\Lambda)})^{1/2}$ and using (3.17).

In fact, combining (3.11) with (3.17) we get an extension of (3.23),

$$\begin{aligned} &\|\mathbf{u}(\cdot,t)\|_q \le \delta(t;\Lambda)t^{-1+\frac{1}{p}}, \quad t > 0, \\ &1 < p \le 2, \quad \frac{1}{q} = \frac{1}{p} - \frac{1}{2}. \end{aligned} \tag{3.24}$$

3.2 Estimates involving $\|\omega_0\|_{L^p(\mathbb{R}^2)}$

In (2.54) we have already encountered a uniform bound for $\mathbf{u}$, based on the use of $\|\omega_0\|_\infty$, which can be written as

$$(3.25) \qquad |\mathbf{u}(\mathbf{x},t)| \le C[\|\omega_0\|_\infty + \|\omega_0\|_1], \quad (\mathbf{x},t) \in \mathbb{R}^2 \times \overline{\mathbb{R}_+}.$$

In order to obtain estimates involving $\|\omega_0\|_p$ instead of $\|\omega_0\|_\infty$, we introduce the Banach space

$$(3.26) \qquad Y_p = L^1(\mathbb{R}^2) \cap L^p(\mathbb{R}^2), \quad \|g\|_{Y_p} = \|g\|_1 + \|g\|_p, \quad p \in (1,\infty).$$

The space Y_∞, where L^∞ is replaced by $D^0(\mathbb{R}^2)$, was already introduced in (2.61).

As in the previous section, we take our initial data $\omega_0 \in C_0^\infty(\mathbb{R}^2)$, but our estimates will be based on its norm in one of the Y_p spaces (and note Remark 3.1).

It follows from (2.3) that $\|\omega(\cdot,t)\|_{Y_p}$ is nonincreasing as a function of t,

$$(3.27) \qquad \|\omega(\cdot,t_2)\|_{Y_p} \le \|\omega(\cdot,t_1)\|_{Y_p}, \quad t_2 \ge t_1, \quad 1 \le p \le \infty.$$

The velocity $\mathbf{u}(\mathbf{x},t)$ is obtained from $\omega(\mathbf{x},t)$ by convolution with the Biot–Savart kernel $\mathbf{K}$ (2.2) and the estimate (3.25) shows that it is bounded from Y_∞ to $D^0(\mathbb{R}^2)$. More generally we have the following estimate.

Lemma 3.5. *Let $2 < p \le \infty$. The convolution operator $\mathbf{K}*$ is bounded from Y_p into $D^0(\mathbb{R}^2)$, so that*

$$(3.28) \qquad \|\mathbf{u}(\cdot,t)\|_\infty \le C_p\|\omega(\cdot,t)\|_{Y_p} \le C_p\|\omega_0\|_{Y_p}, \quad t \ge 0,$$

where $C_p > 0$ depends only on p.

Proof. It is easy to see that $\mathbf{u}(\cdot,t) \in D^0(\mathbb{R}^2)$ by employing a decomposition as in (2.52) and using the Hölder inequality to get

$$\begin{aligned} |\mathbf{u}(\mathbf{x},t)| &\le \int_{|\mathbf{y}|\le a} + \int_{|\mathbf{y}|\ge a} |\mathbf{K}(\mathbf{y})\omega(\mathbf{x}-\mathbf{y},t)|d\mathbf{y} \\ &\le (2\pi)^{\frac{1-q}{q}}\Big(\frac{a^{\frac{2-q}{q}}}{(2-q)^{\frac{1}{q}}}\Big)\Big\{\int_{|\mathbf{y}|\le a} |\omega(\mathbf{x}-\mathbf{y},t)|^p d\mathbf{y}\Big\}^{\frac{1}{p}} \\ &\quad + (2\pi a)^{-1}\int_{|\mathbf{y}|\ge a} |\omega(\mathbf{x}-\mathbf{y},t)|d\mathbf{y}, \end{aligned}$$

where $q^{-1} + p^{-1} = 1$, so that $q \in [1,2)$.

In particular, taking $a^2 = \frac{2-q}{2\pi}$ we get

$$|\mathbf{u}(\mathbf{x},t)| \le (2\pi(2-q))^{-\frac{1}{2}} \|\omega(\cdot,t)\|_{Y_p},$$

which completes the proof in view of the monotonicity (3.27).

■

Observe that the bound (3.12) for $\|\mathbf{u}(\cdot,t)\|_\infty$ in terms of $\|\omega_0\|_1$ blows up as $t \to 0$. In fact, we cannot expect a bound for $\|\mathbf{u}(\cdot,t)\|_\infty$ in terms of $\|\omega(\cdot,t)\|_{Y_p}$ if $p \le 2$. This can be seen by noting that standard interior elliptic estimates applied to (2.1)–(2.2) yield only $\mathbf{u}(\cdot,t) \in W^{1,p}_{loc}(\mathbb{R}^2)$, which is not contained in L^∞ .

Recall the estimate (3.11) of $\|\mathbf{u}(\cdot,t)\|_q$ for $2 < q < \infty$,

$$\|\mathbf{u}(\cdot,t)\|_q \le C_p \|\omega(\cdot,t)\|_p \le \xi_p \|\omega_0\|_p\,, \quad 1 < p < 2, \quad \frac{1}{q} = \frac{1}{p} - \frac{1}{2}.$$

Remark 3.2 asserts that we cannot expect a finite "energy" norm $\|\mathbf{u}(\cdot,t)\|_2$, even for compactly supported (smooth) initial vorticity.

3.3 Estimating derivatives

We again take $\omega_0 \in C_0^\infty(\mathbb{R}^2)$. By Theorem 2.1 the solution functions $\omega(\mathbf{x},t) = S\omega_0(t)$, $\mathbf{u}(\mathbf{x},t) = \mathbf{U}\omega_0(t)$ [see (3.1) and (3.2)] are smooth. We proceed to estimate the derivatives $\nabla\omega$, $\nabla\mathbf{u}$ in terms of the Y_p (or L^1) norm of the initial vorticity ω_0.

Lemma 3.6. *Let $\omega(\mathbf{x},t) = S\omega_0(t)$, $\mathbf{u}(\mathbf{x},t) = \mathbf{U}\omega_0(t)$ be the solution to (2.1)–(2.2). Then,*

(1) For every $p \in (2,\infty]$ there exists $T > 0$, depending on $\|\omega_0\|_{Y_p}$, such that,

$$\|\nabla\omega(\cdot,t)\|_{Y_p} \le Ct^{-1/2}\|\omega_0\|_{Y_p}\,, \quad 0 < t \le T, \tag{3.29}$$

where $C = C(p,\nu)$, but does not depend on ω_0.

(2) For every $\varepsilon > 0$ and $p \in (2,\infty]$ there exists a constant $E > 0$, depending only on ε, p, ν and $\|\omega_0\|_1$, such that

$$\|\nabla\omega(\cdot,t)\|_{Y_p} \le E\,, \quad \varepsilon < t < \infty. \tag{3.30}$$

Proof. Differentiating (3.13) we obtain

(3.31)

$$\nabla_{\mathbf{x}}\omega(\mathbf{x},t) = \int_{\mathbb{R}^2} \nabla_{\mathbf{x}} G_\nu(\mathbf{x}-\mathbf{y},t)\omega_0(\mathbf{y})d\mathbf{y}$$

$$- \int_0^t \int_{\mathbb{R}^2} \nabla_{\mathbf{x}} G_\nu(\mathbf{x}-\mathbf{y},t-s)\cdot(\mathbf{u}(\mathbf{y},s)\cdot\nabla_{\mathbf{y}})\omega(\mathbf{y},s)d\mathbf{y}ds.$$

In order to estimate $\nabla_{\mathbf{x}}\omega(\mathbf{x},t)$ we employ a method already used in obtaining (2.57). However, we insist on using only the Y_p norm of ω_0, and in particular avoiding any reference to $\nabla\omega_0$. Thus, a modification of this method is needed. In view of (3.28) we have, for every $p \in (2,\infty]$,

$$(3.32) \qquad A := \sup_{0\le t<\infty} \|\mathbf{u}(\cdot,t)\|_\infty \le C_p\|\omega_0\|_{Y_p},\ p\in(2,\infty],$$

so that, using (2.26) in (3.31), and denoting $N_p(t) = \sup_{0<\tau\le t}\|\nabla\omega(\cdot,\tau)\|_p$,

$$(3.33) \qquad N_p(t) \le C\Big[t^{-1/2}\|\omega_0\|_p + A\int_0^t (t-s)^{-1/2}N_p(s)ds\Big].$$

The generic constant $C>0$ will only depend on p,ν throughout the rest of the proof.

A similar inequality is obtained for $\|\nabla\omega(\cdot,t)\|_1$, namely

$$N_1(t) \le C\Big[t^{-1/2}\|\omega_0\|_1 + A\int_0^t (t-s)^{-1/2}N_1(s)ds\Big].$$

Thus, with $M_p(t) = N_1(t)+N_p(t)$ we have

$$(3.34) \quad M_p(t) \le C\Big[t^{-1/2}\|\omega_0\|_{Y_p} + A\int_0^t (t-s)^{-1/2}M_p(s)ds\Big], \quad p\in(2,\infty].$$

Defining $Q_p(t) = \sup_{0<\tau\le t}\tau^{\frac12}M_p(\tau)$ we get from this inequality

$$\begin{aligned} Q_p(t) &\le C\Big[\|\omega_0\|_{Y_p} + At^{\frac12}\int_0^t (t-s)^{-1/2}s^{-\frac12}Q_p(s)ds\Big] \\ &\le C\Big[\|\omega_0\|_{Y_p} + At^{\frac12}Q_p(t)\int_0^t (t-s)^{-1/2}s^{-1/2}ds\Big] \\ &= C\Big[\|\omega_0\|_{Y_p} + A\pi t^{\frac12}Q_p(t)\Big], \quad p\in(2,\infty]. \end{aligned}$$

Let $T>0$ be such that $CA\pi T^{\frac12} < \frac12$. Then the final inequality implies,

$$\|\nabla\omega(\cdot,t)\|_{Y_p} \le Ct^{-1/2}\|\omega_0\|_{Y_p}, \quad 0<t\le T,\ p\in(2,\infty],$$

which establishes the first statement of the lemma.

To establish the second statement of the lemma, we shift the initial time to $t=\varepsilon$, and take $T>0$ as in the first statement of the lemma, where now it depends on $\|\omega(\cdot,\varepsilon)\|_{Y_p}$. Note that by (3.27) we have $\|\omega(\cdot,T+\varepsilon)\|_{Y_p} \le \|\omega(\cdot,\varepsilon)\|_{Y_p}$, so the first statement applies also to the time interval $[T+\varepsilon, T+2\varepsilon]$ and so on.

By (3.9)

$$\|\omega(\cdot,\varepsilon)\|_p \le (\nu\eta\varepsilon)^{-1+\frac1p}\|\omega_0\|_1,$$

so that

$$\|\omega(\cdot,\varepsilon)\|_{Y_p} \le [1+(\nu\eta\varepsilon)^{-1+\frac{1}{p}}]\|\omega_0\|_1,$$

which implies that in fact T depends only on $\|\omega_0\|_1$ and ε (in addition to p and ν).

Advancing from $t=\varepsilon$ by steps of size $\theta=\min(\varepsilon,T/2)$, we have for every integer $k=0,1,2...$, by (3.29),

$$\|\nabla\omega(\cdot,t)\|_{Y_p} \le C\big(t-\frac{k}{2}\theta\big)^{-1/2}\|\omega(\cdot,\varepsilon)\|_{Y_p} \le E, \quad \varepsilon+\frac{k+1}{2}\theta < t \le \varepsilon+\frac{k+2}{2}\theta,$$

where $E=C[1+(\nu\eta\varepsilon)^{-1+\frac{1}{p}}]\|\omega_0\|_1\theta^{-\frac{1}{2}}$, which proves (3.30). ∎

We note that the estimate (3.29) gives bounds both for $\|\nabla\omega(\cdot,t)\|_p$ and $\|\nabla\omega(\cdot,t)\|_1$, which are linear in $\|\omega_0\|_{Y_p}$. This will turn out to be most useful when dealing with the extension of the space of initial data in the following chapters.

On the other hand, these estimates are *local* in time, valid only in the interval $(0,T]$. This reflects the fact that (3.29) is not *"scale invariant."* The meaning of this will be clarified in the proof of the following lemma, which gives such a scale invariant estimate.

Lemma 3.7. *Let $\omega(\mathbf{x},t)=S\omega_0(t)$, $\mathbf{u}(\mathbf{x},t)=\mathbf{U}\omega_0(t)$ be the solution to (2.1)–(2.2). Then for every $p\in[1,\infty]$,*

(3.35) $$\|\nabla\omega(\cdot,t)\|_p \le \Upsilon t^{-\frac{3}{2}+\frac{1}{p}}, \quad t>0,\ p\in[1,\infty],$$

where $\Upsilon>0$ depends (nonlinearly) only on $\|\omega_0\|_1$ (and on p, ν).

Proof. In the second statement of Lemma 3.6 we take $\varepsilon=1$ and $p=\infty$, and let Υ be the corresponding value of E in the estimate (3.30), so that

(3.36) $$\|\nabla\omega(\cdot,\tau)\|_{Y_\infty} \le \Upsilon, \quad \tau\ge 1,$$

where Υ depends on $\|\omega_0\|_1$ and on ν.

A very useful tool available for the study of solutions to the vorticity equation (2.1)–(2.2) is the **scaling invariance** of the equation. Specifically, it is easily verified that if $\omega(\mathbf{x},t)$, $\mathbf{u}(\mathbf{x},t)$ is a solution, then so is $\lambda^2\omega(\lambda\mathbf{x},\lambda^2 t)$, $\lambda\mathbf{u}(\lambda\mathbf{x},\lambda^2 t)$, for all $\lambda>0$, subject to the initial condition $\lambda^2\omega(\lambda\mathbf{x},0)$. Observe that the $L^1(\mathbb{R}^2)$ norm of the initial vorticity remains unchanged. In terms of the scaled solution, the estimate (3.36) becomes

$$\lambda^3\|(\nabla\omega)(\lambda\mathbf{x},\lambda^2\tau)\|_1 \le \Upsilon, \quad \tau\ge 1,$$
$$\lambda^3\|(\nabla\omega)(\lambda\mathbf{x},\lambda^2\tau)\|_\infty \le \Upsilon, \quad \tau\ge 1.$$

Note that the above norms are with respect to the $\mathbf{x}$ variable.

We now fix $\tau = 1$ and let $\lambda^2 = t > 0$. The above estimates yield (3.35) for $p = 1, \infty$, and the interpolation inequality $\|g\|_p \le \|g\|_\infty^{1-\frac{1}{p}} \|g\|_1^{\frac{1}{p}}$ implies this estimate for all $1 \le p \le \infty$. Since $\lambda > 0$ is arbitrary we conclude that (3.35) holds for all $t > 0$. ■

Our next step is to estimate $\nabla \mathbf{u}$ (by which we mean the four functions $\partial_{x^i} u^j(\mathbf{x}, t)$, $1 \le i, j \le 2$). Contrary to the estimates in Lemma 3.7, the following estimate is *linear* in $\|\omega_0\|_1$.

Lemma 3.8. *Let $\omega_0 \in C_0^\infty(\mathbb{R}^2)$ and let $\omega(\mathbf{x}, t) = S\omega_0(t)$, $\mathbf{u}(\mathbf{x}, t) = \mathbf{U}\omega_0(t)$ be the solution to* (2.1)–(2.2). *Then*

$$\|\nabla \mathbf{u}(\cdot, t)\|_p \le C t^{-1+\frac{1}{p}} \|\omega_0\|_1 , \quad 1 < p < \infty, \quad t > 0, \tag{3.37}$$

where $C > 0$ depends only on p and ν. In addition

$$\|\nabla \mathbf{u}(\cdot, t)\|_\infty \le \Upsilon t^{-1}, \quad 0 < t < \infty, \tag{3.38}$$

where $\Upsilon > 0$ depends (nonlinearly) only on $\|\omega_0\|_1$ (and on ν).

Proof. When establishing the estimate (3.11) we alluded to the basic ingredient needed in the proof, namely, the fact that $\nabla \mathbf{K}$ is a Calderón–Zygmund kernel (see Appendix A, Section A.1). Differentiation of (2.2) yields

$$\nabla_{\mathbf{x}} \mathbf{u}(\mathbf{x}, t) = (\nabla \mathbf{K} * \omega)(\mathbf{x}, t) = \int_{\mathbb{R}^2} \nabla_{\mathbf{x}} \mathbf{K}(\mathbf{x} - \mathbf{y}) \omega(\mathbf{y}, t) d\mathbf{y},$$

which entails

$$\|\nabla \mathbf{u}(\cdot, t)\|_p \le C \|\omega(\cdot, t)\|_p, \quad 1 < p < \infty.$$

The estimate (3.37) follows by applying (3.9) to the right-hand side.

To prove (3.38) we use $\nabla \mathbf{u}(\cdot, t) = \mathbf{K} * \nabla \omega(\cdot, t)$ and combine the linear estimate (3.28) and (3.35) (with $p = 1$ and $p = \infty$) to get, at $t = 1$,

$$\|\nabla \mathbf{u}(\cdot, 1)\|_\infty \le C \|\nabla \omega(\cdot, 1)\|_{Y_\infty} \le C\Upsilon.$$

We now invoke the scaling argument introduced in the proof of Lemma 3.7. This allows us to replace $\mathbf{u}(\mathbf{x}, 1)$ with $\lambda \mathbf{u}(\lambda \mathbf{x}, \lambda^2)$, for any $\lambda > 0$. We therefore get

$$\lambda^2 \|\nabla \mathbf{u}(\cdot, \lambda^2)\|_\infty \le \Upsilon.$$

Letting $\lambda^2 = t > 0$ we obtain (3.38). ■

We collect the estimates obtained here in the following frame (retaining their equation numbers above), where we employ the solution operators in (3.1) and (3.2).

(3.29) $\|\nabla(S\omega_0)(t)\|_{Y_p} \le Ct^{-1/2}\|\omega_0\|_{Y_p}$, $p > 2, 0 < t \le T = T(\|\omega_0\|_{Y_p})$,
(3.35) $\|\nabla(S\omega_0)(t)\|_p \le \Upsilon(\|\omega_0\|_1)t^{-\frac{3}{2}+\frac{1}{p}}, t > 0,\ p \in [1,\infty]$,
(3.37) $\|\nabla(\mathbf{U}\omega_0)(t)\|_p \le Ct^{-1+\frac{1}{p}}\|\omega_0\|_1, \quad 1 < p < \infty, \quad t > 0$,
(3.38) $\|\nabla(\mathbf{U}\omega_0)(t)\|_\infty \le \Upsilon(\|\omega_0\|_1)t^{-1}, \quad 0 < t < \infty$.

3.4 Notes for Chapter 3

- Note the similarity of the estimates (3.16) for the heat (linear) equation and (3.17) for the vorticity (nonlinear) equation.
- The estimate (3.7) was first derived by Cottet [49]. His method of proof is identical to the one given here, except that the Nash inequality is not explicitly invoked, but basically proved again, using Fourier representation and relying on a similar method introduced by Schonbek [160].
- The L^1-L^∞ decay estimate (3.8) is interesting. It was derived in [9,108], where the constant $\eta > 0$ is the "best constant" in the Nash inequality. As pointed out in [34], $\eta \approx 3.67\pi$ whereas the corresponding estimate for the heat kernel is $\|G_\nu(\cdot,t)\|_\infty = (4\pi\nu t)^{-1}$. This was improved by Carlen and Loss [35], replacing η by 4π. The proof of this improvement [35, Theorem 5] was obtained by using a logarithmic Sobolev inequality instead of the Nash inequality. Thus, quite surprisingly, in spite of the nonlinearity, the L^∞ estimate for $\omega(\cdot,t)$ (in terms of $\|\omega_0\|_1$) is identical to that of the linear heat solution. Observe, however, that *radial* solutions of (2.1) are also solutions of the heat equation, since the nonlinear term vanishes identically. It also follows [35, Theorem 2], that η in (3.10) can be replaced by 4π.
- Observe that only the classical estimates for the heat kernel have been used. A similar approach was used by Kato [108], using also the classical heat kernel but different functional spaces. Kato derives (3.9), but not (3.12), (3.17) and (3.23), which are essential in the uniqueness proof here.
- The estimates in Subsection 3.1.1 are crucial in the proof of the *well-posedness* of (2.1)–(2.2) in $L^1(\mathbb{R}^2)$, hence also in the case of measure-valued initial data, studied in Chapters 4 and 5. They are mostly taken from [9].

- The linear (in terms of $\|\omega_0\|_1$) estimate (3.10) was obtained in [9]. Giga, Miyakawa and Osada [79, Section 3] and Kato [108, Remark in Introduction] derived an estimate of the form $t^{\frac{1}{2}}\|\mathbf{u}(\cdot,t)\|_\infty \leq C(\|\omega_0\|_1)$, where in general C is a nonlinear function.
- Continuing the previous note, the estimate (3.29) is *linear* in ω_0. It was derived in [9]. It applies only to $p > 2$ and only in a time interval $[0,T]$ depending on the initial norm, but gives an estimate also for $\|\nabla\omega(\cdot,t)\|_1$.
  Similarly, the estimate (3.37) for $\|\nabla\mathbf{u}(\cdot,t)\|_p$, $1 < p < \infty$, is linear in $\|\omega_0\|_1$.
- The estimates in Section 3.3 can be extended to all derivatives $\partial_t^k\nabla^{\boldsymbol{\alpha}}\omega(\cdot,t)$ and $\partial_t^k\nabla^{\boldsymbol{\alpha}}\mathbf{u}(\cdot,t)$ by differentiating repeatedly Equation (3.31). This is actually the method used in the proof of Lemma 2.6.
  Kato derived estimates of the form

$$\|\partial_t^k\nabla^{\boldsymbol{\alpha}}\omega(\cdot,t)\|_p \leq C(\|\omega_0\|_1)t^{-k-|\boldsymbol{\alpha}|/2-1+\frac{1}{p}}, \quad t > 0,\ p \in (1,\infty],$$

  and

$$\|\partial_t^k\nabla^{\boldsymbol{\alpha}}\mathbf{u}(\cdot,t)\|_p \leq C(\|\omega_0\|_1)t^{-k-|\boldsymbol{\alpha}|/2-1/2+\frac{1}{p}}, \quad t > 0,\ p \in (1,\infty],$$

  for every integer k and double-index $\boldsymbol{\alpha}$. However C is a nonlinear function [108, Equations (0.4)–(0.5)]. When $k = 0$, $|\boldsymbol{\alpha}| = 1$ these are the estimates (3.35) and (3.38) except that the case $p = 1$ is also included in our estimate.
  The proof given here is very different from Kato's proof, and relies on the scaling properties of the vorticity equation.
  We do not derive here estimates for higher derivatives, since the details of such estimates will not be needed later. On the other hand, we shall use a similar method in the next chapter, showing the smoothness of the solution even when the initial data is "rough."
- The estimates for the gradient of the vorticity, as given in Section 3.3, are also derived in Chapter 2 of the recent book [77] by Giga, Giga and Saal. While our method of proof is based on a scaling argument, their proof uses a representation of the vorticity in terms of the Duhamel integral and utilizes the fact that the velocity field is divergence-free.

Chapter 4

Extension of the Solution Operator

We shall now study the extension of the solution operators introduced in (3.1)–(3.2),

$$S\omega_0(t), \ \mathbf{U}\omega_0(t), \ t \in \overline{\mathbb{R}_+} = [0, \infty),$$

to initial data $\omega_0 \in L^1(\mathbb{R}^2)$. Our goal is to show that the system (2.1)–(2.2) is well-posed in $L^1(\mathbb{R}^2)$.

As in the case of the heat equation, we shall see that the solution "regularizes" for positive time.

Recall that the space

$$Y_\infty = L^1(\mathbb{R}^2) \cap D^0(\mathbb{R}^2)$$

was introduced in (2.61). This space was actually used in the study of vorticity by Marchioro and Pulvirenti [134] in their treatment of "diffusive vortices" (approximation by finite-dimensional diffusion processes). In addition to the interest in Y_∞ as a "persistence" space for vorticity, some basic estimates in this space serve in the study of the "zero viscosity" limit, being independent of $\nu > 0$ [9].

More generally, we shall use the "intermediate" spaces Y_p introduced in (3.26). Recall (Lemma 3.5) that the convolution operator $\mathbf{K}* : Y_p \to D^0(\mathbb{R}^2)$ is bounded, for $p \in (2, \infty]$. This will enable us to prove the well posedness of the system (2.1)–(2.2) in these spaces, as an intermediate step towards the extension to $L^1(\mathbb{R}^2)$.

4.1 An intermediate extension

Let us first consider the situation in Theorem 2.1, with initial data $\omega_0 \in C_0^\infty(\mathbb{R}^2)$. The regularity and decay results stated there, in conjunction with the estimates (2.3) and (3.28) imply that, for every $p \in (2, \infty]$,

$$S\omega_0(t) \in C(\overline{\mathbb{R}_+}; Y_p),$$
$$\mathbf{U}\omega_0(t) \in C(\overline{\mathbb{R}_+}; D^0(\mathbb{R}^2)).$$

Furthermore, in view of Lemma 3.6 [see (3.29)] for every $T > 0$,

$$\nabla(S\omega_0)(t) \in C((0,T]; Y_p) \cap L^q([0,T]; Y_p), \quad q \in [1,2).$$

We note that these continuity and integrability properties involve only the Y_p norm of ω_0. This enables us to extend the solution operators to initial data in Y_p.

Observe that the properties of $S\omega_0(t)$ can be summarized as (see [58, Chapter 5] for basic definitions of Sobolev spaces),

$$\begin{aligned} S\omega_0(t) &\in C(\overline{\mathbb{R}_+}; Y_p) \cap C(\mathbb{R}_+; W^{1,1}(\mathbb{R}^2) \cap W^{1,p}(\mathbb{R}^2)) \\ &\cap L^q_{loc}(\overline{\mathbb{R}_+}; W^{1,1}(\mathbb{R}^2) \cap W^{1,p}(\mathbb{R}^2)), \quad q \in [1,2),\ p \in (2,\infty]. \end{aligned} \tag{4.1}$$

Generally speaking, these extended solutions are singular (at least) near $t = 0$, in exact analogy to the heat equation when the initial data are non-smooth functions in $L^1(\mathbb{R}^2)$. Thus, we must first consider a more general definition of the solution. We encountered a similar situation when **defining** the solution to a convection–diffusion equation by (2.14). Here our basis for the extended solution is the integral form (3.13) of the vorticity equation. We introduce the following definition.

Definition 4.1. *Let $p \in (2, \infty]$ and let*

$$\begin{aligned} \omega(\cdot, t) &\in C(\overline{\mathbb{R}_+}; Y_p) \cap C(\mathbb{R}_+; W^{1,1}(\mathbb{R}^2) \cap W^{1,p}(\mathbb{R}^2)) \\ &\cap L^q_{loc}(\overline{\mathbb{R}_+}; W^{1,1}(\mathbb{R}^2) \cap W^{1,p}(\mathbb{R}^2)), \quad q \in [1,2). \end{aligned}$$

Denote $\mathbf{u}(\cdot,t) = \mathbf{K} * \omega(\cdot,t) \in C(\overline{\mathbb{R}_+}; D^0(\mathbb{R}^2))$. We say that ω is a **mild solution** *of Equation (2.1) if it satisfies, in the sense of $C(\overline{\mathbb{R}_+}; Y_p)$,*

$$\begin{aligned} \omega(\mathbf{x}, t) &= \int_{\mathbb{R}^2} G_\nu(\mathbf{x} - \mathbf{y}, t)\omega(\mathbf{y}, 0)d\mathbf{y} \\ &+ \int_0^t \int_{\mathbb{R}^2} \nabla_{\mathbf{y}} G_\nu(\mathbf{x} - \mathbf{y}, t - s) \cdot \omega(\mathbf{y}, s)\mathbf{u}(\mathbf{y}, s)d\mathbf{y}ds, \quad t \in \mathbb{R}_+. \end{aligned} \tag{4.2}$$

The extension to initial data in Y_p is expressed in the following lemma.

Lemma 4.2. *Let $p \in (2, \infty]$. The operators S, $\mathbf{U}$ can be extended continuously as*

$$\begin{aligned} S &: Y_p \to C(\overline{\mathbb{R}_+}; Y_p) \\ \mathbf{U} &: Y_p \to C(\overline{\mathbb{R}_+}; D^0(\mathbb{R}^2)). \end{aligned} \tag{4.3}$$

In addition, the map ∇S *can be extended continuously as*

$$\nabla S : Y_p \to C(\mathbb{R}_+; Y_p) \cap L^q_{loc}(\mathbb{R}_+; Y_p), \quad q \in [1, 2), \tag{4.4}$$

and it satisfies the estimates of Lemma 3.6.

For any $\omega_0 \in Y_p$*, the function* $S\omega_0(t)$ *is the unique mild solution to* (2.1)*, subject to the initial condition* ω_0*, and it satisfies the contraction estimate* (3.27).

Proof. Given $R > 0$, we consider initial vorticities $\omega_0 \in C_0^\infty(\mathbb{R}^2)$ and $\theta_0 \in C_0^\infty(\mathbb{R}^2)$ such that $\|\omega_0\|_{Y_p} < R$, $\|\theta_0\|_{Y_p} < R$. Let $T = T(R) > 0$ be as in the first statement of Lemma 3.6. Note that the estimate (3.29) is applicable (in $[0, T]$) to all solutions with initial data in the ball of radius R, centered at the origin, in the space Y_p.

We fix R and T, and consider only initial data in the ball of radius R. Here and below the generic constant $C > 0$ may depend on p, ν and R, but not on individual initial data in the ball. As before, we may indicate parameters of dependence (for instance $C(p)$) if needed for clarity.

Let $\omega(\cdot, t) = S\omega_0(t)$ and $\theta(\cdot, t) = S\theta_0(t)$ be solutions to (2.1), with associated velocities $\mathbf{u} = \mathbf{K} * \omega$ and $\mathbf{v} = \mathbf{K} * \theta$. We have as in Equation (2.64),

$$\begin{aligned}\omega(\mathbf{x},t) - \theta(\mathbf{x}, t) &= \int_{\mathbb{R}^2} G_\nu(\mathbf{x} - \mathbf{y}, t)(\omega_0(\mathbf{y}) - \theta_0(\mathbf{y}))d\mathbf{y} \\ &+ \int_0^t \int_{\mathbb{R}^2} \nabla_{\mathbf{y}} G_\nu(\mathbf{x} - \mathbf{y}, t - s) \cdot (\omega(\mathbf{y}, s) - \theta(\mathbf{y}, s))\mathbf{u}(\mathbf{y}, s) d\mathbf{y} ds \\ &+ \int_0^t \int_{\mathbb{R}^2} \nabla_{\mathbf{y}} G_\nu(\mathbf{x} - \mathbf{y}, t - s) \cdot \theta(\mathbf{y}, s)(\mathbf{u}(\mathbf{y}, s) - \mathbf{v}(\mathbf{y}, s)) d\mathbf{y} ds.\end{aligned} \tag{4.5}$$

By Lemma 3.5 we have

$$A := \sup_{0 \le t \le T} \|\mathbf{u}(\cdot, t)\|_\infty + \sup_{0 \le t \le T} \|\mathbf{v}(\cdot, t)\|_\infty \le C(\|\omega_0\|_{Y_p} + \|\theta_0\|_{Y_p}), \tag{4.6}$$

and also

$$\|\mathbf{u}(\cdot, t) - \mathbf{v}(\cdot, t)\|_\infty \le C\|\omega(\cdot, t) - \theta(\cdot, t)\|_{Y_p}. \tag{4.7}$$

Recall that by the monotonicity (3.27) we have $\|\theta(\cdot, t)\|_{Y_p} \le \|\theta_0\|_{Y_p}$. Hence, taking into account (2.15) and (2.26), we obtain from (4.5), using Young's inequality (see Appendix A, Section A.2),

$$
\begin{aligned}
&\|\omega(\cdot,t)-\theta(\cdot,t)\|_{Y_p} \le \|\omega_0-\theta_0\|_{Y_p} \\
&\quad + A\beta_\nu \int_0^t (t-s)^{-\frac{1}{2}} \|\omega(\cdot,s)-\theta(\cdot,s)\|_{Y_p} ds \\
&\quad + C\beta_\nu \|\theta_0\|_{Y_p} \int_0^t (t-s)^{-\frac{1}{2}} \|\omega(\cdot,s)-\theta(\cdot,s)\|_{Y_p} ds \\
&\le \|\omega_0-\theta_0\|_{Y_p} + M \int_0^t (t-s)^{-\frac{1}{2}} \|\omega(\cdot,s)-\theta(\cdot,s)\|_{Y_p} ds,
\end{aligned} \tag{4.8}
$$

where $M = A\beta_\nu + C\beta_\nu \|\theta_0\|_{Y_p}$.
This inequality is similar to (2.57) and we proceed in a similar fashion.

Define

$$N(s) = \sup_{0\le\tau\le s} \|\omega(\cdot,\tau)-\theta(\cdot,\tau)\|_{Y_p},$$

and let $Q(s) = e^{-Ls}N(s)$, with $L > 0$ to be determined. Then (4.8) can be rewritten as

$$Q(t) \le Q(0) + M \int_0^t (t-s)^{-\frac{1}{2}} e^{-L(t-s)} Q(s) ds.$$

If $L = L(M,T)$ is taken so large that $M\int_0^T s^{-\frac{1}{2}} e^{-Ls} ds < \frac{1}{2}$, it follows that $Q(t) \le 2Q(0)$, or

$$\|\omega(\cdot,t)-\theta(\cdot,t)\|_{Y_p} \le 2e^{Lt}\|\omega_0-\theta_0\|_{Y_p}, \quad t\in[0,T]. \tag{4.9}$$

Note that the constant L depends (in terms of M) on $\|\omega_0\|_{Y_p} + \|\theta_0\|_{Y_p}$. Hence the last inequality means that the map

$$S : C_0^\infty(\mathbb{R}^2) \to C([0,T]; Y_p)$$

is locally (namely, in every ball) uniformly continuous with respect to the Y_p norm. Since $C_0^\infty(\mathbb{R}^2)$ is dense in Y_p (note that for $p=\infty$ we defined $Y_\infty = L^1(\mathbb{R}^2)\cap D^0(\mathbb{R}^2)$), it follows that S can indeed be extended to all of Y_p, as a continuous map

$$S : Y_p \to C([0,T]; Y_p).$$

In order to prove the statement concerning ∇S we turn to the estimate of $\nabla\omega(\cdot,t)-\nabla\theta(\cdot,t)$. Recall [see (3.31)] that,

$$\begin{aligned}
&\nabla_{\mathbf{x}}\omega(\mathbf{x},t)-\nabla_{\mathbf{x}}\theta(\mathbf{x},t)\\
&\quad=\int_{\mathbb{R}^2}\nabla_{\mathbf{x}}G_\nu(\mathbf{x}-\mathbf{y},t)(\omega_0(\mathbf{y})-\theta_0(\mathbf{y}))d\mathbf{y}\\
&\quad-\int_0^t\int_{\mathbb{R}^2}\nabla_{\mathbf{x}}G_\nu(\mathbf{x}-\mathbf{y},t-s)\\
&\quad\cdot[(\mathbf{u}(\mathbf{y},s)\cdot\nabla_{\mathbf{y}})\omega(\mathbf{y},s)-(\mathbf{v}(\mathbf{y},s)\cdot\nabla_{\mathbf{y}})\theta(\mathbf{y},s)]d\mathbf{y}ds\\
&\quad=\int_{\mathbb{R}^2}\nabla_{\mathbf{x}}G_\nu(\mathbf{x}-\mathbf{y},t)(\omega_0(\mathbf{y})-\theta_0(\mathbf{y}))d\mathbf{y}\\
&\quad-\int_0^t\int_{\mathbb{R}^2}\nabla_{\mathbf{x}}G_\nu(\mathbf{x}-\mathbf{y},t-s)\\
&\quad\cdot(\mathbf{u}(\mathbf{y},s)\cdot\nabla_{\mathbf{y}})(\omega(\mathbf{y},s)-\theta(\mathbf{y},s))d\mathbf{y}ds\\
&\quad-\int_0^t\int_{\mathbb{R}^2}\nabla_{\mathbf{x}}G_\nu(\mathbf{x}-\mathbf{y},t-s)\\
&\quad\cdot[(\mathbf{u}(\mathbf{y},s)-\mathbf{v}(\mathbf{y},s))\cdot\nabla_{\mathbf{y}})\theta(\mathbf{y},s)]d\mathbf{y}ds.
\end{aligned}\tag{4.10}$$

We apply Young's inequality (see Appendix A, Section A.2). The norm $\|\nabla G_\nu(\cdot,t)\|_1$ is evaluated by (2.26) and the sup-norms of the velocity terms are estimated by (4.6)–(4.7). We have

$$\begin{aligned}
&\|\nabla\omega(\cdot,t)-\nabla\theta(\cdot,t)\|_{Y_p}\\
&\quad\le C\Big\{\|\omega_0-\theta_0\|_{Y_p}t^{-\frac{1}{2}}\\
&\quad+\|\omega_0\|_{Y_p}\int_0^t(t-s)^{-\frac{1}{2}}\|\nabla\omega(\cdot,s)-\nabla\theta(\cdot,s)\|_{Y_p}ds\\
&\quad+\int_0^t(t-s)^{-\frac{1}{2}}\|\omega(\cdot,s)-\theta(\cdot,s)\|_{Y_p}\|\nabla\theta(\cdot,s)\|_{Y_p}ds\Big\}.
\end{aligned}$$

The term $\|\nabla\theta(\cdot,s)\|_{Y_p}$ can be estimated by (3.29) and $\|\omega(\cdot,s)-\theta(\cdot,s)\|_{Y_p}$ can be estimated by (4.9), so that

$$\begin{aligned}\|\nabla\omega(\cdot,t) &-\nabla\theta(\cdot,t)\|_{Y_p}\\ &\le C\Big\{\|\omega_0-\theta_0\|_{Y_p}t^{-\frac12}\\ &\quad+\|\omega_0\|_{Y_p}\int_0^t (t-s)^{-\frac12}\|\nabla\omega(\cdot,s)-\nabla\theta(\cdot,s)\|_{Y_p}ds\\ &\quad+2e^{LT}\|\omega_0-\theta_0\|_{Y_p}\|\theta_0\|_{Y_p}\int_0^t (t-s)^{-\frac12}s^{-\frac12}ds\Big\}.\end{aligned}$$

The third term on the right-hand side can be absorbed into the first if the constant $C>0$ is replaced by $C+C_1$, so that

$$C_1T^{-\frac12}\ge 2Ce^{LT}\|\theta_0\|_{Y_p}\int_0^1 (1-s)^{-\frac12}s^{-\frac12}ds.$$

Retaining the notation C for the generic constant (depending on p,ν and R), we finally get

$$\begin{aligned}\|\nabla\omega(\cdot,t) &-\nabla\theta(\cdot,t)\|_{Y_p}\\ &\le C\Big\{\|\omega_0-\theta_0\|_{Y_p}t^{-\frac12}\\ &\quad+\int_0^t (t-s)^{-\frac12}\|\nabla\omega(\cdot,s)-\nabla\theta(\cdot,s)\|_{Y_p}ds\Big\}.\end{aligned}\tag{4.11}$$

We deal with this inequality by following the procedure employed in the treatment of (3.34). This yields,

$$\|\nabla\omega(\cdot,t)-\nabla\theta(\cdot,t)\|_{Y_p}\le C\|\omega_0-\theta_0\|_{Y_p}t^{-\frac12},\quad 0<t\le T.\tag{4.12}$$

We conclude that the map

$$\nabla S: C_0^\infty(\mathbb{R}^2)\to C([0,T];Y_p)\cap L^q_{loc}([0,T];Y_p),\quad q\in[1,2),$$

is locally uniformly continuous with respect to the Y_p norm, and hence it can be extended to all of Y_p.

In light of Lemma 3.5, the associated velocity $\mathbf{U}\omega(\cdot,t)=\mathbf{K}*\omega(\cdot,t)$ can indeed be extended as

$$\mathbf{U}: Y_p\to C([0,T];D^0(\mathbb{R}^2)).$$

We will now show that the function $S\omega_0(t)$, $\omega_0\in Y_p$, is a mild solution, as stated. We are (still) limited to the time interval $[0,T]$.

Take a sequence $\{\omega_0^{(j)}\}_{j=1}^\infty\subseteq C_0^\infty(\mathbb{R}^2)$ converging to ω_0 in Y_p. By (4.9), (4.12) and Lemma 3.5 we obtain $S\omega_0(t)=\lim_{j\to\infty}S\omega_0^{(j)}(t)$ and $\mathbf{U}\omega_0(t)=\lim_{j\to\infty}\mathbf{U}\omega_0^{(j)}(t)$, in the functional setting (4.3) and (4.4).

In particular, denoting $\omega^{(j)}(t) = S\omega_0^{(j)}(t)$ and $\mathbf{u}^{(j)}(t) = \mathbf{U}\omega_0^{(j)}(t)$, the sequence $\omega^{(j)}(t)\mathbf{u}^{(j)}(t)$ converges to $S\omega_0(t)\mathbf{U}\omega_0(t)$ in $C([0,T];Y_p)$. Since Equation (4.2) is satisfied by the pair $(\omega^{(j)}(t),\ \mathbf{u}^{(j)}(t))$ for $j = 1, 2, ...$ and $t \in [0,T]$, we can pass to the limit (using once again Young's inequality) and get

$$\omega(\mathbf{x},t) = \int_{\mathbb{R}^2} G_\nu(\mathbf{x}-\mathbf{y},t)\omega(\mathbf{y},0)d\mathbf{y} \tag{4.13}$$

$$+ \int_0^t \int_{\mathbb{R}^2} \nabla_{\mathbf{y}} G_\nu(\mathbf{x}-\mathbf{y},t-s)\cdot\omega(\mathbf{y},s)\mathbf{u}(\mathbf{y},s)d\mathbf{y}ds, \quad \mathbf{x}\in\mathbb{R}^2,\ t\in[0,T].$$

To conclude the proof of the lemma, we need to prove the stated uniqueness of the mild solution and replace the interval $[0,T]$ with the full $\overline{\mathbb{R}_+} = [0,\infty)$. In fact, these two goals are closely related, as we have already seen in the conclusion of the proof of Theorem 2.1 (see the discussion preceding (2.75)).

The monotonicity property (3.27) implies that $\|\omega(\cdot,T/2)\|_{Y_p} \le \|\omega_0\|_{Y_p}$. Thus, since $T > 0$ depends only on this norm, we obtain a mild solution also in the interval $[\frac{1}{2}T, \frac{3}{2}T]$. If uniqueness is established, this solution must coincide with the one constructed in $[0,T]$. Thus, the solution can be extended indefinitely to all $t \in \overline{\mathbb{R}_+}$.

To prove the uniqueness assertion, let $\theta(\mathbf{x},t)$, $(\mathbf{x},t)\in\mathbb{R}^2\times[0,T]$, be a mild solution such that $\theta(\mathbf{x},0) = \omega(\mathbf{x},0)$. Let $\mathbf{v} = \mathbf{K}*\theta$ be the associated velocity, so that (4.13) holds, with ω, $\mathbf{u}$ replaced by θ, $\mathbf{v}$. Subtracting the two equations and denoting

$$N(s) = \sup_{0\le\tau\le s} \|\omega(\cdot,\tau)-\theta(\cdot,\tau)\|_{Y_p},$$

we proceed as in the argument leading to (4.9), and conclude that $N(t)\equiv 0$. This establishes the uniqueness and completes the proof of the lemma. ■

4.2 Extension to Initial vorticity in $L^1(\mathbb{R}^2)$

We may now proceed to the main result of this chapter, concerning the extension of the solution operators S, $\mathbf{U}$ to all of $L^1(\mathbb{R}^2)$ (as initial data of the vorticity). We shall show that the vorticity equation (2.1) is well-posed in $L^1(\mathbb{R}^2)$, when properly interpreted in terms of mild solutions. To simplify the notation, we shall mostly use $\omega(t)$, $\mathbf{u}(t)$ instead of $\omega(\cdot,t)$, $\mathbf{u}(\cdot,t)$, respectively.

We first relax the requirements imposed on $\omega(t)$, $\mathbf{u}(t) = \mathbf{K} * \omega(t)$ in Definition 4.1 for mild solutions. We now assume only that

$$\omega(t) \in C(\overline{\mathbb{R}_+}; L^1(\mathbb{R}^2)) \cap C(\mathbb{R}_+; W^{1,1}(\mathbb{R}^2) \cap W^{1,\infty}(\mathbb{R}^2)),$$

(4.14)

$$\mathbf{u}(t) \in C(\mathbb{R}_+; D^0(\mathbb{R}^2)), \quad \mathbf{u}(t) = \mathbf{K} * \omega(t),\ t > 0.$$

Then $\omega(t)$ is a mild solution if Equation (4.2) holds for all $t > 0$, in the sense of $C(\overline{\mathbb{R}_+}; L^1(\mathbb{R}^2))$. In other words, the double integral on the right-hand side of (4.2) converges and is continuous as a function of $t \in \overline{\mathbb{R}_+}$ with values in $L^1(\mathbb{R}^2)$, and the equality (4.2) holds for all $t \geq 0$.

Observe that no continuity assumption is imposed on $\mathbf{u}(t)$ at $t = 0$, hence the convergence of the integral on the right-hand side of (4.2) needs to be verified in the proof.

Our extension theorem is the following.

Theorem 4.3.

(a) (Existence). The operators S, $\mathbf{U}$ can be extended continuously as

$$S : L^1(\mathbb{R}^2) \to C(\overline{\mathbb{R}_+}; L^1(\mathbb{R}^2)) \cap C(\mathbb{R}_+; W^{1,1}(\mathbb{R}^2) \cap W^{1,\infty}(\mathbb{R}^2)),$$

(4.15)

$$\mathbf{U} : L^1(\mathbb{R}^2) \to C(\mathbb{R}_+; D^0(\mathbb{R}^2)).$$

For every $\omega_0 \in L^1(\mathbb{R}^2)$ and $t > 0$ we have

$$(\mathbf{U}\omega_0)(t) = \mathbf{K} * (S\omega_0)(t),$$

and $\omega(t)$ is a mild solution of (2.1). Furthermore, the estimates (3.9) and (3.12) are valid, as well as

$$\|(S\omega_0)(t)\|_1 \leq \|\omega_0\|_1.$$

(b) (Uniqueness). Let $\theta(\mathbf{x}, t)$, $\mathbf{v}(\mathbf{x}, t)$, with $\mathbf{v}(\cdot, t) = \mathbf{K} * \theta(\cdot, t)$, $t > 0$, be such that,

$$\theta(t) \in C(\overline{\mathbb{R}_+}; L^1(\mathbb{R}^2)) \cap C(\mathbb{R}_+; L^\infty(\mathbb{R}^2)),$$

(4.16)

$$\theta(\mathbf{x}, 0) = \omega_0(\mathbf{x}) \in L^1(\mathbb{R}^2).$$

Assume that $\theta(\mathbf{x}, t)$ is a mild solution of (2.1) in $\mathbb{R}^2 \times \mathbb{R}_+$. Then $\theta(t) = (S\omega_0)(t)$ for all $0 \leq t < \infty$.

(c) (Regularity). For every $\omega_0 \in L^1(\mathbb{R}^2)$ the functions $\omega(t) = (S\omega_0)(t)$, $\mathbf{u}(t) = (\mathbf{U}\omega_0)(t)$ are in $C^\infty(\mathbb{R}^2 \times \mathbb{R}_+)$ and Equation (2.1) is satisfied in the

classical sense. Furthermore, for every integer k and double-index $\boldsymbol{\alpha}$, the maps

$$
\begin{aligned}
&\partial_t^k \nabla^{\boldsymbol{\alpha}} S : L^1(\mathbb{R}^2) \to C(\mathbb{R}_+; L^1(\mathbb{R}^2) \cap L^\infty(\mathbb{R}^2)), \\
&\partial_t^k \nabla^{\boldsymbol{\alpha}} \mathbf{U} : L^1(\mathbb{R}^2) \to C(\mathbb{R}_+; D^0(\mathbb{R}^2)),
\end{aligned} \tag{4.17}
$$

are continuous.

Proof. Let $\Lambda \subseteq C_0^\infty(\mathbb{R}^2)$ be precompact (in the L^1 topology). Using the fact that the heat kernel G_ν generates a semigroup [58, Section 7.4] it is clear that the family of maps

$$t \to G_\nu(\cdot, t) * \omega_0, \quad t \geq 0, \ \omega_0 \in \Lambda,$$

is equicontinuous in $L^1(\mathbb{R}^2)$. Recall that this means that, given $\varepsilon > 0$, there exists $\rho > 0$ such that, if $|t_1 - t_2| < \rho$, then

$$\|G_\nu(\cdot, t_2) * \omega_0 - G_\nu(\cdot, t_1) * \omega_0\|_1 < \varepsilon, \quad \text{for all } \omega_0 \in \Lambda.$$

We first establish an analogous statement for the family of vorticity trajectories with initial data in Λ.

Lemma 4.4. *The family*

$$t \to (S\omega_0)(t), \quad t \geq 0, \ \omega_0 \in \Lambda, \tag{4.18}$$

is equicontinuous in $L^1(\mathbb{R}^2)$.

Proof of the Lemma. The crucial element in the proof is the application of the "refined" estimates obtained in Subsection 3.1.1. Our starting point is the integral equation (4.2). Noting the estimates (3.9) and (3.12), combined with Lemma 4.2, the equicontinuity is obvious in every interval $[\varepsilon, \infty)$, $\varepsilon > 0$. We therefore need to study the behavior of the maps near $t = 0$.

In order to estimate the difference $S\omega_0(t) - \omega_0$ we need to estimate the double integral in (4.2). Using (2.3) with $p = 1$, (2.26) and (3.23) we obtain from Young's inequality (see Appendix A, Section A.2),

$$\|(S\omega_0)(t) - \omega_0\|_1 < \|G_\nu(\cdot, t) * \omega_0 - \omega_0\|_1 + \delta(t; \Lambda), \tag{4.19}$$

which converges to 0 (as $t \to 0$) uniformly in $\omega_0 \in \Lambda$. ■

We continue with the proof of the theorem. Let $\{\omega_0^{(n)}(\mathbf{x})\}_{n=1}^\infty \subseteq C_0^\infty(\mathbb{R}^2)$ converge to ω_0 in L^1. Taking $\Lambda = \{\omega_0^{(n)}(\mathbf{x})\}_{n=1}^\infty$ (which is clearly precompact in L^1) the foregoing argument yields the equicontinuity (in L^1)

of the trajectories $\omega^{(n)}(t) = S\omega_0^{(n)}(t)$. In what follows we prove the uniform convergence of these trajectories in $[0, T]$. The generic constant $C > 0$ depends on various parameters, such as p and ν, but not on the initial data.

Writing $\mathbf{u}^{(n)}(t) = \mathbf{U}\omega_0^{(n)}(t)$ we have from (4.2),

$$\omega^{(n)}(t) - \omega^{(m)}(t) = G_\nu(\cdot, t) * (\omega_0^{(n)} - \omega_0^{(m)})$$

$$(4.20)\quad \begin{aligned} &+ \int_0^t \nabla G_\nu(\cdot, t-s) * \mathbf{u}^{(n)}(s)(\omega^{(n)}(s) - \omega^{(m)}(s))ds \\ &+ \int_0^t \nabla G_\nu(\cdot, t-s) * (\mathbf{u}^{(n)}(s) - \mathbf{u}^{(m)}(s))\omega^{(m)}(s)ds \end{aligned}$$

$$= I_1(\cdot, t) + I_2(\cdot, t) + I_3(\cdot, t).$$

Let $p \in (1, 2)$. Since $\|G_\nu(\cdot, t)\|_p = Ct^{-1+\frac{1}{p}}$, Young's inequality entails,

$$(4.21)\qquad \|I_1(\cdot, t)\|_p \le Ct^{-1+\frac{1}{p}} \|\omega_0^{(n)} - \omega_0^{(m)}\|_1.$$

To obtain suitable bounds for the integrals I_2 and I_3, we invoke certain estimates obtained in Chapter 3, as follows.

From (3.23) we have, for $0 < s \le t$,

(4.22)

$$\begin{aligned} \|\mathbf{u}^{(n)}(s)(\omega^{(n)}(s) - \omega^{(m)}(s))\|_p &\le \|\mathbf{u}^{(n)}(s)\|_\infty \|\omega^{(n)}(s) - \omega^{(m)}(s)\|_p \\ &\le \delta(t; \Lambda) \cdot s^{-1/2} \|(\omega^{(n)}(s) - \omega^{(m)}(s))\|_p. \end{aligned}$$

From (3.11) (which is a bound for the linear operator $\mathbf{K}*$) we have

$$(4.23)\qquad \|\mathbf{u}^{(n)}(s) - \mathbf{u}^{(m)}(s)\|_q \le C\|\omega^{(n)}(s) - \omega^{(m)}(s)\|_p, \quad \frac{1}{q} = \frac{1}{p} - \frac{1}{2},$$

so that by Hölder's inequality and (3.17),

(4.24)

$$\begin{aligned} \|(\mathbf{u}^{(n)}(s) - \mathbf{u}^{(m)}(s))\omega^{(m)}(s)\|_p &\le C\|\mathbf{u}^{(n)}(s) - \mathbf{u}^{(m)}(s)\|_q \|\omega^{(m)}(s)\|_2 \\ &\le \delta(t; \Lambda) \cdot s^{-1/2} \|\omega^{(n)}(s) - \omega^{(m)}(s)\|_p. \end{aligned}$$

The estimates for $\|I_2(\cdot, t)\|_p$ and $\|I_3(\cdot, t)\|_p$ are now obtained by combining (4.22) and (4.24) with (2.26) and applying Young's inequality. Inserting these estimates in (4.20) we get, with $p \in (1, 2)$,

$$(4.25)\quad \begin{aligned} \|\omega^{(n)}(t) - \omega^{(m)}(t)\|_p &\le Ct^{-1+\frac{1}{p}} \|\omega_0^{(n)} - \omega_0^{(m)}\|_1 \\ &+ \delta(t; \Lambda) \int_0^t (t-s)^{-1/2} s^{-1/2} \|\omega^{(n)}(s) - \omega^{(m)}(s)\|_p ds. \end{aligned}$$

This inequality is similar to the one we encountered in (4.11) and, as was the case there, the treatment is similar to that accorded to (3.34). In fact, the presence of the vanishing term $\delta(t;\Lambda)$ (as $t \to 0$) makes it simpler and we proceed as follows.

Denoting $N(t) = \sup_{0\le\tau\le t} \tau^{1-\frac{1}{p}} \|\omega^{(n)}(\tau) - \omega^{(m)}(\tau)\|_p$, the inequality (4.25) can be rewritten as

$$N(t) \le C\|\omega_0^{(n)} - \omega_0^{(m)}\|_1 + \delta(t;\Lambda)t^{1-\frac{1}{p}} \int_0^t (t-s)^{-1/2} s^{-3/2+1/p} N(s)ds$$
$$\le C\|\omega_0^{(n)} - \omega_0^{(m)}\|_1 + \delta(t;\Lambda)N(t),$$

which implies, for $0 < t \le t^* = t^*(\Lambda)$,

$$\|\omega^{(n)}(t) - \omega^{(m)}(t)\|_p \le C\|\omega_0^{(n)} - \omega_0^{(m)}\|_1 \cdot t^{-1+\frac{1}{p}}, \quad 1 < p < 2. \tag{4.26}$$

Turning back to (4.20) we now obtain

$$\begin{aligned}\|\omega^{(n)}(t) - \omega^{(m)}(t)\|_1 \le C\Big\{&\|\omega_0^{(n)} - \omega_0^{(m)}\|_1 \\ &+ \int_0^t (t-s)^{-1/2}\|\mathbf{u}^{(n)}(s)(\omega^{(n)}(s) - \omega^{(m)}(s))\|_1 ds \\ &+ \int_0^t (t-s)^{-1/2}\|(\mathbf{u}^{(n)}(s) - \mathbf{u}^{(m)}(s))\omega^{(m)}(s)\|_1 ds\Big\}.\end{aligned} \tag{4.27}$$

To estimate the integrands on the right-hand side, take $p = \frac{4}{3}$ and use (3.9), (3.12), (4.23), and (4.26) in conjunction with Hölder's inequality,

$$\|\mathbf{u}^{(n)}(s)(\omega^{(n)}(s) - \omega^{(m)}(s))\|_1 \le \|\mathbf{u}^{(n)}(s)\|_4 \|\omega^{(n)}(s) - \omega^{(m)}(s)\|_{\frac{4}{3}}$$
$$\le Cs^{-\frac{1}{4}}\|\omega_0^{(n)}\|_1 \cdot s^{-\frac{1}{4}}\|\omega_0^{(n)} - \omega_0^{(m)}\|_1,$$

$$\|(\mathbf{u}^{(n)}(s) - \mathbf{u}^{(m)}(s))\omega^{(m)}(s)\|_1 \le \|\mathbf{u}^{(n)}(s) - \mathbf{u}^{(m)}(s)\|_4 \|\omega^{(m)}(s)\|_{\frac{4}{3}}$$
$$\le Cs^{-\frac{1}{4}}\|\omega_0^{(n)} - \omega_0^{(m)}\|_1 \cdot s^{-\frac{1}{4}}\|\omega_0^{(m)}\|_1.$$

Inserting these inequalities into (4.27) yields, for $0 < t \le t^*$,

$$\|\omega^{(n)}(t) - \omega^{(m)}(t)\|_1 \le C\|\omega_0^{(n)} - \omega_0^{(m)}\|_1, \tag{4.28}$$

where $C > 0$ depends on Λ.

Now fix $\tau \in (0, t^*)$. By (3.9) we have the estimate

$$\|\omega^{(n)}(\tau)\|_\infty \le (\nu\eta\tau)^{-1}\|\omega_0^{(n)}\|_1, \quad n = 1, 2, \ldots$$

Using the interpolation inequality

$$\|g\|_p \le \|g\|_\infty^{1-\frac{1}{p}} \|g\|_1^{\frac{1}{p}}$$

we infer from (4.28) that

$$\|\omega^{(n)}(\tau) - \omega^{(m)}(\tau)\|_p \xrightarrow[n,m\to\infty]{} 0, \quad 1 \le p \le \infty.$$

Thus

$$\|\omega^{(n)}(\tau) - \omega^{(m)}(\tau)\|_{Y_p} \xrightarrow[n,m\to\infty]{} 0, \quad 1 \le p \le \infty. \tag{4.29}$$

Lemma 4.2 implies that, taking $p > 2$, the limit function $\omega(t)$ exists for $t \in [\tau, \infty)$ and

$$\|\omega^{(n)}(t) - \omega(t)\|_{Y_p} \xrightarrow[n\to\infty]{} 0, \quad 2 < p \le \infty,\ t \in [\tau, \infty). \tag{4.30}$$

In particular, $\omega(t) \in C([\tau, \infty); Y_p)$ is a mild solution for all $p > 2$ (with initial data $\omega(\tau)$). Furthermore, the uniqueness result of Lemma 4.2 ensures that this limit is independent of τ; we obtain the same limit function $\omega(t)$ for different values of $\tau \in (0, t^*]$ (in the common half line in t). Hence,

$$\omega(t) \in C(\mathbb{R}_+; Y_p), \quad p > 2. \tag{4.31}$$

We can now complete the proof of the existence part (a) of the theorem.

First observe that since the functions $\left\{\omega^{(n)}(t)\right\}_{n=1}^{\infty}$ are continuous on $[0, t^*]$, the estimate (4.28) implies that the limit function $\omega(t)$ is uniformly continuous on $(0, t^*]$. It can therefore be extended as a continuous function on $[0, t^*]$, with $\omega(0) = \omega_0$. Combining this observation with Lemma 4.2, we conclude that the sequence $\omega^{(n)}(t) = S\omega_0^{(n)}(t)$ converges in $C(\overline{\mathbb{R}_+}; L^1(\mathbb{R}^2))$ to $\omega(t) \in C(\overline{\mathbb{R}_+}; L^1(\mathbb{R}^2))$. The lemma also implies that the sequence converges in

$$C(\mathbb{R}_+; W^{1,1}(\mathbb{R}^2) \cap W^{1,p}(\mathbb{R}^2)),\ p > 2, \tag{4.32}$$

and that the corresponding velocities $\mathbf{u}^{(n)}(t) = \mathbf{U}\omega_0^{(n)}(t)$ converge in

$$C(\mathbb{R}_+; D^0(\mathbb{R}^2)). \tag{4.33}$$

We retain the notation S and $\mathbf{U}$ for the extended operators, so that $\omega(t) = (S\omega_0)(t)$ and $\mathbf{u}(t) = \mathbf{K} * \omega(t) = \mathbf{U}\omega_0(t)$ are now defined for all $\omega_0 \in L^1(\mathbb{R}^2)$.

Of course, this will be justified only after we prove the uniqueness of these extensions.

We note that the convergence (4.29) implies that the estimate (3.17) extends to the limiting function $\omega(t)$. In fact, if $\Lambda \subseteq L^1(\mathbb{R}^2)$ is precompact, we can associate with every function in Λ a converging sequence of functions in $C_0^\infty(\mathbb{R}^2)$. The union of these sequences is again precompact (in $L^1(\mathbb{R}^2)$), so that,

$$\|S\omega_0(t)\|_p \le \delta(t;\Lambda)t^{-1+\frac{1}{p}}, \quad \omega_0 \in \Lambda, \quad 1 < p \le \infty, \tag{4.34}$$

for any $\Lambda \subseteq L^1(\mathbb{R}^2)$ that is precompact.

We see that, as in the case of the heat equation (see Lemma 3.3), the evolution operator S has the property that $t^{1-\frac{1}{p}}\|S\omega_0(t)\|_p$ **vanishes as** $t \to 0$, **uniformly on precompact subsets of** $L^1(\mathbb{R}^2)$. This observation will be crucial in the proof of uniqueness below.

From the above considerations it follows that

$$\boxed{\begin{array}{c}\text{All the estimates of}\\ S\omega_0(t),\ \mathbf{U}\omega_0(t)\\ \text{obtained in Chapter 3 hold true for}\\ \omega_0 \in L^1(\mathbb{R}^2).\end{array}} \tag{4.35}$$

To establish the uniqueness claim in part (b) of the theorem, let $\theta(t)$ satisfy the assumptions (4.16).

Assume first that $\omega_0 = \theta(0) \in L^1(\mathbb{R}^2) \cap L^\infty(\mathbb{R}^2)$. We are then in the framework of Lemma 4.2, where $\omega_0 \in Y_r$, for any $r < \infty$. Then $\theta(t)$ satisfies (4.2) (with ω, ω_0 and $\mathbf{u}$ replaced respectively by θ, ω_0 and $\mathbf{v}$) and repeating the argument used in the proof of (3.17) we get,

$$\|\theta(t)\|_\infty \le \delta(t)t^{-1}, \quad t > 0, \tag{4.36}$$

and as in (3.23),

$$\|\mathbf{v}(t)\|_\infty \le \delta(t)t^{-1/2}, \quad t > 0. \tag{4.37}$$

For the difference of $\omega(t) - (S\omega_0)(t)$ and $\theta(t)$, we get, as in (4.25)

$$\|\omega(t)-\theta(t)\|_p < \delta(t)\int_0^t (t-s)^{-1/2}s^{-1/2}\|\omega(s) - \theta(s)\|_p ds, \ 1 < p < 2. \tag{4.38}$$

We can now repeat the argument employed after (4.25) Define

$$N(t) = \sup_{0\le\tau\le t} \tau^{1-\frac{1}{p}}\|\omega(\tau) - \theta(\tau)\|_p,$$

so that

$$N(t) \le \delta(t)t^{1-\frac{1}{p}}N(t)\int_0^t (t-s)^{-1/2}s^{-3/2+1/p}ds = C\delta(t)N(t).$$

This implies $\omega(t) = \theta(t)$, $0 \le t \le t^*$, for some $t^* > 0$. We can then proceed stepwise in time to obtain $\omega(t) = \theta(t)$ for all $t > 0$.

This completes the proof of uniqueness if $\omega_0 \in L^1(\mathbb{R}^2) \cap L^\infty(\mathbb{R}^2)$.

Dropping the assumption $\omega_0 \in L^\infty(\mathbb{R}^2)$, we still have by hypothesis, for any $s > 0$, that $\theta(s) \in L^\infty(\mathbb{R}^2)$. Invoking the foregoing argument (with $\theta(s)$ as initial data) we obtain

$$(S\theta(s))(t) = \theta(t+s), \quad s > 0,\ t \ge 0. \tag{4.39}$$

Also, since by assumption $\theta(t) \in C(\overline{\mathbb{R}_+}; L^1(\mathbb{R}^2))$, the set $\Lambda = \{\theta(s), 0 < s \le 1\} \subseteq L^1(\mathbb{R}^2)$ is precompact. Hence, the crucial estimate (4.34) yields

$$\|\theta(t+s)\|_p \le \delta(t;\Lambda) t^{-1+\frac{1}{p}}, \quad 0 < s \le 1, \qquad 1 < p \le \infty. \tag{4.40}$$

Letting $s \to 0$ in this inequality (and recalling the continuity assumption on $\theta(t)$ for $t > 0$) we have, with $t > 0$,

$$\|\theta(t)\|_p \le \delta(t;\Lambda) t^{-1+\frac{1}{p}}, \quad 1 < p \le \infty, \tag{4.41}$$

and, in particular, we obtain (4.36) and as in (3.23), we get also (4.37). These estimates enable us to repeat the first part of the uniqueness proof above, starting with the estimate (4.38). This in turn leads to $\theta(t) = (S\omega_0)(t)$, $t > 0$, as above. This concludes the proof of the uniqueness part (b) of the theorem.

The continuity properties of the maps S and $\mathbf{U}$, as stated in part (a), are now immediate consequences of the foregoing considerations. Indeed, let $\omega_0 \in L^1(\mathbb{R}^2)$ and let $\{\omega_0^{(n)}(\mathbf{x})\}_{n=1}^\infty \subseteq L^1(\mathbb{R}^2)$ converge to ω_0 in L^1. Denote $\omega^{(n)}(t) = (S\omega_0^{(n)})(t)$ and $\omega(t) = (S\omega_0)(t)$. Then, as in the derivation of (4.28), noting that all estimates used there are valid due to (4.35), there exist constants $C > 0$, $t^* > 0$, such that,

$$\|\omega^{(n)}(t) - \omega(t)\|_1 \le C\|\omega_0^{(n)} - \omega_0\|_1, \ 0 \le t \le t^*. \tag{4.42}$$

Furthermore, for any $0 < \tau < t^*$ we have, as in (4.29),

$$\|\omega^{(n)}(\tau) - \omega(\tau)\|_{Y_p} \xrightarrow[n\to\infty]{} 0, \quad 1 \le p \le \infty. \tag{4.43}$$

Lemma 4.2 can now be implemented, thus completing the proof of the continuity properties.

We now turn to the proof of the regularity of $\omega = S\omega_0$ and $\mathbf{u} = \mathbf{U}\omega_0$, as stated in part (c) of the theorem.

In fact, the proof can be deduced from the properties of ω and $\mathbf{u}$ as given in (a) and classical results on the regularity of solutions to Navier–Stokes equations [163].

However, we provide a direct proof of the regularity claim, based on the straightforward methodology used in the proof of Lemma 2.6. A similar approach was used in proving the uniform estimates of Lemma 2.11. Unlike the previous parts of this proof, we do not worry too much about the precise blow-up rate of successive derivatives near $t = 0$. Instead, once we gain a certain order of regularity, we advance (in t) by a small $\varepsilon > 0$ in order to establish the existence of derivatives of the next order up.

Observe that if we knew that the velocity $\mathbf{u}(\mathbf{x}, t)$ is smooth, then Equation (2.1) would be identical to the linear equation (2.7) and we could directly apply Lemma 2.6 to obtain the regularity. But since here $\mathbf{u}$ is connected to ω, we need to prove simultaneously the regularity of both functions.

So, let $\omega_0 \in L^1(\mathbb{R}^2)$ and $\omega(t) = (S\omega_0)(t)$, $\mathbf{u}(t) = (\mathbf{U}\omega_0)(t)$, the unique solution obtained above.

Let $\varepsilon > 0$ be sufficiently small and set $\omega_\varepsilon(\mathbf{x}) = \omega(\mathbf{x}, \varepsilon)$ and $\mathbf{u}_\varepsilon(\mathbf{x}) = \mathbf{u}(\mathbf{x}, \varepsilon)$. We have by (4.32)

$$\omega_\varepsilon(\mathbf{x}) \in W^{1,1}(\mathbb{R}^2) \cap W^{1,\infty}(\mathbb{R}^2),$$

and applying (3.28) and (3.38) we get,

$$\mathbf{u}_\varepsilon(\mathbf{x}) \in W^{1,\infty}(\mathbb{R}^2).$$

Lemma 4.2 is thus applicable, yielding in particular that

$$\omega(t) \in C([\varepsilon, \infty); W^{1,1}(\mathbb{R}^2) \cap W^{1,\infty}(\mathbb{R}^2)),$$

where $\|\omega(\cdot, t)\|_{Y_\infty}$ is nonincreasing.

Also, by Lemma 3.7,

$$\sup_{2\varepsilon \le t < \infty} \{\|\nabla \omega(\cdot, t)\|_{Y_\infty}\} \le C, \tag{4.44}$$

where $C = C(\nu, \varepsilon, \|\omega_0\|_1)$. Applying Lemma 3.5 to $\mathbf{u}(t) = \mathbf{K} * \omega(t)$ and $\nabla \mathbf{u}(t) = \mathbf{K} * \nabla \omega(t)$ we have,

$$\mathbf{u}(t) \in C([\varepsilon, \infty); D^0(\mathbb{R}^2)),$$
$$\nabla \mathbf{u}(t) \in C([\varepsilon, \infty); D^0(\mathbb{R}^2)),$$

and from (3.12) and (3.38)

$$\sup_{2\varepsilon \le t < \infty} \{\|\mathbf{u}(\cdot, t)\|_\infty + \|\nabla \mathbf{u}(\cdot, t)\|_\infty\} \le C, \tag{4.45}$$

where $C = C(\nu, \varepsilon, \|\omega_0\|_1)$.

Given indices $j, \ell \in \{1,2\}$ we now consider, as in the linear case (2.29), a pair of equations for $\partial_{x^j}\omega$ and $\partial_{x^\ell}\partial_{x^j}\omega$ as follows (we write full dependence on the coordinates for clarity),

$$\begin{aligned}(4.46)\qquad \partial_{x^j}\omega(\mathbf{x},t) &= \int_{\mathbb{R}^2} G_\nu(\mathbf{x}-\mathbf{y},t-2\varepsilon)\partial_{y^j}\omega_{2\varepsilon}(\mathbf{y})d\mathbf{y} \\ &\quad - \int_{2\varepsilon}^t\int_{\mathbb{R}^2} G_\nu(\mathbf{x}-\mathbf{y},t-s)\partial_{y^j}[(\mathbf{u}(\mathbf{y},s)\cdot\nabla_{\mathbf{y}})\omega(\mathbf{y},s)]d\mathbf{y}ds,\end{aligned}$$

$$\begin{aligned}(4.47)\quad \partial_{x^\ell}\partial_{x^j}\omega(\mathbf{x},t) &= \int_{\mathbb{R}^2} \partial_{x^\ell}G_\nu(\mathbf{x}-\mathbf{y},t-2\varepsilon)\partial_{x^j}\omega_{2\varepsilon}(\mathbf{y})d\mathbf{y} \\ &\quad - \int_{2\varepsilon}^t\int_{\mathbb{R}^2} \partial_{x^\ell}G_\nu(\mathbf{x}-\mathbf{y},t-s)\partial_{x^j}[(\mathbf{u}(\mathbf{y},s)\cdot\nabla_{\mathbf{y}})\omega(\mathbf{y},s)]d\mathbf{y}ds.\end{aligned}$$

Using (4.44) and (4.45) in (4.46)–(4.47), we obtain the existence of $\partial_{x^\ell}\partial_{x^j}\omega(\mathbf{x},t)$ as in the case of (2.29). Furthermore, a derivation analogous to that of (3.29) yields, with $T>0$ (depending on $\varepsilon,\nu,\ ||\omega_\varepsilon||_{Y_\infty}$) and any double-index $\boldsymbol{\alpha}$, $|\boldsymbol{\alpha}|=2$,

$$(4.48)\qquad ||\nabla^{\boldsymbol{\alpha}}\omega(\cdot,t)||_{Y_\infty} \le C(\nu,\varepsilon)\cdot(t-2\varepsilon)^{-\frac{1}{2}},\quad t\in(2\varepsilon,T+2\varepsilon].$$

Also, by Lemma 3.5,

$$(4.49)\qquad \begin{aligned}||\nabla^{\boldsymbol{\alpha}}\mathbf{u}(\cdot,t)||_\infty &= ||\mathbf{K}*\nabla^{\boldsymbol{\alpha}}\omega(\cdot,t)||_\infty \\ &\le C(\nu,\varepsilon)\cdot(t-2\varepsilon)^{-\frac{1}{2}},\quad t\in(2\varepsilon,T+2\varepsilon].\end{aligned}$$

The monotonicity property (3.27) of the Y_∞ norm implies that, with the same T, the argument can be continued to the interval $[T+2\varepsilon, 2T+2\varepsilon]$ and so on, thus establishing the existence of the second-order derivatives for all $t>\varepsilon$.

We can actually get more explicit estimates of the second-order derivatives by noting (3.9), which implies $||\omega_\varepsilon||_{Y_\infty}\le C(\nu,\varepsilon)||\omega_0||_1$. Thus, in fact, $T=T(\varepsilon,\nu,\ ||\omega_0||_1)$ and the preceding argument (combined with (4.48) and (4.49)) yields, for any double-index $\boldsymbol{\alpha}$, $|\boldsymbol{\alpha}|=2$,

$$(4.50)\qquad ||\nabla^{\boldsymbol{\alpha}}\omega(\cdot,t)||_{Y_\infty}\le C(\nu,\varepsilon,||\omega_0||_1),\quad t\in(\varepsilon,\infty),$$

and

$$(4.51)\qquad ||\nabla^{\boldsymbol{\alpha}}\mathbf{u}(\cdot,t)||_\infty\le C(\nu,\varepsilon,||\omega_0||_1),\quad t\in(\varepsilon,\infty).$$

It is now clear how to prove the continuous differentiability of higher orders. Observe that the t-derivatives can be replaced by spatial derivatives using the vorticity equation (2.1) itself.

Finally, we proceed to prove the continuity of the maps (4.17).

Let $\theta_0 \in L^1(\mathbb{R}^2)$ and $\theta(t) = (S\theta_0)(t)$ and $\mathbf{v}(t) = (\mathbf{U}\theta_0)(t)$. Notice that by the regularity result above, in the interval $[\varepsilon, \infty)$ we are in the improved situation of Lemma 4.2; the initial values $\theta_\varepsilon(\mathbf{x}) = \theta(\mathbf{x}, \varepsilon)$ and ω_ε are in Y_∞.

Note in addition that all our estimates hold uniformly for initial data in precompact subsets of $L^1(\mathbb{R}^2)$, as explained in Subsection 3.1.1. Thus, even when we do not make an effort to indicate the precise dependence of various constants on the initial data, these constants are uniformly bounded on precompact subsets. In particular, this is the case when dealing with converging sequences of initial data, as was the case with the extension of the solution operators.

Replacing 2ε by ε in (4.47) and subtracting from it the analogous expression for $\partial_{x^\ell}\partial_{x^j}\theta$ [compare with (4.10)] we get

(4.52)

$$\begin{aligned}
&\partial_{x^\ell}\partial_{x^j}\omega(\mathbf{x},t) - \partial_{x^\ell}\partial_{x^j}\theta(\mathbf{x},t)\\
&= \int_{\mathbb{R}^2} \partial_{x^\ell}G_\nu(\mathbf{x}-\mathbf{y}, t-\varepsilon)\cdot(\partial_{y^j}\omega_\varepsilon(\mathbf{y}) - \partial_{y^j}\theta_\varepsilon(\mathbf{y}))d\mathbf{y}\\
&- \int_\varepsilon^t \int_{\mathbb{R}^2} \partial_{x^\ell}G_\nu(\mathbf{x}-\mathbf{y}, t-s)\partial_{y^j}[(\mathbf{u}(\mathbf{y},s)\cdot\nabla_\mathbf{y})(\omega(\mathbf{y},s)-\theta(\mathbf{y},s))]d\mathbf{y}ds\\
&- \int_\varepsilon^t \int_{\mathbb{R}^2} \partial_{x^\ell}G_\nu(\mathbf{x}-\mathbf{y}, t-s)\cdot\partial_{y^j}[(\mathbf{u}(\mathbf{y},s)-\mathbf{v}(\mathbf{y},s))\cdot\nabla_\mathbf{y}]\theta(\mathbf{y},s)d\mathbf{y}ds.
\end{aligned}$$

For the difference $\|\partial_{y^j}\omega_\varepsilon(\mathbf{y}) - \partial_{y^j}\theta_\varepsilon(\mathbf{y})\|_{Y_\infty}$ we have the estimate (4.12). Using (4.45) and (4.48) we obtain from (4.52), as in the derivation of (4.12), for every double-index $\boldsymbol{\alpha}$, $|\boldsymbol{\alpha}| = 2$, and $\varepsilon < t \le T$,

$$\begin{aligned}
&\|\nabla^{\boldsymbol{\alpha}}\omega(\cdot,t) - \nabla^{\boldsymbol{\alpha}}\theta(\cdot,t)\|_{Y_\infty}\\
&\le C(\nu,\varepsilon,T,\|\omega_\varepsilon\|_{Y_\infty},\|\theta_\varepsilon\|_{Y_\infty})\cdot(t-\varepsilon)^{-\frac{1}{2}}\|\omega_\varepsilon - \theta_\varepsilon\|_{Y_\infty},
\end{aligned} \tag{4.53}$$

hence also, by $\nabla^{\boldsymbol{\alpha}}\mathbf{u} = \mathbf{K} * \nabla^{\boldsymbol{\alpha}}\omega$, $\nabla^{\boldsymbol{\alpha}}\mathbf{v} = \mathbf{K} * \nabla^{\boldsymbol{\alpha}}\theta$ and (3.28), for $\varepsilon < t \le T$ and $|\boldsymbol{\alpha}| = 2$,

$$\begin{aligned}
&\|\nabla^{\boldsymbol{\alpha}}\mathbf{u}(\cdot,t) - \nabla^{\boldsymbol{\alpha}}\mathbf{v}(\cdot,t)\|_\infty\\
&\le C(\nu,\varepsilon,T,\|\omega_\varepsilon\|_{Y_\infty},\|\theta_\varepsilon\|_{Y_\infty})\cdot(t-\varepsilon)^{-\frac{1}{2}}\|\omega_\varepsilon - \theta_\varepsilon\|_{Y_\infty}.
\end{aligned} \tag{4.54}$$

Using Equation (2.1), as well as the estimates obtained in the proof of Lemma 4.2, it follows that, again for $\varepsilon < t \le T$,

$$\begin{aligned}
&\|\partial_t\omega(\cdot,t) - \partial_t\theta(\cdot,t)\|_{Y_\infty}\\
&\le C(\nu,\varepsilon,T,\|\omega_\varepsilon\|_{Y_\infty},\|\theta_\varepsilon\|_{Y_\infty})\cdot(t-\varepsilon)^{-\frac{1}{2}}\|\omega_0 - \theta_0\|_{Y_\infty}.
\end{aligned} \tag{4.55}$$

These estimates directly imply the continuity of the maps (4.17) in the cases $k = 0$, $|\boldsymbol{\alpha}| = 2$ or $k = 1$, $\boldsymbol{\alpha} = 0$. Using exactly the same procedure it is now clear how to establish the continuity for higher order derivatives. Indeed, since $\varepsilon \in (0,T)$ is arbitrary, we obtain the continuity of the maps

$$\partial_t^k \boldsymbol{\nabla}^{\boldsymbol{\alpha}} S : Y_\infty \to C(\mathbb{R}_+; L^1(\mathbb{R}^2) \cap L^\infty(\mathbb{R}^2)),$$

$$\partial_t^k \boldsymbol{\nabla}^{\boldsymbol{\alpha}} \mathbf{U} : Y_\infty \to C(\mathbb{R}_+; D^0(\mathbb{R}^2)).$$

On the other hand, as already proved [see (4.32)], for every $\varepsilon > 0$ the map

$$S : \omega_0 \in L^1(\mathbb{R}^2) \to (S\omega_0)(\varepsilon) \in Y_\infty,$$

is continuous. We therefore obtain the continuity of the maps

$$\partial_t^k \boldsymbol{\nabla}^{\boldsymbol{\alpha}} S : L^1(\mathbb{R}^2) \to C(\mathbb{R}_+; L^1(\mathbb{R}^2) \cap L^\infty(\mathbb{R}^2)),$$

$$\partial_t^k \boldsymbol{\nabla}^{\boldsymbol{\alpha}} \mathbf{U} : L^1(\mathbb{R}^2) \to C(\mathbb{R}_+; D^0(\mathbb{R}^2)).$$

Observe that taking into account the Sobolev embedding theorem [58, Chapter 5] we actually have the continuity of the maps

$$\partial_t^k \boldsymbol{\nabla}^{\boldsymbol{\alpha}} S : L^1(\mathbb{R}^2) \to C(\mathbb{R}_+; C_b(\mathbb{R}^2))$$

for all values of $\boldsymbol{\alpha}$, k (C_b is the space of bounded continuous functions). The proof of Theorem 4.3 is now complete. ∎

4.3 Notes for Chapter 4

- The existence of a solution to the vorticity equation (2.1)–(2.2), when $\omega_0 \in L^1(\mathbb{R}^2)$, was first stated in the note by Cottet [49]. His proof relied on the Nash inequality (see the Notes for Chapter 3) and a regularization of the initial data. The passage to the limit was carried out in velocity spaces of the type $L^q((0,T]; H^1(\Omega))$, $q < 2$, where $\Omega \subseteq \mathbb{R}^2$ is a bounded set (compare Equation (3.37) with $p = 2$).
- At about the same time, the existence of a solution to the vorticity equation (2.1)–(2.2), when $\omega_0 \in L^1(\mathbb{R}^2)$, was proved by Giga, Miyakawa and Osada [79], using a delicate estimate for Green's function of a perturbed heat equation. See also [143]. The constants appearing in their treatment are unspecified and depend nonlinearly on $\|\omega_0\|_1$, in contrast to the linear dependence in (3.9) and (3.12).
- The proof given here follows [9, 10] and the uniqueness part relies also on [26].

- We note in particular the continuous dependence on the initial data, an issue that was not addressed in the above mentioned papers (partly due to the lack of suitable estimates as in Subsection 3.1.1).
- As an "intermediate space" we chose (Lemma 4.2) the spaces Y_p, $p \in (2, \infty]$. In addition to being "persistence spaces" for the vorticity, their associated velocity fields are globally continuous and decaying at infinity, in view of Lemma 3.5. This choice facilitates the treatment, reducing it to a convection–diffusion (nonlinear) equation with globally continuous convection coefficient. Giga, Miyakawa and Osada [79, Theorem 2.5] show the persistence of the (vorticity) solution in the intermediate spaces $L^p(\mathbb{R}^2)$ for $p \in (2, \infty)$.
- We can obtain other persistence spaces of vorticities, where explicit decay assumptions are imposed on the initial data (and then inherited by the solution). In particular, this is true for the weighted-L^2 space $L^{2,s}(\mathbb{R}^2)$, $s > 1$, normed by

$$\|f\|_{L^{2,s}}^2 := \int_{\mathbb{R}^2} (1 + |\mathbf{x}|^2)^s |f(\mathbf{x})|^2 d\mathbf{x},$$

  see Proposition 6.5.
  Note that (for $s > 1$) $L^{2,s}(\mathbb{R}^2) \subseteq L^1(\mathbb{R}^2)$. It plays an important role in the study of the large-time asymptotic behavior of the solution, as will be seen in Chapter 6.
- The "smallness" condition (4.36) has been extensively used in proving uniqueness for solutions of nonlinear parabolic equations (see "note added in proof" in [26]), and is commonly referred to as the "Kato–Fujita condition [109]." In Theorem 4.3 we avoided it by establishing (4.34) for the extension operator. Then, assuming that the solution $\omega(t)$ is in $C(\overline{\mathbb{R}_+}; L^1(\mathbb{R}^2)) \cap C(\mathbb{R}_+, L^\infty(\mathbb{R}^2))$, we actually *derived* (4.36). The requirement $\omega(\cdot, t) \in L^\infty$ for $t > 0$ can be considerably relaxed, still avoiding a "Kato–Fujita condition". We refer to [18, 27] where similar uniqueness arguments have been used in the study of nonlinear parabolic equations.
- The estimates in this chapter deal with the time dependence of L^p norms of the solutions and their derivatives. However, it can be shown that the solution $\omega(\mathbf{x}, t)$ decays as a Gaussian in space. In fact, we have

Theorem 4.5. *For $\omega_0 \in L^1(\mathbb{R}^2)$, let $\omega(\cdot, t) = (S\omega_0)(t)$ be the solution given by Theorem 4.3. Then for any $\kappa \in (0, 1)$ there exists $C_\kappa > 0$,* *depending on* $\|\omega_0\|_1$, *such that*

(4.56)
$$|\omega(\mathbf{x},t)| \le \frac{C_\kappa}{t}\int\limits_{\mathbb{R}^2} \exp\Big(-\kappa\frac{|\mathbf{x}-\mathbf{y}|^2}{4\nu t}\Big)|\omega_0(\mathbf{y})|d\mathbf{y}, \quad \mathbf{x}\in\mathbb{R}^2,\, t>0.$$

For a proof, see [35, Theorem 3].

Chapter 5

Measures as Initial Data

Let $\mathcal{M}$ be the Banach space of finite (signed) measures on $\mathbb{R}^2$, normed by total variation $\|\cdot\|_{\mathcal{M}}$. The reader is referred to Appendix A, Section A.4, for the required background material concerning these measures, including some basic notation (such as the Dirac measure $\delta_{\mathbf{a}}$).

Our aim in this chapter is to establish the existence and uniqueness of solutions to (2.1)–(2.2) for initial vorticity $\omega_0 \in \mathcal{M}$. While this is of course very interesting (and challenging) from the mathematical point of view, it is also of great significance to some basic fluid dynamical models; indeed, it is exactly in this context that one deals with the evolution of a point vortex or a vortex filament. As was observed in the Introduction, the array of "vortex methods" still represents one of the most common approaches in the numerical approximation to (2.1)–(2.2). Interestingly, these methods typically approximate a given initial vorticity (even smooth) by a "collection of point vortices," which are then evolved by a discrete algorithm. It is then expected that the result, at any given time, serves as a good approximation to the analytical solution. Mathematically speaking, the underlying assumption is therefore that (2.1)–(2.2) is "well posed" in $\mathcal{M}$, or, more explicitly, that "small variations" in this space (with respect to $\|\cdot\|_{\mathcal{M}}$) lead to correspondingly small variations of the ensuing solutions.

We note in passing that the approximation methodology expounded in Part II of this monograph is very different from these vortex methods.

Recall (Appendix A, Section A.4) that there is a natural identification (norm preserving) $L^1(\mathbb{R}^2) \subseteq \mathcal{M}$. In fact (Claim A.8 in Appendix A), the subspace $L^1(\mathbb{R}^2)$ (and even $C_0^\infty(\mathbb{R}^2)$) is weak* dense in $\mathcal{M}$. Thus, we need to extend the solution operators S and $\mathbf{U}$ (see Theorem 4.3) to initial data in the larger space $\mathcal{M}$.

Example 5.1. The most elementary (and significant) example of measure initial data is that of a **point vortex** located at the origin. It corresponds to the Dirac measure

$$\omega_0^{pv} = \alpha\delta_{\mathbf{x}=0},$$

where $0 \neq \alpha \in \mathbb{R}$ is the "strength" of the vortex. In fluid dynamical language, it is the circulation of the associated velocity field.

Define, for $\mathbf{x} = (x^1, x^2) \in \mathbb{R}^2$, $t > 0$,

$$\begin{aligned} &\omega^{pv}(\mathbf{x}, t) = \alpha G_\nu(\mathbf{x}, t), \\ &\mathbf{u}^{pv}(\mathbf{x}, t) = (u^{pv,1}, u^{pv,2}) = \frac{\alpha}{2\pi} \frac{1 - e^{-\frac{|\mathbf{x}|^2}{4\nu t}}}{|\mathbf{x}|^2}(-x^2, x^1). \end{aligned} \tag{5.1}$$

Note that the velocity $\mathbf{u}^{pv}(\mathbf{x}, t)$ is orthogonal to $\mathbf{x}$, and since $\omega^{pv}(\mathbf{x}, t)$ is radially symmetric (depending only on $|\mathbf{x}|$) it follows that

$$\mathbf{u}^{pv}(\mathbf{x}, t) \cdot \nabla\omega^{pv}(\mathbf{x}, t) = 0.$$

Thus, the heat solution $\omega^{pv}(\mathbf{x}, t)$ is actually a solution to

$$\partial_t\omega^{pv}(\mathbf{x}, t) + \mathbf{u}^{pv}(\mathbf{x}, t) \cdot \nabla_{\mathbf{x}}\omega^{pv}(\mathbf{x}, t) = \nu\Delta\omega^{pv}(\mathbf{x}, t), \ t > 0.$$

In addition, the reader can verify by a straightforward calculation that

$$\nabla_{\mathbf{x}} \cdot \mathbf{u}^{pv}(\mathbf{x}, t) = 0, \quad \omega^{pv}(\mathbf{x}, t) = \partial_{x^1} u^{pv,2}(\mathbf{x}, t) - \partial_{x^2} u^{pv,1}(\mathbf{x}, t).$$

This implies that $\mathbf{u}^{pv} = \mathbf{K} * \omega^{pv}$, so that $\omega^{pv}(\mathbf{x}, t)$ and $\mathbf{u}^{pv}(\mathbf{x}, t)$ constitute a (classical) solution to the Navier–Stokes system (2.1)–(2.2) for $t > 0$.

In fluid dynamics this solution is usually known as the **Oseen vortex** (of strength α). Of course, if we want to claim that $\omega^{pv}(\cdot, t) = S(t)\omega_0^{pv}$, $t > 0$, we need to address the continuity of this solution as $t \to 0$. In this case, in view of Equation (A.13) in Appendix A we have

$$\omega^{pv}(\cdot, t) \xrightarrow[t\to 0]{weak^*} \alpha\delta_{\mathbf{x}=0},$$

which will motivate our definition of a solution for measure-valued initial data below.

The following definition is a modification of Definition 4.1. We use the space Y_∞ introduced in (2.61), and note that by (3.25) the associated velocity field is continuous and globally bounded for any vorticity in Y_∞.

Definition 5.2. *Let $\omega_0 \in \mathcal{M}$. We say that the pair*

$$\omega(\cdot, t) \in C(\mathbb{R}_+; Y_\infty),$$

$$\mathbf{u}(\cdot, t) = \mathbf{K} * \omega(\cdot, t) \in C(\mathbb{R}_+; D^0(\mathbb{R}^2)),$$

is a **mild solution** *to* (2.1)–(2.2), *subject to the initial condition* $\omega(\cdot, 0) = \omega_0$, *if it satisfies, for all* $0 < t_0 < t < \infty$,

$$\begin{aligned}\omega(\mathbf{x},t) &= \int_{\mathbb{R}^2} G_\nu(\mathbf{x}-\mathbf{y}, t-t_0)\omega(\mathbf{y}, t_0)d\mathbf{y} \\ &+ \int_{t_0}^{t}\int_{\mathbb{R}^2} \nabla_{\mathbf{y}} G_\nu(\mathbf{x}-\mathbf{y}, t-s)\cdot \omega(\mathbf{y}, s)\mathbf{u}(\mathbf{y}, s)d\mathbf{y}ds, \quad \mathbf{x} \in \mathbb{R}^2,\end{aligned} \tag{5.2}$$

and

$$\omega(\cdot, t) \xrightarrow[t\to 0]{weak^*} \omega_0.$$

The equality in (5.2) *is understood in the sense of* $L^1(\mathbb{R}^2)$, *for every* $0 < t_0 < t < \infty$.

The estimates obtained in Chapter 3 will serve as our basic tools in constructing the operators S and $\mathbf{U}$. Recall (Theorem 4.3) that these estimates apply to any initial $\omega_0 \in L^1(\mathbb{R}^2)$.

We can now state the extension theorem for initial data in $\mathcal{M}$.

Theorem 5.3. *Let* $\omega_0 \in \mathcal{M}$. *Then the system* (2.1)–(2.2) *has a mild solution* $\omega(\cdot, t) \in C(\mathbb{R}_+; Y_\infty)$ *and* $\mathbf{u}(\cdot, t) = \mathbf{K} * \omega(\cdot, t)$ *such that,*

$$\omega(\cdot, t) \xrightarrow[t\to 0]{weak^*} \omega_0, \tag{5.3}$$

and

$$\|\omega(\cdot, t)\|_1 \le \|\omega_0\|_{\mathcal{M}}, \; t > 0. \tag{5.4}$$

The conditions (5.3) *and* (5.4) *determine this solution uniquely.*

The properties (5.3) and (5.4) of the solution are essential in establishing its uniqueness. Before proceeding to the proof of the theorem, we show in the following remark and two corollaries how these properties shape the behavior of a mild solution near $t = 0$.

Remark 5.4. Since for $\tau > 0$, $\omega(\cdot, \tau) \in L^1(\mathbb{R}^2)$, Theorem 4.3 can be applied to $t \ge \tau$. Thus $\omega(\mathbf{x}, t) \in C^\infty(\mathbb{R}^2 \times \mathbb{R}_+)$ and we obtain also that for every $1 \le p \le \infty$, $\|\omega(\cdot, t)\|_p$ is a nonincreasing function of $t \in \mathbb{R}_+$, and, by (3.9),

$$\|\omega(\cdot, t)\|_p \le Ct^{-1+\frac{1}{p}}\|\omega(\cdot, t/2)\|_1 \le Ct^{-1+\frac{1}{p}}\|\omega_0\|_{\mathcal{M}}, \tag{5.5}$$

where $C > 0$ depends only on p and ν.

Furthermore, the estimate (3.12) implies similarly that, for any $q \in (2,\infty]$, there exists a constant $C > 0$, depending only on q and ν, so that

$$\|\mathbf{u}(\cdot,t)\|_q \le Ct^{-1/2+1/q}\|\omega_0\|_{\mathcal{M}}, \quad t>0. \tag{5.6}$$

Corollary 5.5. *Let $\omega(\mathbf{x},t)$ be a mild solution satisfying* (5.3) *and* (5.4). *Then it satisfies Equation* (5.2) *in the L^1 sense with $t_0 = 0$. More explicitly,*

$$\begin{aligned}\omega(\mathbf{x},t) &= (G_\nu(\cdot,t) * \omega_0)(\mathbf{x})\\ &\quad + \int_0^t \int_{\mathbb{R}^2} \nabla_{\mathbf{y}} G_\nu(\mathbf{x}-\mathbf{y},t-s)\cdot\omega(\mathbf{y},s)\mathbf{u}(\mathbf{y},s)d\mathbf{y}ds, \quad \mathbf{x}\in\mathbb{R}^2,\end{aligned} \tag{5.7}$$

where the two sides are equal as $L^1(\mathbb{R}^2)$ functions for any $t > 0$.

Refer to Definition A.5 in Appendix A for the definition of a convolution with a measure.

Proof of the corollary. Let $t > 0$ and fix $\mathbf{x}\in\mathbb{R}^2$. Clearly

$$G_\nu(\mathbf{x}-\mathbf{y},t-t_0) \xrightarrow[t_0\to 0]{} G_\nu(\mathbf{x}-\mathbf{y},t) \text{ in } D^0(\mathbb{R}^2_{\mathbf{y}}).$$

Young's inequality [see Equation (A.4) in Appendix A] implies that

$$\begin{aligned}&\left|\int_{\mathbb{R}^2}(G_\nu(\mathbf{x}-\mathbf{y},t-t_0)-G_\nu(\mathbf{x}-\mathbf{y},t))\omega(\mathbf{y},t_0)d\mathbf{y}\right|\\ &\qquad \le \|G_\nu(\mathbf{x}-\cdot,t-t_0)-G_\nu(\mathbf{x}-\cdot,t)\|_\infty\|\omega_0\|_{\mathcal{M}} \xrightarrow[t_0\to 0]{} 0.\end{aligned}$$

Since $\omega(\cdot,t_0) \xrightarrow[t_0\to 0]{weak^*} \omega_0$ we have by definition (of weak* convergence) that

$$\int_{\mathbb{R}^2} G_\nu(\mathbf{x}-\mathbf{y},t)\omega(\mathbf{y},t_0)d\mathbf{y} \xrightarrow[t_0\to 0]{} (G_\nu(\cdot,t)*\omega_0)(\mathbf{x}).$$

We therefore have, for every $\mathbf{x}\in\mathbb{R}^2$,

$$\int_{\mathbb{R}^2} G_\nu(\mathbf{x}-\mathbf{y},t-t_0)\omega(\mathbf{y},t_0)d\mathbf{y} \xrightarrow[t_0\to 0]{} (G_\nu(\cdot,t)*\omega_0)(\mathbf{x}). \tag{5.8}$$

Observe that in view of Young's inequality [see Equation (A.6) in Appendix A] $G_\nu(\cdot,t)*\omega_0 \in L^1(\mathbb{R}^2)$ (for every $t>0$, as a function of $\mathbf{x}$).

Next we claim that, for any $t > 0$,

$$\begin{aligned}&\int_{t_0}^t\int_{\mathbb{R}^2}\nabla_{\mathbf{y}}G_\nu(\mathbf{x}-\mathbf{y},t-s)\cdot\omega(\mathbf{y},s)\mathbf{u}(\mathbf{y},s)d\mathbf{y}ds\\ &\xrightarrow[t_0\to 0]{} \int_0^t\int_{\mathbb{R}^2}\nabla_{\mathbf{y}}G_\nu(\mathbf{x}-\mathbf{y},t-s)\cdot\omega(\mathbf{y},s)\mathbf{u}(\mathbf{y},s)d\mathbf{y}ds,\end{aligned} \tag{5.9}$$

where the convergence is in the sense of $L^1(\mathbb{R}^2_{\mathbf{x}})$.

Indeed, if $0 < t_0^1 < t_0^2 < t$, we estimate the $L^1(\mathbb{R}^2)$ norm by invoking (5.5) (with $p = 1$) and (5.6) (with $q = \infty$), as well as (2.26), so as to obtain by Young's inequality

$$\left\| \int_{t_0^1}^{t_0^2} \int_{\mathbb{R}^2} \nabla_{\mathbf{y}} G_\nu(\mathbf{x}-\mathbf{y}, t-s) \cdot \omega(\mathbf{y}, s)\mathbf{u}(\mathbf{y}, s) d\mathbf{y} ds \right\|_1 \leq C\|\omega_0\|_{\mathcal{M}}^2 \int_{t_0^1}^{t_0^2} s^{-\frac{1}{2}}(t-s)^{-\frac{1}{2}} ds.$$

The right-hand side clearly tends to zero as $t_0^2 \to 0$, so that (5.9) is established. It follows that the convergence in (5.8) is also in L^1 (since the sum is equal to the fixed function $\omega(\mathbf{x}, t)$) and the proof is complete. ∎

Lemma 3.4 states that if the initial data ω_0 is an L^1 function (and, in particular, has no atomic part), then $t^{1-\frac{1}{p}}\|\omega(\cdot, t)\|_p$ vanishes as $t \to 0$, for all $p > 1$. The same is true for solutions of the heat equation, by Claim A.9 in Appendix A. We now show that such a "smallness" condition implies uniqueness. This result will be useful when dealing with the uniqueness of the solution, when the initial atomic part is small.

Corollary 5.6. *For any $p \in [\frac{4}{3}, 2)$ there exists a constant $\delta_p > 0$ (for notational simplicity we omit the dependence on ν), so that the following uniqueness claim holds:*

Fix any $t_p > 0$. Then there is at most one mild solution $\omega(\mathbf{x}, t)$ satisfying (5.3) *and* (5.4), *as well as the "smallness" condition*

$$\|\omega(\cdot, t)\|_p < \delta_p t^{-1+\frac{1}{p}}, \quad 0 < t < t_p. \tag{5.10}$$

Proof of the corollary. We prove the corollary for the important case $p = \frac{4}{3}$ and refer to [108, Proposition 3.2] for other values of p.

Let $\omega(\cdot, t)$ and $\theta(\cdot, t)$ be two mild solutions to (2.1)–(2.2), subject to the same initial condition $\omega_0 \in \mathcal{M}$, and satisfying (5.3) and (5.4).

Let $\mathbf{u}(\cdot, t) = \mathbf{K} * \omega(\cdot, t)$ and $\mathbf{v}(\cdot, t) = \mathbf{K} * \theta(\cdot, t)$ be their associated velocities. Corollary 5.5 can be applied, so that $\omega(\mathbf{x}, t)$ satisfies (5.7), and also, in the L^1 sense,

$$\begin{aligned} \theta(\mathbf{x}, t) &= (G_\nu(\cdot, t) * \omega_0)(\mathbf{x}) \\ &\quad + \int_0^t \int_{\mathbb{R}^2} \nabla_{\mathbf{y}} G_\nu(\mathbf{x}-\mathbf{y}, t-s) \cdot \theta(\mathbf{y}, s)\mathbf{v}(\mathbf{y}, s) d\mathbf{y} ds, \quad \mathbf{x} \in \mathbb{R}^2. \end{aligned} \tag{5.11}$$

For the difference of the two mild solutions we have [compare (4.20)]

$$
\begin{aligned}
&\omega(\mathbf{x},t)-\theta(\mathbf{x},t)\\
(5.12)\qquad &=\int_0^t \nabla_{\mathbf{x}} G_\nu(\mathbf{x}-\mathbf{y},t-s)*\mathbf{u}(\mathbf{y},s)(\omega(\mathbf{y},s)-\theta(\mathbf{y},s))d\mathbf{y}ds\\
&+\int_0^t \nabla_{\mathbf{x}} G_\nu(\mathbf{x}-\mathbf{y},t-s)*(\mathbf{u}(\mathbf{y},s)-\mathbf{v}(\mathbf{y},s))\theta(\mathbf{y},s)d\mathbf{y}ds.
\end{aligned}
$$

Take $p=\frac{4}{3}$ and $q=4$ in (3.11) to estimate by Hölder's inequality [compare the treatment of (4.27)]

$$
\begin{aligned}
\|\mathbf{u}(\cdot,s)(\omega(\cdot,s)-\theta(\cdot,s))\|_1 &\le \|\mathbf{u}(\cdot,s)\|_4\|\omega(\cdot,s)-\theta(\cdot,s)\|_{\frac{4}{3}}\\
&\le \xi_{\frac{4}{3}}\|\omega(\cdot,s)\|_{\frac{4}{3}}\cdot\|\omega(\cdot,s)-\theta(\cdot,s)\|_{\frac{4}{3}},\\
\|(\mathbf{u}(\cdot,s)-\mathbf{v}(\cdot,s))\theta(\cdot,s)\|_1 &\le \|\mathbf{u}(\cdot,s)-\mathbf{v}(\cdot,s)\|_4\|\theta(\cdot,s)\|_{\frac{4}{3}}\\
&\le \xi_{\frac{4}{3}}\|\omega(\cdot,s)-\theta(\cdot,s)\|_{\frac{4}{3}}\cdot\|\theta(\cdot,s)\|_{\frac{4}{3}}.
\end{aligned}
$$

Inserting these inequalities in (5.12) yields, by Young's inequality,

$$
\begin{aligned}
&\|\omega(\cdot,t)-\theta(\cdot,t)\|_{\frac{4}{3}}\\
&\le \xi_{\frac{4}{3}}\int_0^t \|\nabla G_\nu(\cdot,t-s)\|_{\frac{4}{3}}(\|\omega(\cdot,s)\|_{\frac{4}{3}}+\|\theta(\cdot,s)\|_{\frac{4}{3}})\|\omega(\cdot,s)-\theta(\cdot,s)\|_{\frac{4}{3}}ds.
\end{aligned}
$$

Recall that by (3.18)

$$
\|\nabla G_\nu(\cdot,t)\|_r = C(r,\nu)t^{-\frac{3}{2}+\frac{1}{r}},\quad t>0,
$$

so that we have

$$
\begin{aligned}
(5.13)\qquad &\|\omega(\cdot,t)-\theta(\cdot,t)\|_{\frac{4}{3}}\\
&\le C(\frac{4}{3},\nu)\xi_{\frac{4}{3}}\int_0^t (t-s)^{-\frac{3}{4}}(\|\omega(\cdot,s)\|_{\frac{4}{3}}+\|\theta(\cdot,s)\|_{\frac{4}{3}})\|\omega(\cdot,s)-\theta(\cdot,s)\|_{\frac{4}{3}}ds.
\end{aligned}
$$

[Compare with the estimate (4.38) in the uniqueness proof for L^1 initial data].

We now invoke the assumption that the solutions $\omega(\mathbf{x},t)$ and $\theta(\mathbf{x},t)$ satisfy (5.10), with $\delta_{\frac{4}{3}}$ yet to be determined (and any fixed $t_{\frac{4}{3}}>0$).

Thus, the last inequality yields, for $t<t_{\frac{4}{3}}$,

$$
\begin{aligned}
&\|\omega(\cdot,t)-\theta(\cdot,t)\|_{\frac{4}{3}}\\
(5.14)\qquad &\le 2C(\frac{4}{3},\nu)\xi_{\frac{4}{3}}\delta_{\frac{4}{3}}\int_0^t (t-s)^{-\frac{3}{4}}s^{-\frac{1}{4}}\|\omega(\cdot,s)-\theta(\cdot,s)\|_{\frac{4}{3}}ds.
\end{aligned}
$$

Define

$$N_{\frac{4}{3}}(t) = \sup_{0<\tau\le t} \|\omega(\cdot,\tau) - \theta(\cdot,\tau)\|_{\frac{4}{3}},$$

so that

$$N_{\frac{4}{3}}(t) \le 2C(\frac{4}{3},\nu)\xi_{\frac{4}{3}}\delta_{\frac{4}{3}}N_{\frac{4}{3}}(t)\int_0^1 (1-s)^{-\frac{3}{4}}s^{-\frac{1}{4}}ds.$$

We conclude that if we take

$$\delta_{\frac{4}{3}} < \left[2C(\frac{4}{3},\nu)\xi_{\frac{4}{3}}\int_0^1 (1-s)^{-\frac{3}{4}}s^{-\frac{1}{4}}ds\right]^{-1},$$

then

$$\omega(\cdot,t) = \theta(\cdot,t), \quad t < t_{\frac{4}{3}}.$$

Since $\omega(\cdot,t),\ \theta(\cdot,t) \in L^1(\mathbb{R}^2)$, they are equal for $t \in [t_{\frac{4}{3}},\infty)$ by the uniqueness claim of Theorem 4.3. This concludes the proof of the corollary. ■

Proof of Theorem 5.3. Let $\omega_{0,j}$ be the restriction of ω_0 to the disk $\{|\mathbf{x}| < j\}$. Using Claim A.8 in Appendix A (and its proof), we construct a sequence $\{\omega_0^{(j)} = \eta_j * \omega_{0,j}\}_{j=1}^{\infty} \subseteq C_0^\infty(\mathbb{R}^2)$, so that

$$\omega_0^{(j)} \xrightarrow[j\to\infty]{weak^*} \omega_0, \tag{5.15}$$

and $\|\omega_0^{(j)}\|_1 \le \|\omega_0\|_{\mathcal{M}}$. Let $\omega^{(j)}(\mathbf{x},t)$ be the solution given by Theorem 2.1, subject to the initial condition $\omega^{(j)}(\cdot,0) = \omega_0^{(j)}$, and let $\mathbf{u}^{(j)}(\cdot,t) = \mathbf{K} * \omega^{(j)}(\cdot,t)$ be the associated velocity.

The estimates (3.9) and (3.12) imply that, given $p \in [1,\infty]$ and $q \in (2,\infty]$, there exist constants C_1, $C_2 > 0$, depending only on p, q, respectively, so that

$$\begin{aligned} \|\omega^{(j)}(\cdot,t)\|_p &\le C_1 t^{-1+1/p}, \quad t>0,\ j=1,2,\ldots \\ \|\mathbf{u}^{(j)}(\cdot,t)\|_q &\le C_2 t^{-1/2+1/q}, \quad t>0,\ j=1,2,\ldots \end{aligned} \tag{5.16}$$

The obvious dependence on ν and $\|\omega_0\|_{\mathcal{M}}$ will not be mentioned here and below.

Fix $\tau > 0$. In view of Lemmas 3.7 and 3.8, there exist positive constants C_3, $C_4 > 0$, depending on τ, such that

$$\begin{aligned} \|\nabla\omega^{(j)}(\cdot,t)\|_{Y_\infty} &\le C_3, \quad \tau < t < \infty,\ j=1,2,\ldots \\ \|\nabla\mathbf{u}^{(j)}(\cdot,t)\|_\infty &\le C_4, \quad \tau < t < \infty,\ j=1,2,\ldots \end{aligned} \tag{5.17}$$

In addition, by (4.50), there exists a constant $C_5 > 0$, depending only on τ, such that

(5.18) $$\|\Delta\omega^{(j)}(\cdot,t)\|_{Y_\infty} \le C_5, \quad \tau < t < \infty,\ j = 1,2,\dots$$

Since $\omega^{(j)}$ is a solution to (2.1), the above estimates also yield

(5.19) $$\|\partial_t\omega^{(j)}(\cdot,t)\|_p \le C_6, \quad 1 \le p \le \infty,\ \tau < t < \infty,\ j = 1,2,\dots$$

where C_6 depends on p and τ. Also, as in (3.10)–(3.11),

(5.20)
$$\|\partial_t\mathbf{u}^{(j)}(\cdot,t)\|_q = \|\mathbf{K}* \partial_t\omega^{(j)}(\cdot,t)\|_q \le C\|\partial_t\omega^{(j)}(\cdot,t)\|_p \le CC_6, \quad 2 < q \le \infty,$$

where $\frac{1}{q} = \frac{1}{p} - \frac{1}{2}$.

In particular, both sequences $\left\{\omega^{(j)}(\mathbf{x},t)\right\}_{j-1}^{\infty}$ and $\left\{\mathbf{u}^{(j)}(\mathbf{x},t)\right\}_{j-1}^{\infty}$ are uniformly bounded and uniformly Lipschitz continuous, with respect to $(\mathbf{x},t) \in \mathbb{R}^2 \times [\tau,\infty)$ for every $\tau > 0$.

Invoking the Arzelà–Ascoli theorem [58, Appendix C.7] we obtain subsequences, which we relabel as $\{\omega^{(j)}\}_{j=1}^{\infty}$, and $\{\mathbf{u}^{(j)}\}_{j=1}^{\infty}$ and functions $\omega(\mathbf{x},t)$ and $\mathbf{u}(\mathbf{x},t)$, such that

(5.21) $$\omega^{(j)}(\mathbf{x},t) \to \omega(\mathbf{x},t),\ \mathbf{u}^{(j)}(\mathbf{x},t) \to \mathbf{u}(\mathbf{x},t), \quad (\mathbf{x},t) \in \mathbb{R}^2 \times \mathbb{R}_+,$$

uniformly in every compact subset of $\mathbb{R}^2 \times \mathbb{R}_+$ (in fact, with all derivatives, see Lemma 4.2).

In particular, for every $t > 0$,

$$\omega^{(j)}(\cdot,t) \xrightarrow[j\to\infty]{weak^*} \omega(\cdot,t),$$

and

$$\omega^{(j)}(\cdot,t)\mathbf{u}^{(j)}(\cdot,t) \xrightarrow[j\to\infty]{weak^*} \omega(\cdot,t)\mathbf{u}(\cdot,t),$$

and by Fatou's lemma [58, Appendix E.3], for every $t > 0$,

(5.22) $$\|\omega(\cdot,t)\|_1 \le \liminf_{j\to\infty} \|\omega^{(j)}(\cdot,t)\|_1 \le \|\omega_0\|_{\mathcal{M}}.$$

Using a decomposition as in (2.52) we infer that $\mathbf{u}(\cdot,t) = \mathbf{K} * \omega(\cdot,t)$.

Notice that from (5.16) we get, again using Fatou's lemma,

(5.23) $$\begin{aligned} \|\omega(\cdot,t)\|_p &\le C_1 t^{-1+1/p}, \quad t > 0,\ p \in [1,\infty], \\ \|\mathbf{u}(\cdot,t)\|_q &\le C_2 t^{-1/2+1/q}, \quad t > 0,\ q \in (2,\infty], \end{aligned}$$

and (5.17) yields the L^∞ bounds,

(5.24) $$\begin{aligned} \|\nabla\omega(\cdot,t)\|_\infty &\le C_3, \quad \tau < t < \infty, \\ \|\nabla\mathbf{u}(\cdot,t)\|_\infty &\le C_4, \quad \tau < t < \infty. \end{aligned}$$

We proceed to show that the pair $(\omega, \mathbf{u})$ is a mild solution.

Fix $t_0 > 0$ and note that each function $\omega^{(j)}(\mathbf{x}, t)$ satisfies, for $\mathbf{x} \in \mathbb{R}^2$ and all $t > t_0 > 0$,

(5.25)
$$\omega^{(j)}(\mathbf{x}, t) = \int_{\mathbb{R}^2} G_\nu(\mathbf{x} - \mathbf{y}, t - t_0)\omega^{(j)}(\mathbf{y}, t_0)d\mathbf{y} + \int_{t_0}^{t}\int_{\mathbb{R}^2} \nabla_{\mathbf{y}} G_\nu(\mathbf{x} - \mathbf{y}, t - s) \cdot \omega^{(j)}(\mathbf{y}, s)\mathbf{u}^{(j)}(\mathbf{y}, s)d\mathbf{y}ds.$$

We claim that we can pass to the limit as $j \to \infty$, concluding that the limit functions $\omega(\mathbf{x}, t)$ and $\mathbf{u}(\mathbf{x}, t)$ satisfy (5.2). Indeed, if $\eta > 0$ then both $G_\nu(\mathbf{x} - \mathbf{y}, \eta)$ and $\nabla_{\mathbf{y}} G_\nu(\mathbf{x} - \mathbf{y}, \eta)$ are in $D^0(\mathbb{R}^2)$ as functions of $\mathbf{y}$, for any fixed $\mathbf{x} \in \mathbb{R}^2$. The weak convergence now yields, for $t > t_0$,

$$(5.26)\qquad \lim_{j\to\infty} \int_{\mathbb{R}^2} G_\nu(\mathbf{x} - \mathbf{y}, t - t_0)\omega^{(j)}(\mathbf{y}, t_0)d\mathbf{y} = \int_{\mathbb{R}^2} G_\nu(\mathbf{x} - \mathbf{y}, t - t_0)\omega(\mathbf{y}, t_0)d\mathbf{y},$$

and, for $t_0 \le s < t$,

$$\lim_{j\to\infty} f^{(j)}(\mathbf{x}, s; t) = \int_{\mathbb{R}^2} \nabla_{\mathbf{y}} G_\nu(\mathbf{x} - \mathbf{y}, t - s) \cdot \omega(\mathbf{y}, s)\mathbf{u}(\mathbf{y}, s)d\mathbf{y},$$

where

$$f^{(j)}(\mathbf{x}, s; t) = \int_{\mathbb{R}^2} \nabla_{\mathbf{y}} G_\nu(\mathbf{x} - \mathbf{y}, t - s) \cdot \omega^{(j)}(\mathbf{y}, s)\mathbf{u}^{(j)}(\mathbf{y}, s)d\mathbf{y}.$$

In (5.16) take $p = q = \infty$. We then have,

$$(5.27)\qquad \|\omega^{(j)}(\cdot, s)\mathbf{u}^{(j)}(\cdot, s)\|_\infty \le C_1 C_2 s^{-\frac{3}{2}},$$

and in conjunction with (2.26) Young's inequality yields

$$|f^{(j)}(\mathbf{x}, s; t)| \le \beta_\nu C_1 C_2 (t - s)^{-\frac{1}{2}} s^{-\frac{3}{2}}.$$

The dominated convergence theorem [58, Appendix E.3] now yields, for fixed $\mathbf{x}$ and t,

(5.28)
$$\int_{t_0}^{t} f^{(j)}(\mathbf{x}, s; t)ds \xrightarrow[j\to\infty]{} \int_{t_0}^{t}\int_{\mathbb{R}^2} \nabla_{\mathbf{y}} G_\nu(\mathbf{x} - \mathbf{y}, t - s) \cdot \omega(\mathbf{y}, s)\mathbf{u}(\mathbf{y}, s)d\mathbf{y}ds.$$

Combining the limits (5.26) and (5.28) we conclude that the equality (5.2) is satisfied for fixed $\mathbf{x}$ and t.

However, Young's inequality implies that both sides of the equality are in $L^1(\mathbb{R}^2)$, for any fixed $t > t_0$, so that it holds also in the sense of L^1, as required by Definition 5.2.

For the rest of the proof, the reader is referred to Section A.4 in Appendix A concerning basic definitions and facts of measure theory. In particular, $\langle\, , \,\rangle$ is the $(\mathcal{M}, D^0(\mathbb{R}^2))$ pairing.

To prove the weak convergence (5.3) we need to prove that, given any $\psi \in C_0^\infty(\mathbb{R}^2)$, the function

$$g(t) = \begin{cases} \int_{\mathbb{R}^2} \omega(\mathbf{x}, t)\psi(\mathbf{x})d\mathbf{x}, & t > 0, \\ \\ \langle \omega_0, \psi \rangle, & t = 0, \end{cases}$$

is continuous at $t = 0$. This will follow if we prove that the sequence of continuous functions

$$g_j(t) = \int_{\mathbb{R}^2} \omega^{(j)}(\mathbf{x}, t)\psi(\mathbf{x})d\mathbf{x}, \quad j = 1, 2, \ldots, t \geq 0,$$

converges to $g(t)$ uniformly in every interval $[0, T]$.

Since $g_j(t) \to g(t)$ as $j \to \infty$, for any fixed $t \geq 0$, the uniform convergence will follow by invoking the Arzelà–Ascoli theorem [58, Appendix C.7]. Thus it suffices to show (compare with Lemma 4.4) that the family $\{g_j(t)\}_{j=1}^\infty$ is equicontinuous in every interval $[0, T]$.

Taking the scalar product of Equation (2.1) (where ω is replaced by $\omega^{(j)}$) with ψ, we get

(5.29)
$$\frac{d}{dt} g_j(t) = \nu \int_{\mathbb{R}^2} \omega^{(j)}(\mathbf{x}, t)\Delta\psi(\mathbf{x})d\mathbf{x} + \int_{\mathbb{R}^2} \omega^{(j)}(\mathbf{x}, t)\mathbf{u}^{(j)}(\mathbf{x}, t) \cdot \nabla\psi(\mathbf{x})d\mathbf{x}.$$

The estimate

$$\|\mathbf{u}^{(j)}(\cdot, t)\omega^{(j)}(\cdot, t)\|_1 \leq C_1 C_2 t^{-1/2}$$

(which follows from (5.16) with $p = 1$ and $q = \infty$) implies the uniform integrability of the derivatives in $[0, T]$, hence the equicontinuity of the family $\{g_j(t)\}_{j=1}^\infty$, as claimed.

Finally, it remains to prove the uniqueness claim in the theorem. In Section 5.1 below we outline the proof in the general case.

Here we content ourselves with the case of *initial measure with small atomic part.* In this case, the estimates obtained so far enable us to provide a simple, short proof.

Refer to Claim A.6 in Appendix A for the definition of the atomic part of a measure.

We prove the uniqueness of the solution in the case where the atomic part of the initial measure ω_0 is small. More explicitly, there exists a constant $\kappa = \kappa(\nu) > 0$, such that if $\|(\omega_0)_{atom}\|_{\mathcal{M}} < \kappa$ then:

- **Our constructed solution satisfies**

$$\limsup_{t\to 0} t^{\frac{1}{4}}\|\omega(\cdot,t)\|_{\frac{4}{3}} \le 2\kappa. \tag{5.30}$$

- **Any other solution $\theta(\mathbf{x},t)$ satisfying**

$$\limsup_{t\to 0} t^{\frac{1}{4}}\|\theta(\cdot,t)\|_{\frac{4}{3}} \le 2\kappa \tag{5.31}$$

is identically equal to $\omega(\mathbf{x},t)$.

Recall that by Claim A.9 in Appendix A, taking $p = \frac{4}{3}$, there exists $t_{\frac{4}{3}} > 0$, depending on ν, such that if $\|(\omega_0)_{atom}\|_{\mathcal{M}} < \kappa$, then

$$t^{\frac{1}{4}}\|G_\nu(\cdot,t) * \omega_0\|_{\frac{4}{3}} < \kappa, \quad t < t_{\frac{4}{3}}. \tag{5.32}$$

By the definition (5.15) (and Claim A.8 in Appendix A) of the sequence $\left\{\omega_0^{(j)}\right\}_{j=1}^{\infty}$, combined with Young's inequality (A.6), we also have, for $t < t_{\frac{4}{3}}$,

$$t^{\frac{1}{4}}\|G_\nu(\cdot,t) * \omega_0^{(j)}\|_{\frac{4}{3}} < \kappa, \quad j = 1,2\ldots$$

The solution $\omega^{(j)}(\mathbf{x},t)$ constructed at the beginning of the proof is smooth and certainly satisfies Equation (5.7). Also by (3.11),

$$\|\omega^{(j)}(\cdot,t)\mathbf{u}^{(j)}(\cdot,t)\|_1 \le \|\omega^{(j)}(\cdot,t)\|_{\frac{4}{3}}\|\mathbf{u}^{(j)}(\cdot,t)\|_4 \le \xi_{\frac{4}{3}}\|\omega^{(j)}(\cdot,t)\|_{\frac{4}{3}}^2.$$

Thus, with the constants as in (5.13), we obtain [compare the similar estimate in (3.20)],

$$\|\omega^{(j)}(\cdot,t)\|_{\frac{4}{3}} \le \kappa t^{-\frac{1}{4}} + C(\frac{4}{3},\nu)\xi_{\frac{4}{3}}\int_0^t (t-s)^{-\frac{3}{4}}\|\omega^{(j)}(\cdot,s)\|_{\frac{4}{3}}^2 ds. \tag{5.33}$$

We can now repeat the proof of Lemma 3.4.

For $j = 1,2\ldots$ let $M_{\frac{4}{3}}^j(t) = \sup_{0\le\tau\le t}\tau^{\frac{1}{4}}\|\omega^{(j)}(\cdot,\tau)\|_{\frac{4}{3}}$. Note that $M_{\frac{4}{3}}^j(t)$ is continuous (recall Theorem 2.1) and $M_{\frac{4}{3}}^j(0) = 0$. We infer from the above inequality that,

$$M_{\frac{4}{3}}^j(t) \le \kappa + C(\frac{4}{3},\nu)\xi_{\frac{4}{3}}t^{\frac{1}{4}}\int_0^t (t-s)^{-\frac{3}{4}}s^{-\frac{1}{2}}s^{\frac{1}{2}}\|\omega^{(j)}(\cdot,s)\|_{\frac{4}{3}}^2 ds$$

$$\le \kappa + C(\frac{4}{3},\nu)\xi_{\frac{4}{3}}(M_{\frac{4}{3}}^j(t))^2\int_0^1 (1-s)^{-\frac{3}{4}}s^{-\frac{1}{2}}ds.$$

We denote $F = C(\frac{4}{3},\nu)\xi_{\frac{4}{3}}\int_0^1(1-s)^{-\frac{3}{4}}s^{-\frac{1}{2}}ds$ and restrict $\kappa > 0$, so that

$$4F\kappa < 1. \tag{5.34}$$

Define

$$t_0 = \sup\left\{\tau \in [0, t_{\frac{4}{3}}), \quad M^j_{\frac{4}{3}}(\tau) < 2\kappa\right\}.$$

Then

$$M^j_{\frac{4}{3}}(t_0) \le \kappa + F(M^j_{\frac{4}{3}}(t_0))^2 \le \kappa + 4F\kappa^2 < 2\kappa.$$

It follows that $t_0 = t_{\frac{4}{3}}$, since otherwise there exists $\tau > t_0$ for which $M^j_{\frac{4}{3}}(\tau) < 2\kappa$.

We conclude that for all $j = 1, 2, \ldots$

$$t^{\frac{1}{4}} \|\omega^{(j)}(\cdot, t)\|_{\frac{4}{3}} \le 2\kappa, \quad 0 < t < t_{\frac{4}{3}}.$$

Passing to the limit as $j \to \infty$ we obtain, as in (5.22),

$$t^{\frac{1}{4}} \|\omega(\cdot, t)\|_{\frac{4}{3}} \le 2\kappa, \quad 0 < t < t_{\frac{4}{3}}. \tag{5.35}$$

Let $\delta_{\frac{4}{3}} > 0$ be the constant in Corollary 5.6. As we saw at the end of the proof of the corollary, we could take any $\delta_{\frac{4}{3}} < (2F)^{-1}$. In particular, condition (5.34) also implies

$$2\kappa < \delta_{\frac{4}{3}}.$$

Thus, in view of Corollary 5.6, this solution is unique in the class of solutions satisfying (5.31).

This concludes the proof of uniqueness for a sufficiently small atomic part of the initial measure ω_0. Observe that all the constants appearing in the smallness requirement imposed on $\|(\omega_0)_{atom}\|_{\mathcal{M}}$ can be explicitly evaluated.

■

For initial data in $L^1(\mathbb{R}^2)$, we saw in Theorem 4.3 that the ensuing "vorticity trajectories," expressed by the evolution operator $S(t)$, depend continuously on the initial data, in the sense of the L^1 norm. This is what is generally meant by saying that Equation (2.1) is "well posed" in $L^1(\mathbb{R}^2)$. On the other hand, for initial data $\omega_0 \in \mathcal{M}$ that is not in $L^1(\mathbb{R}^2)$, the solution $\omega(t)$ converges to ω_0, as $t \to 0$, only in the weak* sense. Indeed, since $\omega(t) \in L^1(\mathbb{R}^2)$ for $t > 0$, convergence in the norm sense of $\mathcal{M}$ is identical to convergence in the norm sense of L^1, and would therefore imply that $\omega_0 \in L^1(\mathbb{R}^2)$. Thus, the space $\mathcal{M}$ is not a "persistence space" for the vorticity equation, as discussed in the Notes for Chapter 4.

However, it can still be shown that the solution $\omega(t)$ varies continuously with respect to the initial data $\omega_0 \in \mathcal{M}$, in a suitable functional setting. Refer to the Notes at the end of this chapter for more detail, while here we establish the continuity in the case of a small atomic part of the initial measure. Our proof is obtained by inspecting the proof of uniqueness in Theorem 5.3 above.

Corollary 5.7. *Let $\kappa > 0$ be as in (5.30). Let $\omega_0 \in \mathcal{M}$ satisfy $\|(\omega_0)_{atom}\|_{\mathcal{M}} < \kappa$, and let $\left\{\omega_0^{(j)}\right\}_{j=1}^{\infty} \subseteq \mathcal{M}$ be a sequence such that*

$$\lim_{j\to\infty} \|\omega_0 - \omega_0^{(j)}\|_{\mathcal{M}} = 0. \tag{5.36}$$

Let $\omega(\cdot,t)$ (respectively $\omega^{(j)}(\cdot,t)$) be the mild solution with initial data ω_0 (respectively $\omega_0^{(j)}$).

Then, for every $T > 0$,

$$\sup_{0<t<T} \|\omega(\cdot,t) - \omega^{(j)}(\cdot,t)\|_1 \xrightarrow[j\to\infty]{} 0. \tag{5.37}$$

Proof. The solutions $\omega(\mathbf{x},t)$ and $\omega^{(j)}(\mathbf{x},t)$ satisfy (5.7), subject to the initial conditions ω_0 and $\omega_0^{(j)}$, respectively. Hence, as in (5.13),

$$\begin{aligned}\|\omega(\cdot,t) - \omega^{(j)}(\cdot,t)\|_{\frac{4}{3}} &\le \|G_\nu(\cdot,t) * (\omega_0 - \omega_0^{(j)})\|_{\frac{4}{3}} \\ &+ C(\frac{4}{3},\nu)\xi_{\frac{4}{3}} \int_0^t (t-s)^{-\frac{3}{4}}(\|\omega(\cdot,s)\|_{\frac{4}{3}} + \|\omega^{(j)}(\cdot,s))\|_{\frac{4}{3}})\|\omega(\cdot,s) \\ &- \omega^{(j)}(\cdot,s)\|_{\frac{4}{3}} ds.\end{aligned} \tag{5.38}$$

By (A.12) in Appendix A we have

$$t^{\frac{1}{4}}\|G_\nu(\cdot,t) * (\omega_0 - \omega_0^{(j)})\|_{\frac{4}{3}} \le (4\pi\nu)^{-\frac{1}{4}}\|\omega_0 - \omega_0^{(j)}\|_{\mathcal{M}}, \quad t > 0.$$

We designate $Q_j(t) = \sup_{0<\tau\le t} \tau^{\frac{1}{4}}\|\omega(\cdot,\tau) - \omega^{(j)}(\cdot,\tau)\|_{\frac{4}{3}}$.

For j sufficiently large (which we relabel as $j = 1$) we have $\|(\omega_0^{(j)})_{atom}\|_{\mathcal{M}} < \kappa$. The convergence assumption (5.36) implies that $\omega(\mathbf{x},t)$, as well as all the functions $\left\{\omega^{(j)}(\mathbf{x},t)\right\}_{j=1}^{\infty}$, satisfy the smallness condition (5.35), with the same $t_{\frac{4}{3}}$. Incorporating this in (5.38) we get, with $F = C(\frac{4}{3},\nu)\xi_{\frac{4}{3}} \int_0^1 (1-s)^{-\frac{3}{4}} s^{-\frac{1}{2}} ds$,

$$Q_j(t) \le (4\pi\nu)^{-\frac{1}{4}}\|\omega_0 - \omega_0^{(j)}\|_{\mathcal{M}} + 4F\kappa Q_j(t), \quad 0 < t < t_{\frac{4}{3}}. \tag{5.39}$$

Using condition (5.34) we get

$$t^{\frac{1}{4}}\|\omega(\cdot,t) - \omega^{(j)}(\cdot,t)\|_{\frac{4}{3}} \le (1-4F\kappa)^{-1}(4\pi\nu)^{-\frac{1}{4}}\|\omega_0 - \omega_0^{(j)}\|_{\mathcal{M}}, \quad 0 < t < t_{\frac{4}{3}}. \tag{5.40}$$

Now, as in the estimate (5.38), we can estimate the L^1 norm of the difference, for $t \in (0, t_{\frac{4}{3}})$, as,

$$\begin{aligned}\|\omega(\cdot,t) - \omega^{(j)}(\cdot,t)\|_1 &\le \|G_\nu(\cdot,t) * (\omega_0 - \omega_0^{(j)})\|_1 \\ &+ \beta_\nu \xi_{\frac{4}{3}} \int_0^t (t-s)^{-\frac{1}{2}}(\|\omega(\cdot,s)\|_{\frac{4}{3}} + \|\omega^{(j)}(\cdot,s)\|_{\frac{4}{3}})\|\omega(\cdot,s) - \omega^{(j)}(\cdot,s)\|_{\frac{4}{3}} ds\end{aligned} \tag{5.41}$$

$$\leq \|\omega_0 - \omega_0^{(j)}\|_{\mathcal{M}} + \beta_\nu \xi_{\frac{4}{3}} \int_0^t (t-s)^{-\frac{1}{2}} s^{-\frac{1}{2}} s^{\frac{1}{4}} (\|\omega(\cdot,s)\|_{\frac{4}{3}} + \|\omega^{(j)}(\cdot,s)\|_{\frac{4}{3}})$$
$$\cdot s^{\frac{1}{4}} \|\omega(\cdot,s) - \omega^{(j)}(\cdot,s)\|_{\frac{4}{3}} ds$$
$$\leq \|\omega_0 - \omega_0^{(j)}\|_{\mathcal{M}} + 4\kappa\pi\beta_\nu\xi_{\frac{4}{3}} Q_j(t)$$
$$\leq \left(1 + (4\pi)^{\frac{3}{4}} \nu^{-\frac{1}{4}} \kappa \beta_\nu \xi_{\frac{4}{3}} (1 - 4F\kappa)^{-1}\right) \|\omega_0 - \omega_0^{(j)}\|_{\mathcal{M}}.$$

We used Young's inequality (A.12) in Appendix A, the equality $\int_0^1 (1 - s)^{-\frac{1}{2}} s^{-\frac{1}{2}} ds = \pi$, as well as $C(1,\nu) = \beta_\nu$, see (3.18).

For $t_0 \in (0, t_{\frac{4}{3}})$ we therefore get $\lim_{j\to\infty} \|\omega(\cdot,t_0) - \omega^{(j)}(\cdot,t_0)\|_1 = 0$, so that the continuity in $[t_0,\infty)$ follows from Theorem 4.3. ■

Condition (5.36) is clearly a very strong condition. On the other hand, the weak* convergence of a sequence of initial data is too weak to guarantee convergence of the associated solutions. However, sometimes an intermediate assumption can be useful. Take for example a sequence of points $\left\{\mathbf{x}_0^{(j)}\right\}_{j=1}^{\infty} \subseteq \mathbb{R}^2$ converging to a point $\mathbf{x}_0$. Consider the sequence of point vortices $\left\{\omega_0^{(j)} = \delta_{\mathbf{x}_0^{(j)}}\right\}_{j=1}^{\infty} \subseteq \mathcal{M}$. It converges, in the weak* sense, to $\delta_{\mathbf{x}_0}$, but clearly not in the norm sense (5.36). The following corollary is applicable to this case.

Corollary 5.8. *Let $\omega_0 \in \mathcal{M}$ and let $\left\{\omega_0^{(j)}\right\}_{j=1}^{\infty} \subseteq \mathcal{M}$ be a sequence such that*

$$\omega_0^{(j)} \xrightarrow[j\to\infty]{weak^*} \omega_0, \tag{5.42}$$

and

$$\lim_{j\to\infty} \|\omega_0^{(j)}\|_{\mathcal{M}} = \|\omega_0\|_{\mathcal{M}}. \tag{5.43}$$

Let $\omega(\cdot,t)$ (respectively $\omega^{(j)}(\cdot,t)$) be the mild solution with initial data ω_0 (respectively $\omega_0^{(j)}$).

Then, for any $T_2 > T_1 > 0$,

$$\sup_{T_1<t<T_2} \|\omega(\cdot,t) - \omega^{(j)}(\cdot,t)\|_1 \xrightarrow[j\to\infty]{} 0. \tag{5.44}$$

Proof. The proof is a verbatim repetition of the first (existence) part of the proof of Theorem 5.3. Notice that the estimate (5.22), which is crucial in the proof, is satisfied due to the hypothesis (5.43). ■

5.1 Uniqueness for general initial measures

The question of the uniqueness of the solution $\omega(\cdot, t)$ given in Theorem 5.3, for any initial measure $\omega_0 \in \mathcal{M}$, was settled by Gallagher and Gallay [70]. Here we outline briefly some of the ideas of their proof.

We start by reviewing the setting in Subsection 2.1.1. Thus, we consider the equation

$$(5.45) \quad \partial_t \phi(\mathbf{x}, t) - \nu \Delta_{\mathbf{x}} \phi(\mathbf{x}, t) = \mathbf{a}(\mathbf{x}, t) \cdot \nabla_{\mathbf{x}} \phi(\mathbf{x}, t), \qquad (\mathbf{x}, t) \in \mathbb{R}^2 \times (0, \infty),$$

subject to the initial condition $\phi_0 \in \mathcal{M}$, which is attained in the weak* sense (2.34). The divergence-free coefficient $\mathbf{a}(\mathbf{x}, t)$ can blow up at $t = 0$ but is limited by the growth condition (2.35).

The existence, uniqueness and regularity of the solution are stated in Lemma 2.9. In particular, shifting the initial time to $t_0 > 0$, we have $\phi(\cdot, t_0) \in L^1(\mathbb{R}^2)$.

We can now repeat verbatim the argument employed in Section 3.1 (based on the Nash inequality). Letting $t_0 \to 0$ we get, as in (3.9),

$$(5.46) \qquad \|\phi(\cdot, t)\|_p \leq (\nu\eta t)^{-1+\frac{1}{p}} \|\phi_0\|_{\mathcal{M}}, \quad t > 0, \ p \in [1, \infty].$$

This estimate is an important ingredient in the proof, as it applies to the general equation (5.45), where $\mathbf{a}(\mathbf{x}, t)$ is not necessarily the velocity field associated (via the Biot–Savart kernel) with $\phi(\mathbf{x}, t)$. Notice that it is independent of $\mathbf{a}(\mathbf{x}, t)$.

Turning back to the vorticity solution $\omega(\mathbf{x}, t)$, it follows from Remark 5.4 that the associated velocity field $\mathbf{u}(x, t)$ satisfies the conditions imposed on $\mathbf{a}(\mathbf{x}, t)$.

The first step in the uniqueness proof is to separate the (finite) large atomic part of ω_0 from the rest of the measure, so that with some small $\kappa > 0$ yet to be determined,

$$(5.47) \qquad \omega_0 = \widetilde{\omega_0} + \sum_{l=1}^{N} \alpha_l \delta_{\mathbf{x}=\mathbf{x}_l}, \quad \|(\widetilde{\omega_0})_{atom}\|_{\mathcal{M}} < \kappa.$$

Refer to Section A.4 in Appendix A for the notation relating to measures.

Let the pair $\omega(\mathbf{x},t)$ and $\mathbf{u}(\mathbf{x},t)$ be a solution satisfying the conditions of Theorem 5.3. We need to prove the uniqueness of this solution. Lemma 2.9 can be used, with $\mathbf{a}(\mathbf{x},t) = -\mathbf{u}(\mathbf{x},t)$, yielding unique global solutions (for $(\mathbf{x},t) \in \mathbb{R}^2 \times (0,\infty)$) to the following *linear* equations:

$$\begin{aligned} &\partial_t \zeta_0(\mathbf{x},t) + \mathbf{u}(\mathbf{x},t) \cdot \nabla_{\mathbf{x}} \zeta_0(\mathbf{x},t) = \nu \Delta_{\mathbf{x}} \zeta_0(\mathbf{x},t), \\ &\zeta_0(\mathbf{x},0) = \widetilde{\omega_0}, \end{aligned} \tag{5.48}$$

$$\begin{aligned} &\partial_t \zeta_l(\mathbf{x},t) + \mathbf{u}(\mathbf{x},t) \cdot \nabla_{\mathbf{x}} \zeta_l(\mathbf{x},t) = \nu \Delta_{\mathbf{x}} \zeta_l(\mathbf{x},t), \\ &\zeta_l(\mathbf{x},0) = \alpha_l \delta_{\mathbf{x}=\mathbf{x}_l}, \quad l = 1,2,\dots,N. \end{aligned} \tag{5.49}$$

Clearly,

$$\omega(\mathbf{x},t) = \zeta_0(\mathbf{x},t) + \sum_{l=1}^{N} \alpha_l \zeta_l(\mathbf{x},t), \quad (\mathbf{x},t) \in \mathbb{R}^2 \times (0,\infty), \tag{5.50}$$

and

$$\mathbf{u}(\mathbf{x},t) = \mathbf{v}_0(\mathbf{x},t) + \sum_{l=1}^{N} \alpha_l \mathbf{v}_l(\mathbf{x},t) = \mathbf{v}_0(\mathbf{x},t) + \mathbf{U}(\mathbf{x},t), \tag{5.51}$$

where

$$\mathbf{v}_l(\cdot,t) = \mathbf{K} * \zeta_l(\cdot,t), \quad l = 0,1,2,\dots,N, \ t > 0. \tag{5.52}$$

Note that if in Equation (5.49) the coefficient $\mathbf{u}(\mathbf{x},t)$ is replaced by $\mathbf{v}_l(\mathbf{x},t)$, then the solution is given by the Oseen vortex $\omega^{pv}(\mathbf{x}-\mathbf{x}_l,t)$, introduced in (5.1). The difference between $\zeta_l(\mathbf{x},t)$ and $\omega^{pv}(\mathbf{x}-\mathbf{x}_l,t)$ is measured in terms of weighted norms that will be introduced in Section 6.2 below.

From (5.47) we know that the atomic part of $\widetilde{\omega_0}$ is small. This can be used in order to estimate ζ_0 near $t=0$. As in (5.7), we have

$$\begin{aligned} \zeta_0(\mathbf{x},t) &= (G_\nu(\cdot,t) * \widetilde{\omega_0})(\mathbf{x}) \\ &+ \int_0^t \int_{\mathbb{R}^2} \nabla_{\mathbf{y}} G_\nu(\mathbf{x}-\mathbf{y},t-s) \cdot \zeta_0(\mathbf{y},s) \mathbf{v}_0(\mathbf{y},s) d\mathbf{y} ds \\ &+ \int_0^t \int_{\mathbb{R}^2} \nabla_{\mathbf{y}} G_\nu(\mathbf{x}-\mathbf{y},t-s) \cdot \zeta_0(\mathbf{y},s) \mathbf{U}(\mathbf{y},s) d\mathbf{y} ds, \quad \mathbf{x} \in \mathbb{R}^2. \end{aligned} \tag{5.53}$$

The coefficient $\mathbf{v}_0$ is associated with ζ_0 by the Biot–Savart kernel, so we can estimate as in (5.33),

$$\begin{aligned} \|\zeta_0(\cdot,t)\|_{\frac{4}{3}} &\le \kappa t^{-\frac{1}{4}} + C(\tfrac{4}{3},\nu) \xi_{\frac{4}{3}} \int_0^t (t-s)^{-\frac{3}{4}} \|\zeta_0(\cdot,s)\|_{\frac{4}{3}}^2 ds \\ &+ C(\tfrac{4}{3},\nu) \int_0^t (t-s)^{-\frac{3}{4}} \|\zeta_0(\cdot,s) \mathbf{U}(\cdot,s)\|_1 ds. \end{aligned} \tag{5.54}$$

The measure $\widetilde{\omega_0}$ has no atomic part at the points $\{\mathbf{x}_l\}_{l=1}^N$. On the other hand, by (5.51) the velocity field $\mathbf{U}(\mathbf{x},t)$ is obtained from a convection–diffusion equation with initial data supported precisely at these points. The Gaussian bounds given in Theorem 2.10 imply that $\|\zeta_0(\cdot,s)\mathbf{U}(\cdot,s)\|_1$ vanishes as $s \to 0$, and we obtain, as in (5.35),

$$\limsup_{t\to 0} t^{\frac{1}{4}} \|\zeta_0(\cdot,t)\|_{\frac{4}{3}} \le 2\kappa. \tag{5.55}$$

Notice that this estimate can be obtained directly by using the Gaussian bounds of the fundamental solution [70, Lemma 4.1].

The strategy of the uniqueness proof can be summarized as follows: given two possible (mild) solutions subject to the same initial data, they are both decomposed as in (5.50). The parts associated with $\widetilde{\omega_0}$ remain small by (5.55), while the parts associated with $\sum_{l=1}^{N} \alpha_l \delta_{\mathbf{x}=\mathbf{x}_l}$ are estimated in terms of their distances to the respective Oseen vortices. The difference of the two solutions is then estimated by using the $L^{\frac{4}{3}}$ norm for the first part and weighted norms for the second. This leads, as in the proof of Corollary 5.6, to a Gronwall inequality that forces the identity of the two solutions.

5.2 Notes for Chapter 5

- The existence of solutions to (2.1)–(2.2) with measure initial data was proved, using different methods, by Cottet [49], Giga, Miyakawa and Osada [79]. See also Notes for Chapter 4. Kato [108] used estimates for all derivatives, depending only on $\|\omega_0\|_1$, in order to obtain the case of measure initial data by approximation. Our treatment of the existence statement in Theorem 5.3 follows Kato's proof.
- Well posedness of the Navier–Stokes system, in terms of the velocity–pressure formulation, is established in [75] (for the two-dimensional case) and [116] (in the n-dimensional version, $n \ge 2$). They deal with initial (velocity) data beyond the "energy space" $L^2(\mathbb{R}^2)$. However, when recast in the framework of the vorticity formulation, they require at least sufficiently small initial data (even in $L^1(\mathbb{R}^2)$).
- Our interest in this chapter is in the case where the initial velocity is not square-integrable. We obtained the well-posedness of (1.10)–(1.11) in the full plane. However, the same problem in bounded domains, is still open, when "natural" boundary conditions (such as the "no-slip"

condition) are imposed. See [140] for the case of the "non-physical" requirement of vanishing vorticity on the boundary.

- Corollary 5.7 for general initial measures was proved by Gallagher and Gallay [70, Section 5.3]. Note that for sufficiently small $t > 0$ the estimate (4.3) yields

$$\|\omega(\cdot,t) - \omega^{(j)}(\cdot,t)\|_1 \leq C\|\omega_0 - \omega_0^{(j)})\|_{\mathcal{M}}.$$

However, the map $\omega_o \to \omega(\cdot,t)$ is not locally Lipschitz. The basic reason is that in general we cannot find $t_{\frac{4}{3}} > 0$ so that the smallness condition (5.35) is satisfied for all initial measures ω_0 in a small ball in $\mathcal{M}$.
- See [19] for a treatment of well-posedness for vorticity initial data in functional spaces beyond finite measures. In fact, these functional spaces are defined by suitable restrictions of the action of the heat kernel on the initial data.
- The uniqueness proof for $\omega_0 = \alpha\delta_{\mathbf{x}=0}$, for any $\alpha \in \mathbb{R}$, was given in [71].

Chapter 6

Asymptotic Behavior for Large Time

In the previous chapters we have dealt with the global (in time) well-posedness of the vorticity equation (2.1)–(2.2). We have gradually relaxed the assumptions imposed on the initial data; first very smooth and finally measure-valued.

The irregularity of the initial data forced us to focus on the behavior of solutions near the initial time $t = 0$. In fact, we have seen that the solution becomes smooth for $t > 0$ (see Theorem 4.3) and (restricting to the vorticity) belongs to $Y_p(\mathbb{R}^2)$ for any $p \in [1, \infty]$ [see (3.26) for the definition of these spaces].

In this chapter we turn our attention to the asymptotic behavior of the solution for large time. By the above comments we may certainly assume that $\omega_0 \in L^1(\mathbb{R}^2)$, and, wherever needed, $\omega_0 \in Y_p$, $p \in [1, \infty]$.

To simplify notation, we now set the viscosity coefficient in Equation (2.1) equal to one,

$$\nu = 1.$$

It is clear that in the context of asymptotic behavior, this involves no loss of generality, changing t to νt in (2.1) and replacing the velocity $\mathbf{u}$ by $\nu^{-1}\mathbf{u}$.

As in the previous chapters, our main object is the vorticity (in a two-dimensional setting), which is a solution of the scalar equation (2.1) (with $\nu = 1$). As pointed out in the Notes to this chapter, the study of the asymptotic properties of such solutions has much in common with similar properties for other types of nonlinear "convection–diffusion" equations. In our case, the asymptotic behavior of the velocity is obtained from (2.2).

In Section 6.1 we study the decay of the vorticity in the L^p norm (including the important case $p = 1$).

In Section 6.2 we follow the work of Gallay and Wayne [73] and establish the fact that, following an appropriate scaling, the solution (for *any* initial measure-valued data) approaches the Oseen vortex (5.1).

Finally, in Section 6.3 we discuss the global stability of the Oseen vortex and refer to the challenging problem of stability in bounded domains. These comments, concluding Part I of this monograph, are strongly connected to the results of the numerical simulations described in the last chapter of Part II.

6.1 Decay estimates for large time

For any $t > 0$, the solution $\omega(\mathbf{x}, t)$ to (2.1)–(2.2) satisfies

$$\int_{\mathbb{R}^2} \omega(\mathbf{x}, t) d\mathbf{x} = \int_{\mathbb{R}^2} \omega_0(\mathbf{x}) d\mathbf{x}.$$

Thus, in general, there is no decay (for large time) in the $L^1(\mathbb{R}^2)$ norm. On the other hand, by (3.9), the vorticity decays (as $t \to \infty$) in all L^p norms, $p \in (1, \infty]$.

Observe that as mentioned in the Notes for Chapter 3, the constant η in (3.9) (and the subsequent inequalities) can be replaced by 4π, thus equalizing the L^1–L^p estimates for vorticity with those of the heat equation.

A solution to the heat equation (in any $\mathbb{R}^n$) decays (as $t \to \infty$) in the L^1 norm if the initial function has zero mean value. In addition, its rate of decay in L^∞ is faster than $t^{-\frac{n}{2}}$. For the reader's convenience, we include a short proof of these facts.

Claim 6.1. *Consider the solution $\phi(\mathbf{x}, t)$ to the heat equation* (2.13). *Assume that $\phi_0 \in L^1(\mathbb{R}^n)$ and*

$$\int_{\mathbb{R}^n} \phi_0(\mathbf{x}) d\mathbf{x} = 0. \tag{6.1}$$

Then

$$\lim_{t \to \infty} \|\phi(\cdot, t)\|_1 = 0 \tag{6.2}$$

and

$$\lim_{t \to \infty} t^{\frac{n}{2}} \|\phi(\cdot, t)\|_\infty = 0. \tag{6.3}$$

Proof. Under condition (6.1) the expression (2.13) can be rewritten as

$$\begin{aligned}\phi(\mathbf{x},t) &= \int_{\mathbb{R}^n} (G(\mathbf{x}-\mathbf{y},t) - G(\mathbf{x},t))\phi_0(\mathbf{y})d\mathbf{y} \\ &= \int_{|\mathbf{y}|\geq R} (G(\mathbf{x}-\mathbf{y},t) - G(\mathbf{x},t))\phi_0(\mathbf{y})d\mathbf{y} \\ &\quad + \int_{|\mathbf{y}|\leq R} (G(\mathbf{x}-\mathbf{y},t) - G(\mathbf{x},t))\phi_0(\mathbf{y})d\mathbf{y} \\ &= I_R^1(\mathbf{x},t) + I_R^2(\mathbf{x},t).\end{aligned}$$

Fix $\varepsilon > 0$ and take $R > 0$ sufficiently large, so that $\int_{|\mathbf{y}|\geq R} |\phi_0(\mathbf{y})|d\mathbf{y} < \varepsilon$.

Then, using Fubini's theorem and (2.15),

$$\int_{\mathbb{R}^n} |I_R^1(\mathbf{x},t)|d\mathbf{x} \leq 2 \int_{|\mathbf{y}|\geq R} |\phi_0(\mathbf{y})|d\mathbf{y} < 2\varepsilon, \quad t > 0.$$

On the other hand

$$G(\mathbf{x}-\mathbf{y},t) - G(\mathbf{x},t) = -\int_0^1 \boldsymbol{\nabla}_{\mathbf{x}} G(\mathbf{x}-\lambda\mathbf{y},t)\cdot\mathbf{y}d\lambda,$$

so that, from (2.16) and Fubini's theorem

$$\int_{\mathbb{R}^n} |I_R^2(\mathbf{x},t)|d\mathbf{x} \leq R\beta t^{-\frac{1}{2}} \int_{|\mathbf{y}|\leq R} |\phi_0(\mathbf{y})|d\mathbf{y} \leq R\beta t^{-\frac{1}{2}}\|\phi_0\|_1, \quad t > 0.$$

The two estimates above imply

$$\limsup_{t\to\infty} \|\phi(\cdot,t)\|_1 < 2\varepsilon,$$

which proves (6.2) since $\varepsilon > 0$ is arbitrary.

Next, we note that $\|G(\cdot,t)\|_\infty = C_1 t^{-\frac{n}{2}}$ and $\|\boldsymbol{\nabla} G(\cdot,t)\|_\infty = C_2 t^{-\frac{n+1}{2}}$. Decomposing the solution as before, it follows that

$$\sup_{\mathbf{x}\in\mathbb{R}^n} |I_R^1(\mathbf{x},t)| \leq 2C_1 t^{-\frac{n}{2}} \int_{|\mathbf{y}|\geq R} |\phi_0(\mathbf{y})|d\mathbf{y} < 2C_1 t^{-\frac{n}{2}}\varepsilon, \quad t > 0.$$

On the other hand

$$\sup_{\mathbf{x}\in\mathbb{R}^n} |I_R^2(\mathbf{x},t)| \leq RC_2 t^{-\frac{n+1}{2}} \int_{|\mathbf{y}|\leq R} |\phi_0(\mathbf{y})|d\mathbf{y} \leq RC_2 t^{-\frac{n+1}{2}}\|\phi_0\|_1, \quad t > 0.$$

Adding these two estimates we obtain (6.3).

∎

It is remarkable that a similar fact holds for vorticity in the two-dimensional case. In what follows, the solution $\omega(\mathbf{x}, t)$ is that obtained in Theorem 4.3.

Proposition 6.2. *Consider the system* (2.1)–(2.2) *and assume that* $\omega_0 \in L^1(\mathbb{R}^2)$ *and*

$$\int_{\mathbb{R}^2} \omega_0(\mathbf{x}) d\mathbf{x} = 0. \tag{6.4}$$

Then

$$\lim_{t\to\infty} \|\omega(\cdot, t)\|_1 = 0. \tag{6.5}$$

Proof. By Hölder's inequality and (3.11) we get,

$$\|\omega(\cdot, s)\mathbf{u}(\cdot, s)\|_1 \le \xi_{\frac{4}{3}} \|\omega(\cdot, s)\|^2_{\frac{4}{3}}. \tag{6.6}$$

From the integral representation (5.7) of the mild solution we obtain [as in (3.20) or (4.3)] the estimate

$$\begin{aligned} \|\omega(\cdot, t)\|_1 &\le \|G(\cdot, t) * \omega_0\|_1 + \beta\xi_{\frac{4}{3}} \int_0^t (t-s)^{-\frac{1}{2}} \|\omega(\cdot, s)\|^2_{\frac{4}{3}} ds \\ &\le \|G(\cdot, t) * \omega_0\|_1 + \beta\xi_{\frac{4}{3}}\eta^{-\frac{1}{2}} \|\omega_0\|_1^2 \int_0^t (t-s)^{-\frac{1}{2}} s^{-\frac{1}{2}} ds \\ &= \|G(\cdot, t) * \omega_0\|_1 + F\|\omega_0\|_1^2, \end{aligned} \tag{6.7}$$

where $F = \beta\pi\xi_{\frac{4}{3}}\eta^{-\frac{1}{2}}$ and we used (3.9) to estimate $\|\omega(\cdot, s)\|_{\frac{4}{3}}$.

Let $\varepsilon > 0$. On account of Claim 6.1 we have $T > 0$ such that

$$\|G(\cdot, T) * \omega_0\|_1 < \varepsilon.$$

Shifting time by T in (6.7), so that the initial data is $\omega(\cdot, T)$, and noting that the heat kernel (2.13) contracts the L^1 norm, we get,

$$\|\omega(\cdot, 2T)\|_1 \le \varepsilon + F\|\omega(\cdot, T)\|_1^2.$$

Continuing in the same fashion we have,

$$\|\omega(\cdot, 2^{n+1}T)\|_1 \le \varepsilon + F\|\omega(\cdot, 2^n T)\|_1^2, \quad n = 1, 2, \ldots \tag{6.8}$$

Recall from (2.3) that $\|\omega(\cdot, t)\|_1$ is non increasing as function of t.

If for an infinite subsequence of indices n we have

$$\|\omega(\cdot, 2^{n+1}T)\|_1 \le \frac{1}{2}\|\omega(\cdot, 2^n T)\|_1,$$

then clearly (6.5) holds true. Otherwise, this can happen only finitely many times, so that, shifting T to some 2^kT and retaining notation, we can rewrite (6.8) as

$$\|\omega(\cdot, 2^nT)\|_1 \leq 2\varepsilon + 2F\|\omega(\cdot, 2^nT)\|_1^2, \quad n = 1, 2, \ldots \tag{6.9}$$

Suppose now that

$$\|\omega_0\|_1 < \kappa, \tag{6.10}$$

where $\kappa > 0$ satisfies

$$2F\kappa < 1. \tag{6.11}$$

It follows that $\|\omega(\cdot, 2^nT)\|_1 \leq \|\omega_0\|_1 < \kappa$, hence

$$\|\omega(\cdot, 2T)\|_1 \leq 2(1 - 2F\kappa)^{-1}\varepsilon. \tag{6.12}$$

Since F is a universal constant, this estimate holds for any $\varepsilon > 0$ (with a suitable T), and we obtain the limit (6.5) under condition (6.10).

The reader is advised to compare the above argument to the last part of the proof of Theorem 5.3, establishing the uniqueness of the solution for the case of a sufficiently small atomic part of the initial data.

To remove the restriction (6.10) we take $\omega_0 \in Y_\infty$ (see the opening comments to this chapter). The condition (6.10) certainly implies the validity of the limit (6.5) if $\|\omega_0\|_{Y_\infty} < \kappa$.

Define

$$R = \sup\{r > 0, \text{ such that (6.5) holds if } \|\omega_0\|_{Y_\infty} < r\}.$$

In particular, $R \geq \kappa$. Suppose that $R < \infty$. Let $\theta_0 \in Y_\infty$ satisfy

$$\|\theta_0\|_{Y_\infty} < R + r,$$

where $0 < r < R$ is yet to be determined. Let $\omega_0 = \frac{R}{R+r}\theta_0$, and let $\theta(\cdot, t) = (S\theta_0)(t)$ and $\omega(\cdot, t) = (S\omega_0)(t)$. By the definition of R, there exists $T > 0$ such that

$$\|\omega(\cdot, t)\|_1 < \frac{1}{2}\kappa, \quad t \geq T. \tag{6.13}$$

Due to (4.9) there exists $C_T > 0$, depending on R, so that

$$\|\omega(\cdot, t) - \theta(\cdot, t)\|_1 \leq C_T\|\omega_0 - \theta_0\|_{Y_\infty} \leq C_T\frac{2Rr}{R + r}.$$

We now take $r > 0$ sufficiently small, so that

$$C_T\frac{2Rr}{R + r} < \frac{1}{2}\kappa,$$

hence $\|\theta(\cdot,T)\|_1 < \kappa$ and by the first part of the proof $\|\theta(\cdot,t)\|_1 \to 0$ as $t \to \infty$. This contradicts the definition of R, so that $R = \infty$ and the proof is complete.

■

In Lemma 3.4 we saw that $t^{1-\frac{1}{p}}\|\omega(\cdot,t)\|_p$, $1 < p \le \infty$, vanishes as $t \to 0$. We now state an analogous result for large time.

Proposition 6.3. *Let ω_0 satisfy the zero mean value assumption* (6.4). *Then*

$$\lim_{t\to\infty} t^{1-\frac{1}{p}}\|\omega(\cdot,t)\|_p = 0, \qquad p \in [1,\infty]. \tag{6.14}$$

Proof. The case $p = 1$ is just Proposition 6.2.

On the other hand, it follows from (3.9) that $t\|\omega(\cdot,t)\|_\infty < C$. The interpolation inequality $\|g\|_p \le \|g\|_\infty^{1-\frac{1}{p}}\|g\|_1^{\frac{1}{p}}$ finishes the proof for $p \in (1,\infty)$.

It remains to deal with the case $p = \infty$. In view of (6.3) it is clear that we need to deal only with [see (5.7)],

$$\Omega(\mathbf{x},t) = \int_0^t \int_{\mathbb{R}^2} \nabla_{\mathbf{y}} G_\nu(\mathbf{x}-\mathbf{y},t-s)\cdot\omega(\mathbf{y},s)\mathbf{u}(\mathbf{y},s)d\mathbf{y}ds.$$

Let $\varepsilon > 0$. Taking into account the decay already established (Proposition 6.2) for $p = 1$, there exists $T > 0$ such that $\|\omega(\cdot,T)\|_1 < \varepsilon$, and by shifting time we may assume $\|\omega_0\|_1 < \varepsilon$.

We estimate separately

$$\Omega_1(\mathbf{x},t) = \int_0^{\frac{t}{2}} \int_{\mathbb{R}^2} \nabla_{\mathbf{y}} G_\nu(\mathbf{x}-\mathbf{y},t-s)\cdot\omega(\mathbf{y},s)\mathbf{u}(\mathbf{y},s)d\mathbf{y}ds$$

and

$$\Omega_2(\mathbf{x},t) = \int_{\frac{t}{2}}^t \int_{\mathbb{R}^2} \nabla_{\mathbf{y}} G_\nu(\mathbf{x}-\mathbf{y},t-s)\cdot\omega(\mathbf{y},s)\mathbf{u}(\mathbf{y},s)d\mathbf{y}ds.$$

Using $\|\nabla G(\cdot,t)\|_\infty = C_2 t^{-\frac{3}{2}}$ (see the proof of Claim 6.1) we can estimate, as in (6.7),

$$\begin{aligned} \|\Omega_1(\cdot,t)\|_\infty &\le C_2\xi_{\frac{4}{3}} \int_0^{\frac{t}{2}} (t-s)^{-\frac{3}{2}}\|\omega(\cdot,s)\|_{\frac{4}{3}}^2 ds \\ &\le C_2\xi_{\frac{4}{3}}\eta^{-\frac{1}{2}}\|\omega_0\|_1^2 \int_0^{\frac{t}{2}} (t-s)^{-\frac{3}{2}}s^{-\frac{1}{2}}ds \\ &\le C_2\xi_{\frac{4}{3}}\eta^{-\frac{1}{2}}\varepsilon^2 t^{-1} \int_0^{\frac{1}{2}} (1-s)^{-\frac{3}{2}}s^{-\frac{1}{2}}ds. \end{aligned} \tag{6.15}$$

Hence,

$$\limsup_{t\to\infty} t\|\Omega_1(\cdot,t)\|_\infty < C\varepsilon^2. \tag{6.16}$$

Next, using (2.16), as well as (3.9) and (3.12) (with $p = q = \infty$), we get

$$\begin{aligned}\|\Omega_2(\cdot,t)\|_\infty &\le \beta\xi_2 \int_{\frac{t}{2}}^{t} (t-s)^{-\frac{1}{2}}\|\omega(\cdot,s)\|_\infty\|\mathbf{u}(\cdot,s)\|_\infty ds\\ &\le \beta\xi_2\eta^{-\frac{3}{2}}\|\omega_0\|_1^2 \int_{\frac{t}{2}}^{t}(t-s)^{-\frac{1}{2}}s^{-\frac{3}{2}}ds \qquad (6.17)\\ &\le \beta\xi_2\eta^{-\frac{3}{2}}\varepsilon^2 t^{-1}\int_{\frac{1}{2}}^{1}(1-s)^{-\frac{1}{2}}s^{-\frac{3}{2}}ds,\end{aligned}$$

so that,

$$\limsup_{t\to\infty} t\|\Omega_2(\cdot,t)\|_\infty < C\varepsilon^2. \tag{6.18}$$

The proof is complete in view of (6.16) and (6.18). ■

Remark 6.4. The result of Proposition 6.2 can be improved if it is also assumed that ω_0 is compactly supported (in fact, only exponential decay of ω_0 is required). In this case we have

$$\sup_{0\le t<\infty} t^{\frac{1}{2}}\|\omega(\cdot,t)\|_1 < \infty,$$

and combining this with (3.8) we obtain

$$\sup_{0<t<\infty} t^{\frac{3}{2}}\|\omega(\cdot,t)\|_\infty < \infty.$$

These estimates are identical to the ones obtained for the heat equation (in terms of decay rate), for initial data satisfying (6.4). We refer to [35, Theorem 8] for details and sharp constants.

6.2 Initial data with stronger spatial decay

The L^p decay results of Section 6.1 were obtained assuming zero mean value of the initial data. However, if this is not the case, we still know, by (3.9) and (3.12), that $\|\omega(\cdot,t)\|_p$ (respectively $\|\mathbf{u}(\cdot,t)\|_q$) vanishes, as $t\to\infty$, if $1 < p \le \infty$ (respectively $2 < q \le \infty$).

The nature of this decay is of interest. Specifically, we would like to know if, under appropriate scaling, the solutions converge to one "typical" state or else, if several such "asymptotic states" are possible.

A similar question has been addressed with regard to various (nonlinear) evolution equations. Recall in particular the case of the hyperbolic (scalar) conservation law, where all solutions emanating from compactly supported initial data converge to a unique profile of an N-wave [51, Section 11.6].

It turns out that an analogous phenomenon occurs for the vorticity equation (2.1); all solutions converge, subject to suitable scaling, to the Oseen vortex, introduced in (5.1).

In this section we outline the proof of this result, following the work of Gallay and Wayne [73], for the case of sufficiently small initial data. The "smallness" of the initial data will be stated in terms of weighted-L^2 spaces, to be defined below. In fact, in this case the evolution of the vorticity is determined, to any order of decay, by finite-dimensional invariant manifolds. The global stability result will be stated in Section 6.3.

For any $m \geq 0$ we introduce the Hilbert space $L^{2,m}(\mathbb{R}^2)$ by

$$L^{2,m}(\mathbb{R}^2) = \Big\{ f, \qquad \|f\|^2_{L^{2,m}} := \int_{\mathbb{R}^2} (1+|\mathbf{x}|^2)^m |f(\mathbf{x})|^2 d\mathbf{x} \Big\}. \tag{6.19}$$

We infer from the Cauchy–Schwarz inequality that $L^{2,m}(\mathbb{R}^2) \subseteq L^1(\mathbb{R}^2)$ if $m > 1$. As was mentioned in the Notes for Chapter 4, in this case the space $L^{2,m}(\mathbb{R}^2)$ is a "persistence space" for the vorticity equation (2.1). This observation is an essential tool in the treatment.

Proposition 6.5. *The system* (2.1)–(2.2) *is well posed in* $L^{2,m}(\mathbb{R}^2)$ *for any* $m > 1$. *More explicitly, for any* $\omega_0 \in L^{2,m}(\mathbb{R}^2)$ *there exists a unique global solution*

$$\omega(\cdot, t) \in C(\overline{\mathbb{R}_+}; L^{2,m}(\mathbb{R}^2)).$$

Proof. As in the case of (the more general) L^1 or measure initial vorticity (Chapters 4 and 5), this means that the integral equation (5.7) is solvable in this space. This can be accomplished either by approximating the initial data by regular functions (as we did in the preceding chapters), or by obtaining the solution as a fixed point of a contraction map, as we did in the proof of Theorem 2.2. In either case, the first step of the proof (and really the "heart of the matter") is to show that initial values in $L^{2,m}$ evolve in the same space. This requires suitable estimates for the map

(6.20)

$$\begin{aligned}(\Phi\omega)(\mathbf{x}, t) &= G(\cdot, t) * \omega_0 + \int_0^t \int_{\mathbb{R}^2} \nabla_{\mathbf{y}} G(\mathbf{x} - \mathbf{y}, t - s) \cdot \omega(\mathbf{y}, s)\mathbf{u}(\mathbf{y}, s) d\mathbf{y} ds \\ &= \phi(\mathbf{x}, t) + \int_0^t \Omega(\mathbf{x}, s; t) ds,\end{aligned}$$

where $\omega_0 \in L^{2,m}(\mathbb{R}^2)$, $m > 1$, and $\omega(\cdot, t) \in C(\overline{\mathbb{R}_+}; L^{2,m}(\mathbb{R}^2))$ with $\mathbf{u}(\cdot, t) = \mathbf{K} * \omega(\cdot, t)$ being the associated velocity field.

Thus, while we do not give a full proof of this proposition here [see [73, Section 3]], we derive the required estimates of the two terms on the right-hand side of (6.20).

Fix $T > 0$. We need to estimate $\phi(\mathbf{x}, t)$ and $\Omega(\mathbf{x}, s; t)$. The estimates below make use of "Peetre's inequality" (whose simple proof is left to the reader),

$$(1 + |\mathbf{x}|^2)^s \le C(1 + |\mathbf{y}|^2)^s (1 + |\mathbf{x} - \mathbf{y}|^2)^{|s|}, \quad \mathbf{x},\, \mathbf{y} \in \mathbb{R}^2,\ s \in \mathbb{R}, \tag{6.21}$$

where $C > 0$ depends only on s (actually one can take $C = 2^{|s|}$, valid in any dimension).

In what follows we designate $C > 0$ as a generic constant depending only on m and T, unless dependence on other parameters is explicitly indicated.

We shall make use of the following obvious extensions of (3.14) and (3.18), valid for $s \in (0, T)$,

$$\left\{ \int_{\mathbb{R}^2} (1 + |\mathbf{y}|^2)^m G(\mathbf{y}, s)^p d\mathbf{y} \right\}^{\frac{1}{p}} \le C(T, m, p) s^{-1+\frac{1}{p}}, \ p \in [1, \infty],$$

$$\left\{ \int_{\mathbb{R}^2} (1 + |\mathbf{y}|^2)^m |\nabla_{\mathbf{y}} G(\mathbf{y}, s)|^q d\mathbf{y} \right\}^{\frac{1}{q}} \le C(T, m, q) s^{-\frac{3}{2}+\frac{1}{q}}, \ q \in [1, \infty].$$

By Peetre's inequality

$$(1+|\mathbf{x}|^2)^{\frac{m}{2}} |\phi(\mathbf{x}, t)| \le C \int_{\mathbb{R}^2} (1+|\mathbf{x}-\mathbf{y}|^2)^{\frac{m}{2}} G(\mathbf{x}-\mathbf{y}, t)(1+|\mathbf{y}|^2)^{\frac{m}{2}} |\omega_0(\mathbf{y})| d\mathbf{y},$$

so from Young's inequality [see (A.4) in Appendix A] we infer that, for $t \in (0, T]$,

$$\begin{aligned} &\|\phi(\cdot, t)\|_{L^{2,m}} = \left\{ \int_{\mathbb{R}^2} (1 + |\mathbf{x}|^2)^m |\phi(\mathbf{x}, t)|^2 d\mathbf{x} \right\}^{\frac{1}{2}} \\ &\le C \int_{\mathbb{R}^2} (1 + |\mathbf{y}|^2)^{\frac{m}{2}} G(\mathbf{y}, t) d\mathbf{y} \cdot \left\{ \int_{\mathbb{R}^2} (1 + |\mathbf{y}|^2)^m \omega_0(\mathbf{y})^2 d\mathbf{y} \right\}^{\frac{1}{2}} \\ &\le C \|\omega_0\|_{L^{2,m}}. \end{aligned} \tag{6.22}$$

To estimate the $L^{2,m}$ norm of $\Omega(\mathbf{x}, s; t)$ we again resort to Peetre's inequality and Young's inequality, as above,

$$\tag{6.23}\begin{aligned}&\Big\{\int_{\mathbb{R}^2}(1+|\mathbf{x}|^2)^m\Omega(\mathbf{x},s;t)^2d\mathbf{x}\Big\}^{\frac12}\\&\quad\le C\Big\{\int_{\mathbb{R}^2}\big[(1+|\mathbf{y}|^2)^{\frac m2}|\omega(\mathbf{y},s)\mathbf{u}(\mathbf{y},s)|\big]^{\frac43}d\mathbf{y}\Big\}^{\frac34}\\&\qquad\cdot\Big\{\int_{\mathbb{R}^2}\big|(1+|\mathbf{y}|^2)^{\frac m2}\nabla_{\mathbf{y}}G(\mathbf{y},t-s)\big|^{\frac43}d\mathbf{y}\Big\}^{\frac34}\\&\quad\le C(t-s)^{-\frac34}\Big\{\int_{\mathbb{R}^2}\big[(1+|\mathbf{y}|^2)^{\frac m2}|\omega(\mathbf{y},s)\mathbf{u}(\mathbf{y},s)|\big]^{\frac43}d\mathbf{y}\Big\}^{\frac34}.\end{aligned}$$

By Hölder's inequality

$$\|\omega(\cdot,s)\|_{\frac43}\le\Big\{\int_{\mathbb{R}^2}[(1+|\mathbf{y}|^2)^m\omega(\mathbf{y},s)^2d\mathbf{y}\Big\}^{\frac12}\Big\{\int_{\mathbb{R}^2}[(1+|\mathbf{y}|^2)^{-2m}d\mathbf{y}\Big\}^{\frac14}.$$

Thus, from (3.11) (and once again Hölder's inequality) we get

$$\tag{6.24}\begin{aligned}&\Big\{\int_{\mathbb{R}^2}\big[(1+|\mathbf{y}|^2)^{\frac m2}|\omega(\mathbf{y},s)\mathbf{u}(\mathbf{y},s)|\big]^{\frac43}d\mathbf{y}\Big\}^{\frac34}\\&\quad\le\Big\{\int_{\mathbb{R}^2}\big[|(1+|\mathbf{y}|^2)^m\omega(\mathbf{y},s)\big]^2d\mathbf{y}\Big\}^{\frac12}\Big\{\int_{\mathbb{R}^2}|\mathbf{u}(\mathbf{y},s)|^4d\mathbf{y}\Big\}^{\frac14}\\&\quad\le\xi_{\frac43}\|\omega(\cdot,s)\|_{L^{2,m}}\|\omega(\cdot,s)\|_{\frac43}\le C\xi_{\frac43}\|\omega(\cdot,s)\|^2_{L^{2,m}},\end{aligned}$$

and inserting this in (6.23) yields

$$\|\Omega(\cdot,s;t)\|_{L^{2,m}}\le C\|\omega(\cdot,s)\|^2_{L^{2,m}}(t-s)^{-\frac34}.$$

Turning back to (6.20), we infer from the last estimate and (6.22) ,

$$\|(\Phi\omega)(\cdot,t)\|_{L^{2,m}}\le C\Big(\|\omega_0\|_{L^{2,m}}+\int_0^t(t-s)^{-\frac34}\|\omega(\cdot,s)\|^2_{L^{2,m}}ds\Big),$$

with $C=C(T,m)>0$.

An immediate consequence of this estimate is that, for $0\le t\le T$,

$$\tag{6.25}\sup_{0\le\tau\le t}\|(\Phi\omega)(\cdot,\tau)\|_{L^{2,m}}\le C\Big(\|\omega_0\|_{L^{2,m}}+t^{\frac14}\sup_{0\le\tau\le t}\|\omega(\cdot,\tau)\|^2_{L^{2,m}}\Big).$$

Given $R>0$, we take $T=T(R)>0$ so that $4C^2RT^{\frac14}<1$. Let $\|\omega_0\|_{L^{2,m}}<R$. The last estimate implies that if $\sup\limits_{0\le\tau\le T}\|\omega(\cdot,\tau)\|_{L^{2,m}}<2CR$, then also $\sup\limits_{0\le\tau\le t}\|(\Phi\omega)(\cdot,\tau)\|_{L^{2,m}}<2CR$.

The basic observation in the existence proof is that we thus obtain a fixed point $\Phi\omega=\omega$ in $[0,T]$, and the solution satisfies

$$\tag{6.26}\|\omega(\cdot,t)\|_{L^{2,m}}\le 2C\|\omega_0\|_{L^{2,m}},\quad 0\le t\le T.$$

■

As has already been pointed out (the Claim in the Introduction) the zero mean value condition (6.4) is necessary for finite energy ($\|\mathbf{u}_0\|_2 < \infty$). We now show that it is also sufficient if $\omega_0 \in L^{2,m}(\mathbb{R}^2)$, $m > 1$. In particular, the classical Leray theory is applicable for such initial data.

Proposition 6.6. *Let* $\omega_0 \in L^{2,m}(\mathbb{R}^2)$, $m > 1$, *and assume also that it satisfies the zero mean value condition* (6.4).

Then the associated velocity field satisfies $\|\mathbf{u}_0\|_2 < \infty$.

Proof. Since $\omega_0 \in L^1(\mathbb{R}^2) \cap L^2(\mathbb{R}^2)$, we have $\omega_0 \in L^p(\mathbb{R}^2)$, $p \in [1, 2]$. It follows by (3.11) that $\mathbf{u}_0 \in L^q(\mathbb{R}^2)$, $q \in (2, \infty)$, hence $\mathbf{u}_0$ is square-integrable in every bounded set and it suffices to restrict our consideration to $E_R = \{|\mathbf{x}| > R\}$ for any $R > 1$.

In our estimates below we designate by $C > 0$ a generic constant that may depend on R.

The assumption $\int_{\mathbb{R}^2} \omega_0(\mathbf{x})d\mathbf{x} = 0$, enables us to rewrite (2.2) as

$$\mathbf{u}_0(\mathbf{x}) = \int_{\mathbb{R}^2} (\mathbf{K}(\mathbf{x}-\mathbf{y}) - \mathbf{K}(\mathbf{x}))\omega_0(\mathbf{y})d\mathbf{y}. \tag{6.27}$$

We may assume $1 < m < 2$ and pick $2 - m < \gamma < 1$. To estimate the integral in (6.27), for $|\mathbf{x}| > R > 1$, we break it up into three integrals, as follows.

$$I_1(\mathbf{x}) = \int_{|\mathbf{y}| \le |\mathbf{x}|/2} (\mathbf{K}(\mathbf{x}-\mathbf{y}) - \mathbf{K}(\mathbf{x}))\omega_0(\mathbf{y})d\mathbf{y},$$

$$I_2(\mathbf{x}) = \int_{\{|\mathbf{x}|/2 \le |\mathbf{y}|\} \cap \{|\mathbf{y}-\mathbf{x}| \ge |\mathbf{x}|^\gamma\}} (\mathbf{K}(\mathbf{x}-\mathbf{y}) - \mathbf{K}(\mathbf{x}))\omega_0(\mathbf{y})d\mathbf{y},$$

$$I_3(\mathbf{x}) = \int_{|\mathbf{y}-\mathbf{x}| \le |\mathbf{x}|^\gamma} (\mathbf{K}(\mathbf{x}-\mathbf{y}) - \mathbf{K}(\mathbf{x}))\omega_0(\mathbf{y})d\mathbf{y}.$$

For I_1, we note that, using an obvious estimate for $\nabla\mathbf{K}$,

$$|\mathbf{K}(\mathbf{x}-\mathbf{y}) - \mathbf{K}(\mathbf{x})| \le C|\mathbf{x}|^{-2}|\mathbf{y}|,$$

hence, by the Cauchy–Schwarz inequality,

$$\begin{aligned}
|I_1(\mathbf{x})| &\le C|\mathbf{x}|^{-2} \int_{|\mathbf{y}| \le |\mathbf{x}|/2} |\mathbf{y}||\omega_0(\mathbf{y})|d\mathbf{y} \\
&\le C|\mathbf{x}|^{-2} \Big\{ \int_{|\mathbf{y}| \le |\mathbf{x}|/2} (1 + |\mathbf{y}|^2)^{-(m-1)} d\mathbf{y} \Big\}^{\frac{1}{2}} \|\omega_0\|_{L^{2,m}} \\
&\le C|\mathbf{x}|^{-m} \|\omega_0\|_{L^{2,m}}.
\end{aligned}$$

For I_2, we note that $|\mathbf{K}(\mathbf{x}-\mathbf{y})|+|\mathbf{K}(\mathbf{x})| \le C|\mathbf{x}|^{-\gamma}$, so as in the previous estimate,

$$|I_2(\mathbf{x})| \le C|\mathbf{x}|^{-\gamma}\Big\{\int_{|\mathbf{y}|\ge|\mathbf{x}|/2}(1+|\mathbf{y}|^2)^{-m}d\mathbf{y}\Big\}^{\frac{1}{2}}\cdot\|\omega_0\|_{L^{2,m}}$$
$$\le C|\mathbf{x}|^{-\gamma-m+1}\|\omega_0\|_{L^{2,m}}.$$

To estimate I_3, we split it as

$$I_3(\mathbf{x}) = -\int_{|\mathbf{y}-\mathbf{x}|\le|\mathbf{x}|^\gamma}\mathbf{K}(\mathbf{x})\omega_0(\mathbf{y})d\mathbf{y} + \int_{|\mathbf{y}-\mathbf{x}|\le|\mathbf{x}|^\gamma}\mathbf{K}(\mathbf{x}-\mathbf{y})\omega_0(\mathbf{y})d\mathbf{y}$$
$$= I_{3,1}(\mathbf{x}) + I_{3,2}(\mathbf{x}).$$

A bound for $I_{3,1}(\mathbf{x})$ is obtained as in the case of I_2,

$$|I_{3,1}(\mathbf{x})| \le C|\mathbf{x}|^{-1-m+1}\|\omega_0\|_{L^{2,m}} = C|\mathbf{x}|^{-m}\|\omega_0\|_{L^{2,m}}.$$

Finally, by the Cauchy–Schwarz inequality,

$$|I_{3,2}(\mathbf{x})|^2 \le \int_{|\mathbf{y}-\mathbf{x}|\le|\mathbf{x}|^\gamma}(1+|\mathbf{y}|^2)^{-m}|\mathbf{K}(\mathbf{x}-\mathbf{y})|d\mathbf{y}$$
$$\cdot\int_{|\mathbf{y}-\mathbf{x}|\le|\mathbf{x}|^\gamma}|\mathbf{K}(\mathbf{x}-\mathbf{y})|(1+|\mathbf{y}|^2)^m\omega_0(\mathbf{y})^2d\mathbf{y}$$
$$\le C|\mathbf{x}|^{\gamma-2m}\int_{|\mathbf{y}-\mathbf{x}|\le|\mathbf{x}|^\gamma}|\mathbf{K}(\mathbf{x}-\mathbf{y})|(1+|\mathbf{y}|^2)^m\omega_0(\mathbf{y})^2d\mathbf{y},$$

hence, noting that $C^{-1}|\mathbf{x}| \le |\mathbf{y}| \le C|\mathbf{x}|$ over the domain of integration, we get by Fubini's theorem,

$$\int_{E_R}|I_{3,2}(\mathbf{x})|^2d\mathbf{x} \le C\int_{|\mathbf{y}|\ge\frac{R}{2}}|\mathbf{y}|^{2\gamma-2m}(1+|\mathbf{y}|^2)^m\omega_0(\mathbf{y})^2d\mathbf{y} \le C\|\omega_0\|^2_{L^{2,m}}.$$

The above estimates (and the assumption on γ) imply the square-integrability of I_1, I_2, and I_3 in E_R, so the same is true for $\mathbf{u}_0$. ■

The study of the asymptotic properties of the solution heavily depends on "scaling arguments." Before introducing "rescaling" of space and time, we recall the scaling of the vorticity as introduced in the proof of Lemma 3.7. Thus, if $\omega_0 \in L^{2,m}$, $m > 1$, with $\omega(\mathbf{x},t)$ the corresponding solution (of the vorticity equation), then for any $\lambda > 0$ the function $\lambda^2\omega(\lambda\mathbf{x},\lambda^2 t)$ is a solution of the vorticity equation, with initial data $\lambda^2\omega_0(\lambda\mathbf{x})$. The two initial functions have the same L^1 norm. The following corollary exploits this fact, and gives a rate of growth of the $L^{2,m}$ norm of the solution.

Corollary 6.7. *Let $\omega_0 \in L^{2,m}(\mathbb{R}^2)$, $m > 1$, and let $\omega(\cdot,t) = (S\omega_0)(\cdot,t)$ be the corresponding solution. For any $R > 0$ there exists a constant $C = C(R,m)$ such that, if $\|\omega_0\|_{L^{2,m}} < R$ then*

$$\|\omega(\cdot,t)\|_{L^{2,m}} \le C\|\omega_0\|_{L^{2,m}}(1+t)^{\frac{m}{2}}, \quad t > 0. \tag{6.28}$$

Proof. By (2.3) we know that $\|\omega(\cdot,t)\|_2$ is nonincreasing, so we need only to establish the estimate for $\int_{\mathbb{R}^2}|\mathbf{x}|^{2m}\omega(\mathbf{x},t)^2d\mathbf{x}$.

The preceding comments imply that for any $\lambda>0$, and for some $T>0$ independent of λ, the estimate (6.26) yields, for $\delta\in(0,T)$,

$$\int_{\mathbb{R}^2}|\mathbf{x}|^{2m}\omega(\lambda\mathbf{x},\lambda^2\delta)^2d\mathbf{x}\le C\int_{\mathbb{R}^2}(1+|\mathbf{x}|^2)^m\omega_0(\lambda\mathbf{x})^2d\mathbf{x}.$$

Given any $t>0$, let $\lambda^2=\delta^{-1}t$. Substituting $\mathbf{y}=\lambda\mathbf{x}$, this inequality becomes

$$\int_{\mathbb{R}^2}|\mathbf{y}|^{2m}\omega(\mathbf{y},t)^2d\mathbf{y}\le C\int_{\mathbb{R}^2}(\delta^{-1}t+|\mathbf{y}|^2)^m\omega_0(\mathbf{y})^2d\mathbf{y}.$$

Since $(\delta^{-1}t+|\mathbf{y}|^2)^m<C(1+|\mathbf{y}|^2)^m(1+t)^m$, we obtain (6.28). ■

6.2.1 *Scaling variables and invariant manifolds*

The decay (3.9) of the vorticity in $L^p(\mathbb{R}^2)$, $p\in(1,\infty]$, as $t\to\infty$, means that the zero solution is globally asymptotically stable in all these spaces (but only stable in $L^1(\mathbb{R}^2)$). However, as has already been noted, a suitable scaling of the solution can reveal finer details of the asymptotic behavior. It turns out that the weighted spaces $L^{2,m}(\mathbb{R}^2)$ introduced in (6.19) provide a good framework for this study. However, instead of the "classical" scaling used in the proof of (6.28), which involves the arbitrary parameter $\lambda>0$, and preserves the L^1 norm of the vorticity, we now introduce a fixed transformation, as follows.

$$\boldsymbol{\xi}=(1+t)^{-\frac{1}{2}}\mathbf{x},\quad \tau=\log(1+t),\ (\mathbf{x},t)\in\mathbb{R}^2\times\overline{\mathbb{R}_+}. \tag{6.29}$$

The vorticity $\omega(\mathbf{x},t)$ and velocity $\mathbf{u}(\mathbf{x},t)$ are transformed into new functions $\theta(\boldsymbol{\xi},\tau)$, $\mathbf{v}(\boldsymbol{\xi},\tau)$, respectively, by

$$\theta(\boldsymbol{\xi},\tau)=(1+t)\omega(\mathbf{x},t),\qquad \mathbf{v}(\boldsymbol{\xi},\tau)=(1+t)^{\frac{1}{2}}\mathbf{u}(\mathbf{x},t). \tag{6.30}$$

It can readily be seen that the relation $\mathbf{v}(\cdot,\tau)=\mathbf{K}*\theta(\cdot,\tau)$ is still valid in terms of the new coordinates, namely, with $\mathbf{x}$, $\mathbf{y}$ replaced by $\boldsymbol{\xi}=(1+t)^{-\frac{1}{2}}\mathbf{x}$, $\boldsymbol{\eta}=(1+t)^{-\frac{1}{2}}\mathbf{y}$, respectively, in (2.2).

The new variables $\boldsymbol{\xi}$, τ are referred to as "scaling variables" for the vorticity equation (2.1), with $\nu=1$ (see Notes for Chapter 6).

Equation (2.1) (with $\nu=1$) is transformed into

$$\begin{aligned}&\partial_\tau\theta=\mathcal{L}\theta-(\mathbf{v}\cdot\nabla)\theta, &&\theta(\boldsymbol{\xi},0)=\theta_0(\boldsymbol{\xi}),\\ &\mathcal{L}\theta=\Delta\theta+\tfrac{1}{2}(\boldsymbol{\xi}\cdot\nabla)\theta+\theta\,. &&\end{aligned} \tag{6.31}$$

Notice that the spatial derivatives in Δ and ∇ are with respect to $\boldsymbol{\xi}$.

The following example demonstrates the effect of this transformation.

Example 6.8. The Gaussian

$$G(\boldsymbol{\xi}) = (4\pi)^{-1} \exp\Big(- \frac{|\boldsymbol{\xi}|^2}{4} \Big), \qquad \boldsymbol{\xi} \in \mathbb{R}^2$$

is a stationary solution of (6.31) and an eigenfunction of $\mathcal{L}$ (with zero eigenvalue).

In terms of the original equation (2.1) (with $\nu = 1$), it is the "Oseen vortex" (5.1), with initial data shifted to $t = 1$. Thus, in this framework, the Oseen vortex is a stationary solution and the question of asymptotic convergence for large time is transformed into the question of convergence to this solution.

Throughout the rest of this section we refer to the weighted-L^2 spaces as defined in (6.19), but substituting $\boldsymbol{\xi}$ for $\mathbf{x}$. Thus,

$$\|f\|^2_{L^{2,m}} := \int_{\mathbb{R}^2} (1 + |\boldsymbol{\xi}|^2)^m |f(\boldsymbol{\xi})|^2 d\boldsymbol{\xi}.$$

Observe that a function f is in $L^{2,m}$ with respect to the $\mathbf{x}$ variable if and only if it is in $L^{2,m}$ with respect to the $\boldsymbol{\xi}$ variable.

It follows readily from Proposition 6.5 that Equation (6.31) is well posed in $L^{2,m}$, $m > 1$. For $\theta_0 \in L^{2,m}$ we denote by $(S\theta_0)(\tau) = \theta(\cdot, \tau)$ the solution.

Furthermore, we can restate the result of Corollary 6.7 as:

Corollary 6.9. *Let $\theta_0 \in L^{2,m}(\mathbb{R}^2)$, $m > 1$, and let $\theta(\cdot,\tau) = (S\theta_0)(\cdot,\tau)$ be the corresponding solution. For any $R > 0$ there exists a constant $C = C(R,m)$ such that if $\|\theta_0\|_{L^{2,m}} < R$ then*

$$\|\theta(\cdot,\tau)\|_{L^{2,m}} \le C\|\theta_0\|_{L^{2,m}}, \quad \tau > 0. \tag{6.32}$$

Remark 6.10. Observe that $\int_{\mathbb{R}^2} \theta(\boldsymbol{\xi},\tau))d\boldsymbol{\xi} = \int_{\mathbb{R}^2} \theta_0(\boldsymbol{\xi})d\boldsymbol{\xi}$ for all $\tau > 0$, since this conservation property holds for $\omega(\mathbf{x},t)$. Hence, a necessary condition for $\lim\limits_{\tau\to\infty} \|\theta(\cdot,\tau)\|_{L^{2,m}} = 0$, is that $\int_{\mathbb{R}^2} \theta_0(\boldsymbol{\xi})d\boldsymbol{\xi} = 0$.

In analogy with Propositions 6.2 and 6.3, it can be shown that this condition is also sufficient [73, Theorem 3.2]. Note that in this case the velocity field is square-integrable (assuming $m > 1$) by Proposition 6.6.

The fundamental reason for dealing with the spaces $L^{2,m}$ in the context of asymptotic behavior is the fact that, as m increases, the continuous part of the spectrum is "pushed towards" $-\infty$, exposing a finite set of non-positive eigenvalues. The finite-dimensional eigenspaces associated with those eigenvalues are consequently mapped onto finite-dimensional invariant manifolds for the semiflow of Equation (6.31). The latter determine the full asymptotic behavior for solutions having small initial data. We now discuss this statement in more detail.

The spectrum $\sigma(\mathcal{L})$ of $\mathcal{L}$ in $L^{2,m}$, for any $m > 1$, is given by [73, Appendix A]

$$\sigma(\mathcal{L}) = \Big\{\lambda \in \mathbb{C}, \quad Re(\lambda) \le -\frac{m-1}{2}\Big\} \bigcup \Big\{ -\frac{k}{2}, \quad k = 0, 1, 2, \dots\Big\}.$$

In particular, for a fixed $m > 1$, the finite set of real nonpositive numbers

$$\Lambda(k) = \Big\{\lambda_j = -\frac{j}{2}, \; j = 0, 1, \dots, k\Big\} \tag{6.33}$$

consists of isolated eigenvalues of $\mathcal{L}$ if $k < m - 1$ (in the space $L^{2,m}$).

Given $k \in \mathbb{N}$ (nonnegative integers), take $m \ge k + 2$. Let $\mathcal{H}_k \subseteq L^{2,m}$ be the finite-dimensional subspace spanned by the eigenvectors associated with $\Lambda(k)$, and let $\mathcal{J}_k = L^{2,m} \ominus \mathcal{H}_k$ be its orthogonal complement.

For $r > 0$ we denote by $\mathcal{B}_r$ the ball of radius r in $L^{2,m}$ (centered at 0).

The subspaces $\mathcal{H}_k$ can be mapped onto finite-dimensional invariant manifolds for the semiflow of Equation (6.31), for sufficiently small initial data. This is described in the following theorem. Refer to [73, Section 3] for its proof.

Theorem 6.11. *Let $k \in \mathbb{N}$ and $m > k + 2$. Then there exist $r_0 > 0$ and a globally Lipschitz C^1 map*

$$g : \mathcal{H}_k \to \mathcal{J}_k, \quad g(0) = 0, \; Dg(0) = 0,$$

such that,

- *The manifold*

  $$\mathcal{T}_g = \{w + g(w), \quad w \in \mathcal{H}_k\}$$

  is locally invariant in the following sense.
  There exists $0 < r_1 < r_0$ such that the semiflow $(S\theta_0)(\tau)$ associated with (6.31), for any point $\theta_0 \in \mathcal{T}_g \cap \mathcal{B}_{r_1}$ stays in $\mathcal{T}_g \cap \mathcal{B}_{r_0}$ for all $\tau \ge 0$.

- *This invariant manifold "attracts" all trajectories having small initial data. More explicitly, for every $\theta_0 \in \mathcal{T}_g \cap \mathcal{B}_{r_0}$ there exists a manifold $\mathcal{S}_{\theta_0}$ having the following property.*
  All trajectories $\phi(\cdot, \tau) = (S\phi_0)(\tau)$, for $\phi_0 \in \mathcal{S}_{\theta_0} \cap \mathcal{B}_{r_1}$ (with θ_0 also restricted to $\mathcal{B}_{r_1}$) approach the trajectory $\theta(\cdot, \tau) = (S\theta_0)(\tau) \subseteq \mathcal{T}_g$. In fact, the rate of approach is exponential,

  $$\limsup_{\tau \to \infty} \tau^{-1} \log \|\phi(\cdot, \tau) - \theta(\cdot, \tau)\|_{L^{2,m}} < -\frac{k+1}{2}. \tag{6.34}$$

- *The manifold* $\mathcal{S}_{\theta_0}$ *is a continuous map of* $\mathcal{J}_k$. *It intersects* $\mathcal{T}_g \cap \mathcal{B}_{r_0}$ *only at* θ_0 *and the family*
$$\{\mathcal{S}_{\theta_0}, \qquad \theta_0 \in \mathcal{T}_g \cap \mathcal{B}_{r_0}\}$$
is a foliation of $\mathcal{B}_{r_1}$.
- *The manifold* $\mathcal{S}_0$ *is tangent to* $\mathcal{J}_k$ *at the origin* $\theta_0 = 0$. *It is the (strong) stable manifold for the semiflow of Equation* (6.31) *in* $L^{2,m}(\mathbb{R}^2)$.

Remark 6.12. Observe that the decay rate in Equation (6.34) corresponds to a decay rate of $t^{-\frac{k+1}{2}}$ for solutions of the vorticity equation (2.1). Thus, for sufficiently small initial data in weighted-L^2 spaces, the asymptotic behavior of the vorticity is determined, to any order, by "finite-dimensional dynamics."

Remark 6.13. Continuing Remark 6.10, if $\int_{\mathbb{R}^2} \theta_0(\boldsymbol{\xi})d\boldsymbol{\xi} = 0$, then $\|\theta(\cdot,\tau)\|_{L^{2,m}} \to 0$ as $\tau \to \infty$, and Theorem 6.11 can be applied to determine its asymptotic behavior. Since in this case the velocity field is square-integrable (assuming $m > 1$) by Proposition 6.6, the asymptotic behavior can also be studied in terms of the velocity field [139].

In Example 6.8 we saw that the Oseen vortex is transformed into a steady-state solution of (6.31). In this framework we can also address its stability, as follows.

Proposition 6.14 (Stability of the Oseen vortex). *Fix* $0 < \mu < \frac{1}{2}$. *There exists* $r > 0$ *such that if* $\theta(\boldsymbol{\xi},\tau)$ *is a solution to* (6.31) *with* $\|\theta_0\|_{L^{2,2}} < r$ *and* $\int_{\mathbb{R}^2} \theta_0(\boldsymbol{\xi})d\boldsymbol{\xi} = a$ *then*

$$\|\theta(\cdot,\tau) - aG(\cdot)\|_{L^{2,2}} \le Ce^{-\mu\tau}, \qquad \tau \ge 0. \tag{6.35}$$

Proof. Taking $k = 0$ and $m = 2$ in Theorem 6.11, it is easily seen that $\mathcal{H}_0$ is the one-dimensional subspace spanned by G and coincides with the invariant manifold $\mathcal{T}$ (namely , $g \equiv 0$). Thus, combining Theorem 6.11 and the conservation of $\int_{\mathbb{R}^2} \theta(\boldsymbol{\xi},\tau)d\boldsymbol{\xi}$ finishes the proof. ■

Refer to [73, Section 4] for a detailed analysis of this convergence.

6.3 Stability of steady states

The stability of a given flow or, more specifically, that of a steady state, is a fundamental issue in fluid dynamics. It is nicely described in the following quotation [124, Chapter III], in the introductory comments to the study of *turbulence*.

"Yet not every solution of the equations of motion, even if it is exact, can actually occur in Nature. The flows that occur in Nature must not only obey the equations of fluid dynamics, but also be stable."

A mathematical treatment of stability by the linearization method can be found in [179], and an approach to nonlinear stability by the energy method is described in [54, Chapter 2]. However, the problem of the stability or instability of steady states remains, from the mathematical point of view, a difficult challenge.

Recall that in Proposition 6.14 we saw the stability of the Oseen vortex (up to scaling) for sufficiently small initial vorticities in $L^{2,2}(\mathbb{R}^2)$. However, it is important to note that in the case of the full plane, the zero solution is the only steady state, since all solutions decay to zero in the L^p norm for $p \in (2, \infty]$. Thus the scaled Oseen vortex displays the asymptotic profile of this decay, at least for small initial data in the weighted space. In fact, the (scaled) Oseen vortex is globally asymptotically stable as expressed in the following theorem of Gallay and Wayne [73].

Theorem 6.15. *Let $\omega_0 \in L^1(\mathbb{R}^2)$ and let $\omega(\cdot, t) = (S\omega_0)(t)$ be the solution given by Theorem 4.3. Let $G(\boldsymbol{\xi})$ be the Oseen vortex of Example 6.8. Then, for every $p \in [1, \infty]$,*

$$\lim_{t\to\infty} t^{1-\frac{1}{p}} \left\| \omega(\cdot, t) - \frac{\alpha}{t} G\left(\frac{\cdot}{\sqrt{t}}\right) \right\|_p = 0, \tag{6.36}$$

where $\alpha = \int_{\mathbb{R}^2} \omega_0(\mathbf{x}) d\mathbf{x}$.

While for small initial data (Proposition 6.14) the asymptotic behavior could be inferred from the approach to the one-dimensional invariant subspace spanned by G, the proof of Theorem 6.15 relies on the construction of a suitable Lyapunov function. This function is a "relative entropy" of the solution $\omega(\mathbf{x}, t)$ with respect to the Oseen vortex.

The only key element of the proof that we present here is the following compactness result, which states that the trajectory of the scaled vorticities (6.30) is relatively compact in $L^1(\mathbb{R}^2)$. The proof of the theorem is then achieved by showing that all possible limit points of the trajectory are scalar multiples of G.

Proposition 6.16. *Given the solution $\omega(\mathbf{x}, t)$, let*

$$\theta(\boldsymbol{\xi}, t) = t\omega(\sqrt{t}\,\boldsymbol{\xi}, t), \quad t \geq 1.$$

Then the set $\{\theta(\cdot, t)\}_{t\geq 1}$ is relatively compact in $L^1(\mathbb{R}^2)$.

Proof. Fix $R > 0$. Recall that by Theorem 4.5 we have, for $\kappa < 1$,

$$
\begin{aligned}
|\theta(\boldsymbol{\xi},t)| \le C_\kappa\Big[\int\limits_{|\mathbf{y}|\le R} \exp\Big(-\frac{\kappa}{4}(\boldsymbol{\xi} - t^{-\frac{1}{2}}\mathbf{y})^2\Big)|\omega_0(\mathbf{y})|d\mathbf{y} \\
+ \int\limits_{|\mathbf{y}|\ge R} \exp\Big(-\frac{\kappa}{4}(\boldsymbol{\xi} - t^{-\frac{1}{2}}\mathbf{y})^2\Big)|\omega_0(\mathbf{y})|d\mathbf{y}\Big] = I_{R,1}(\boldsymbol{\xi},t) + I_{R,2}(\boldsymbol{\xi},t).
\end{aligned} \tag{6.37}
$$

We now estimate, using Fubini's theorem,

$$
\int\limits_{|\boldsymbol{\xi}|\ge 2R} |I_{R,1}(\boldsymbol{\xi},t)|d\boldsymbol{\xi} \le C_\kappa \int\limits_{|\boldsymbol{\xi}|\ge 2R} \exp\Big(-\kappa\frac{\boldsymbol{\xi}^2}{16}\Big)d\boldsymbol{\xi}\cdot \|\omega_0\|_1
$$

and

$$
\int\limits_{|\boldsymbol{\xi}|\ge 2R} |I_{R,2}(\boldsymbol{\xi},t)|d\boldsymbol{\xi} \le C_\kappa \int\limits_{\mathbb{R}^2} \exp\Big(-\kappa\frac{\boldsymbol{\xi}^2}{4}\Big)d\boldsymbol{\xi}\cdot \int\limits_{|\mathbf{y}|\ge R} |\omega_0(\mathbf{y})|d\mathbf{y}.
$$

Combining these two estimates it follows that

$$
\lim_{R\to\infty} \int\limits_{|\boldsymbol{\xi}|\ge 2R} |\theta(\boldsymbol{\xi},t)|d\boldsymbol{\xi} = 0, \tag{6.38}
$$

uniformly in $t \ge 1$.

Writing

$$
\theta(\boldsymbol{\xi}+\widetilde{\boldsymbol{\xi}},t) - \theta(\boldsymbol{\xi},t) = \int_0^1 \nabla_{\boldsymbol{\xi}}\theta(\boldsymbol{\xi}+\lambda\widetilde{\boldsymbol{\xi}},t)d\lambda\cdot\widetilde{\boldsymbol{\xi}}
$$

and using (3.35) we get,

$$
\int\limits_{|\boldsymbol{\xi}|\ge 2R} |\theta(\boldsymbol{\xi}+\widetilde{\boldsymbol{\xi}},t) - \theta(\boldsymbol{\xi},t)|d\boldsymbol{\xi} \le \Upsilon|\widetilde{\boldsymbol{\xi}}|, \tag{6.39}
$$

uniformly in $t \ge 1$.

We now invoke the Riesz criterion for compactness in L^1 [155, Vol. IV, Theorem XIII.66]; the two conditions (6.38) and (6.39) imply the (relative) compactness of $\{\theta(\cdot,t)\}_{t\ge 1}$. ∎

In contrast to the global stability in the full plane (or periodic domains), for the Navier–Stokes equations in bounded domains the stability of steady states depends in general on the viscosity coefficient ν. As it decreases (so that the Reynolds number $Re = \nu^{-1}$ increases), small perturbations of the steady state may evolve in very complicated ways. The treatment of such perturbations has been carried out, so far, mostly by numerical simulations.

In Chapter 14 of this monograph we describe in detail the numerical treatment of the "driven cavity" flow. The careful simulations show that for sufficiently high Reynolds number (meaning a small viscosity coefficient $\nu > 0$), the steady-state solution becomes unstable and bifurcates into a time-periodic solution. Yet, there exists no rigorous proof of this fact.

6.4 Notes for Chapter 6

- The study of the large time behavior for solutions of the Navier–Stokes equations has been the topic of many studies. They deal essentially with the decay (in some L^p norm) of the solution as time increases. Typically the decay for large time is obtained by restriction to initial data with sufficient decay as $|\mathbf{x}| \to \infty$. Most of these studies deal with the velocity (assuming at least finite energy, $\|\mathbf{u}_0\|_2 < \infty$) ([72, 132, 139, 160] and references therein). The use of vorticity in connection with the asymptotic behavior is more recent [35, 37, 73, 74, 78]. Recall that even if the initial vorticity is smooth, the associated energy is in general infinite.
- The review paper of Wayne [177] gives a nice exposition of the asymptotic behavior of the vorticity. In addition, it contains nice illustrations of the evolution of vorticity.
- The proof of Proposition 6.2 for large initial data uses the idea of "pulling" the solution to a ball of small radius centered at the origin. This idea was first used by Serre [162] for the decay of solutions to viscous conservation laws.
- Refer to [35, Theorem 4] for a different proof of the decay result in Proposition 6.2. The proof there applies also to the viscous conservation law, but it is much more involved, relying on stochastic integration and properties of the Ornstein–Uhlenbeck semigroup.
- There are strong similarities between the decay estimates (especially the L^1 decay) for the vorticity (in the two-dimensional case) and those of solutions to "viscous" Hamilton–Jacobi and conservation equations [17, 35, 162].

  In particular, since we can take the initial data in Y_∞, it follows from (3.28) that the velocity is uniformly bounded, $\|\mathbf{u}(\cdot, t)\|_\infty \le C$, $t > 0$. Thus, the vorticity equation (2.1) is majorized by the viscous Hamilton–Jacobi equation

$$\partial_t \phi + C|\nabla \phi| = \nu \Delta \phi.$$

However, in this case $\|\phi(\cdot,t)\|_1$ vanishes as $t \to \infty$ only if $C > 0$ (assuming nonnegative initial data, see [17]), whereas it is impossible to impose an analogous sign condition on the convection term $\mathbf{u} \cdot \nabla\omega$.

- The decay result (6.14) was first proved in [37], assuming finite initial energy, $\|\mathbf{u}_0\|_2 < \infty$, where $\mathbf{u}_0$ is the associated velocity. Recall (Remark 3.2) that the zero mean value condition (6.4) is necessary in this case. Another proof is given in [73, Theorem 2.4], also assuming finite energy.
- Another proof of Proposition 6.6, based on a weighted Hardy–Littlewood–Sobolev inequality, can be found in [73, Appendix B].
- For the use of the scaling variables (6.29), see [37, 73, 78]).
- Issues concerning the stability of steady states were considered in [73, 74, 78].
- Refer to [68, 176] and references therein for the study of nonlinear instability, using the velocity–pressure classical formulation.

Appendix A

Some Theorems from Functional Analysis

A.1 The Calderón–Zygmund Theorem

Let $Z(x)$ be a smooth function defined in $\mathbb{R}^n \setminus \{0\}$ and assume that:

(1) $Z(\mathbf{x})$ is homogeneous of degree $-n$, namely,

$$Z(\mathbf{x}) = |\mathbf{x}|^{-n}\Omega\Big(\frac{\mathbf{x}}{|\mathbf{x}|}\Big), \quad 0 \neq \mathbf{x} \in \mathbb{R}^n,$$

where $\Omega(\mathbf{y})$ is a smooth function on the sphere $S^{n-1} = \{\mathbf{y}, |\mathbf{y}| = 1\}$.

(2) The function Ω has mean value zero:

$$\int\limits_{S^{n-1}} \Omega(\mathbf{y})d\sigma(\mathbf{y}) = 0,$$

where $d\sigma$ is the (Lebesgue) surface measure.

Consider the *singular integral*

(A.1) $$\psi(\mathbf{x}) = \int\limits_{\mathbb{R}^n} Z(\mathbf{y})\phi(\mathbf{x}-\mathbf{y})d\mathbf{y}, \quad \phi \in C_0^\infty(\mathbb{R}^n).$$

It is singular because $Z(\mathbf{y})$ is not integrable near the origin (if it is not identically zero). Thus, it should be understood as *principal value*,

(A.2) $$\psi(\mathbf{x}) = \lim_{\varepsilon \to 0} \int\limits_{|\mathbf{y}| \geq \varepsilon} Z(\mathbf{y})\phi(\mathbf{x}-\mathbf{y})d\mathbf{y}, \quad \phi \in C_0^\infty(\mathbb{R}^n).$$

Theorem A.1 (Calderón–Zygmund). *Let $1 < p < \infty$. The function $\psi(\mathbf{x})$ is well defined and satisfies*

$$\|\psi\|_p \leq C\|\phi\|_p,$$

where $C > 0$ depends only on p and n.

In particular, the singular integral (A.1) *can be extended as a bounded operator in $L^p(\mathbb{R}^n)$.*

A proof of the theorem can be found in:

E.M. Stein, *Singular Integrals and Differentiability Properties of Functions*, Princeton University Press (1970), Chapter II.4.

The Biot–Savart Kernel: Let $\mathbf{K}(\mathbf{x})$ be the Biot–Savart kernel (1.12). It can easily be verified that for $j = 1, 2$, the two components of $\partial_{x^j}\mathbf{K}(\mathbf{x})$ satisfy the conditions imposed on Z [133, Section 2.4].

A.2 Young's and the Hardy–Littlewood–Sobolev Inequalities

For $\phi, \psi \in C_0^\infty(\mathbb{R}^n)$ the *convolution function* $\phi * \psi$ is defined as

$$\phi * \psi(\mathbf{x}) = \int_{\mathbb{R}^n} \phi(\mathbf{y})\psi(\mathbf{x}-\mathbf{y})d\mathbf{y}, \quad \mathbf{x} \in \mathbb{R}^n. \tag{A.3}$$

Young's Inequality:

Let $1 \le p, q, r \le \infty$ be such that $\frac{1}{p} + \frac{1}{q} = 1 + \frac{1}{r}$. Then

$$\|\phi * \psi\|_r \le C\|\phi\|_p \|\psi\|_q, \tag{A.4}$$

where $C > 0$ depends only on p, q and n.

In particular, it follows that the convolution operation can be extended as a continuous bilinear operator from $L^p(\mathbb{R}^n) \times L^q(\mathbb{R}^n)$ into $L^r(\mathbb{R}^n)$.

Consider now the case

$$\psi(\mathbf{x}) = |\mathbf{x}|^{-\beta}, \quad 0 < \beta < n.$$

Let $q = \frac{n}{\beta}$. It is easily checked that $\psi \notin L^q(\mathbb{R}^n)$. However, Young's inequality is still valid in this case. In other words, in the convolution operation *this particular function* ψ can be considered "as if" it belongs to $L^q(\mathbb{R}^n)$ as follows.

Hardy–Littlewood–Sobolev Inequality:

Let $1 < p, r < \infty$ and $0 < \beta < n$, so that $\frac{1}{p} + \frac{\beta}{n} = 1 + \frac{1}{r}$. Consider the convolution

$$g(\mathbf{x}) = \int_{\mathbb{R}^n} \phi(\mathbf{y})|\mathbf{x}-\mathbf{y}|^{-\beta}d\mathbf{y}, \quad \phi \in C_0^\infty(\mathbb{R}^n).$$

Then

$$\|g\|_r \le C\|\phi\|_p,$$

where $C > 0$ depends only on p, β and n. In some cases there are "optimal" values for this constant [127].

A proof of these inequalities can be found in:

E.H. Lieb and M. Loss, *Analysis*, American Mathematical Society (1997), Section 4.3.

The Biot–Savart Kernel: Let $\mathbf{K}(\mathbf{x})$ be the Biot–Savart kernel (1.12). It can easily be verified that $|\mathbf{K}(\mathbf{x})| \leq \frac{1}{|\mathbf{x}|}$, hence the convolution

$$\mathbf{u}(\mathbf{x},t) = (\mathbf{K} * \omega)(\mathbf{x},t) = \int_{\mathbb{R}^2} \mathbf{K}(\mathbf{x}-\mathbf{y})\omega(\mathbf{y},t)d\mathbf{y}$$

can be estimated in terms of the Hardy–Littlewood–Sobolev inequality (where $\beta = 1$, $n = 2$, so that $\frac{1}{p} = \frac{1}{2} + \frac{1}{r}$).

A.3 The Riesz–Thorin Interpolation Theorem

The following "interpolation inequality" in $L^p(\mathbb{R}^n)$ is well-known (and holds in fact in any measure space [58, Appendix B.2]).

$$\|g\|_{L^p(\mathbb{R}^n)} \leq \|g\|_{L^\infty(\mathbb{R}^n)}^{1-\frac{1}{p}} \|g\|_{L^1(\mathbb{R}^n)}^{\frac{1}{p}}.$$

It can be generalized as follows.

Let $1 \leq p_1 \leq p_2 \leq \infty$ and $1 \leq q_1 \leq q_2 \leq \infty$. Let $V \subseteq L^{p_1}(\mathbb{R}^n) \cap L^{p_2}(\mathbb{R}^n)$ be a linear subspace that is dense in both spaces. For example, if $p_2 < \infty$, we can take $V = C_0^\infty(\mathbb{R}^n)$.

Theorem A.2 (Riesz–Thorin). *Let $T : V \to L^{q_1}(\mathbb{R}^n) \cap L^{q_2}(\mathbb{R}^n)$ be a linear map and suppose that there exist constants M_1, M_2, such that, for all $v \in V$,*

$$\|Tv\|_{q_1} \leq M_1 \|v\|_{p_1},$$

$$\|Tv\|_{q_2} \leq M_2 \|v\|_{p_2}.$$

Let $\lambda \in [0,1]$, and let $p_\lambda \in [p_1, p_2]$ and $q_\lambda \in [q_1, q_2]$ be defined by

$$p_\lambda^{-1} = \lambda p_1^{-1} + (1-\lambda)p_2^{-1}, \quad q_\lambda^{-1} = \lambda q_1^{-1} + (1-\lambda)q_2^{-1}.$$

Then

$$\|Tv\|_{q_\lambda} \leq M_1^\lambda M_2^{1-\lambda} \|v\|_{p_\lambda}, \quad v \in V.$$

In particular, T can be extended (from its initial domain V) as a continuous linear map from $L^{p_\lambda}(\mathbb{R}^n)$ to $L^{q_\lambda}(\mathbb{R}^n)$, for every $\lambda \in [0,1]$.

In the context of the convection–diffusion equation this theorem serves to interpolate between the "L^1 *contraction*" and the "*Maximum principle*" (see the proof of Lemma 2.7). It is then applied to estimates of the vorticity (see Equation (2.48)).

A proof of this theorem can be found in:

M. Reed and B.Simon, *Methods of Modern Mathematical Physics*, Vol. II, Academic Press (1975), Section IX.4.

A.4 Finite Borel measures in $\mathbb{R}^2$ and the heat kernel

For the basic facts on measure theory, the reader is referred to the books:

[1] P. R. Halmos, *Measure Theory*, Springer-Verlag (1974),

[2] W. Rudin, *Real and Complex Analysis*, McGraw-Hill (1966), Chapters 1–2,

We collect here some facts concerning finite Borel measures in $\mathbb{R}^2$, and the action of the heat kernel on such measures, as needed in Chapter 5. We provide proofs or specific references for claims whose proofs are not readily available in the aforementioned books.

We restrict ourselves to the two-dimensional setting pertinent to this monograph.

The space $\mathcal{M} = \mathcal{M}(\mathbb{R}^2)$ is the space of (finite) Borel measures, which is (in view of the Riesz–Markoff theorem) the dual space of $D^0(\mathbb{R}^2)$, continuous functions decaying as $|\mathbf{x}| \to \infty$.

We denote the pairing between $\mathcal{M}$ and $D^0(\mathbb{R}^2)$ as

$$\langle \mu, \phi \rangle = \int_{\mathbb{R}^2} \phi d\mu.$$

Thus, the norm of $\mu \in \mathcal{M}$ is given by

$$\|\mu\|_{\mathcal{M}} = \sup_{\phi \in D^0(\mathbb{R}^2),\, \|\phi\|_\infty \le 1} |\langle \mu, \phi \rangle|.$$

The space $L^1(\mathbb{R}^2)$ can be identified as a subspace of $\mathcal{M}$. Indeed, for $f \in L^1(\mathbb{R}^2)$ we define the measure μ_f by

$$\langle \mu_f, \phi \rangle = \int_{\mathbb{R}^2} f(\mathbf{x})\phi(\mathbf{x}) d\mathbf{x},$$

where $d\mathbf{x}$ is the Lebesgue measure.

It is easy to see that this is a one-to-one linear and norm-preserving map,

$$\|\mu_f\|_{\mathcal{M}} = \|f\|_1.$$

For any (measurable) subset $\Omega \subseteq \mathbb{R}^2$ and $\mu \in \mathcal{M}$ we define the "truncated" measure μ^Ω by

$$\langle \mu^\Omega, \phi \rangle = \int_{\Omega} \phi d\mu.$$

Definition A.3. *The* ***support*** *of $\mu \in \mathcal{M}$ is the smallest closed set Ω such that $\mu^\Omega = \mu$. It is denoted by* **supp**(μ).

The convolution of a function $g \in L^p(\mathbb{R}^2)$, $p \in [1, \infty]$, with a measure $\kappa \in \mathcal{M}$ is defined as an extension of (A.3),

$$g * \kappa(\mathbf{x}) = \int_{\mathbb{R}^2} g(\mathbf{x} - \mathbf{y}) d\kappa(\mathbf{y}), \quad \mathbf{x} \in \mathbb{R}^2. \tag{A.5}$$

An easy extension of Young's inequality (A.4) gives,

$$\|g * \kappa\|_p \leq C \|g\|_p \|\kappa\|_{\mathcal{M}}, \quad p \in [1, \infty], \tag{A.6}$$

where $C > 0$ depends only on p.

In particular, if $g \in L^1(\mathbb{R}^2)$ then $g * \kappa \in L^1(\mathbb{R}^2)$ with associated measure $\mu_{g*\kappa}$.

If also $h \in L^1(\mathbb{R}^2)$ then we have the following associative rule (analogous to the same rule where three L^1 functions are involved),

$$h * (g * \kappa) = (h * g) * \kappa.$$

Claim A.4. *If $g \in C_0^\infty(\mathbb{R}^2)$ and $\mu \in \mathcal{M}$ then $g * \mu \in C^\infty(\mathbb{R}^2)$ with all derivatives decaying as $|\mathbf{x}| \to \infty$ (namely, in $D^0(\mathbb{R}^2)$).*

Furthermore, if $\mathbf{supp}(\mu)$ *is compact then $g * \mu \in C_0^\infty(\mathbb{R}^2)$ and*

(A.7)

$$\mathbf{supp}(g * \mu) \subseteq \mathbf{supp}(\mu) + \mathbf{supp}(g) = \{\mathbf{x} + \mathbf{y},\, \mathbf{x} \in \mathbf{supp}(\mu),\, \mathbf{y} \in \mathbf{supp}(g)\}.$$

Given a point $\mathbf{a} \in \mathbb{R}^2$, we denote by $\delta_{\mathbf{a}} \in \mathcal{M}$ the (Dirac) measure defined by

$$\langle \delta_{\mathbf{a}}, \phi \rangle = \phi(\mathbf{a}).$$

Definition A.5. *(a) An* ***atomic measure*** *μ is a measure having the form*

$$\mu = \sum_{n=1}^{\infty} \lambda_n \delta_{\mathbf{a}_n}, \quad \{\lambda_n\}_{n=1}^{\infty} \subseteq \mathbb{R}, \quad \{\mathbf{a}_n\}_{n=1}^{\infty} \subseteq \mathbb{R}^2. \tag{A.8}$$

(b) A measure $\mu \in \mathcal{M}$ is ***continuous*** *if $\mu(\mathbf{a}) = 0$ for every $\mathbf{a} \in \mathbb{R}^2$.*

The norm of the atomic measure is readily seen to be

$$\|\mu\|_{\mathcal{M}} = \sum_{n=1}^{\infty} |\lambda_n|.$$

Observe in particular that if $f \in L^1(\mathbb{R}^2)$ then the corresponding measure μ_f is continuous.

Claim A.6. *Every measure $\mu \in \mathcal{M}$ can be decomposed in a unique way as*

$$\mu = \mu_c + \mu_{atom}, \tag{A.9}$$

where μ_c is continuous and μ_{atom} is atomic. Furthermore

$$\|\mu\|_{\mathcal{M}} = \|\mu_c\|_{\mathcal{M}} + \|\mu_{atom}\|_{\mathcal{M}}.$$

In addition to the usual strong convergence with respect to $\|\cdot\|_{\mathcal{M}}$, we have the following important concept of "weak* convergence."

Definition A.7. *Let $\{\mu_n\}_{n=1}^{\infty} \subseteq \mathcal{M}$. We say that the sequence converges in the* **weak* topology** *of $\mathcal{M}$ to $\mu \in \mathcal{M}$ if for every $\phi \in D^0(\mathbb{R}^2)$,*

$$\lim_{n\to\infty} \langle \mu_n, \phi\rangle = \langle \mu, \phi\rangle.$$

We denote this convergence by $\mu_n \xrightarrow[n\to\infty]{weak^*} \mu$.

It is obvious how to extend this definition to the case of a family of measures depending on a continuous parameter.

Claim A.8. *$C_0^\infty(\mathbb{R}^2)$ is weak* dense in $\mathcal{M}$. More explicitly, if $\mu \in \mathcal{M}$ then there exists a sequence $\{\psi_n\}_{n=1}^{\infty} \subseteq C_0^\infty(\mathbb{R}^2)$ such that*

$$\mu_{\psi_n} \xrightarrow[n\to\infty]{weak^*} \mu \tag{A.10}$$

and

$$\|\psi_n\|_1 \le \|\mu\|_{\mathcal{M}}. \tag{A.11}$$

Proof. We can assume without loss of generality that μ is compactly supported. Indeed, if μ^R is the truncation of μ to the ball $B_R = \{|\mathbf{x}| \le R\}$, then clearly $\mu^R \xrightarrow[R\to\infty]{weak^*} \mu$ and it suffices to prove the claim for μ replaced by μ^R.

Let $0 \le \eta(x) \in C_0^\infty(\mathbb{R}^2)$ be a mollifier [58, Appendix C], namely, a smooth function with support in the unit ball $\{|\mathbf{x}| \le 1\}$ and such that $\int_{\mathbb{R}^2} \eta(\mathbf{x})d\mathbf{x} = 1$. We set $\eta_n(\mathbf{x}) = n^2\eta(n\mathbf{x}),\ n = 1, 2, 3...$ and define

$$\psi_n(\mathbf{x}) = \eta_n * \mu = \int_{\mathbb{R}^2} \eta_n(\mathbf{x}-\mathbf{y})d\mu(\mathbf{y}).$$

It is left to the reader to verify that $\{\psi_n\}_{n=1}^{\infty}$ satisfies (A.10). ■

The action of the heat kernel on measures: Consider the heat kernel [see (2.25) with $n = 2$],

$$G_\nu(\mathbf{x}, t) = (4\pi\nu t)^{-1} e^{-\frac{|\mathbf{x}|^2}{4\nu t}}.$$

Since for every fixed $\mathbf{x} \in \mathbb{R}^2$ (and $t > 0$) we have $G_\nu(\mathbf{x} - \cdot, t) \in D^0(\mathbb{R}^2)$, we can define, for $\mu \in \mathcal{M}$,

$$(G_\nu(\cdot, t) * \mu)(\mathbf{x}) = \langle \mu, G_\nu(\mathbf{x} - \cdot, t)\rangle = \int_{\mathbb{R}^2} G_\nu(\mathbf{x}-\mathbf{y}, t)d\mu(\mathbf{y}).$$

Clearly, if $f \in L^1(\mathbb{R}^2)$, then

$$G_\nu(\cdot, t) * \mu_f = G_\nu(\cdot, t) * f, \quad t > 0.$$

In view of (3.14) and Young's inequality (A.6) we have

$$\|G_\nu(\cdot, t) * \mu\|_p \leq (4\pi\nu t)^{-1+\frac{1}{p}} \|\mu\|_{\mathcal{M}}, \quad p \in [1, \infty]. \tag{A.12}$$

For an atomic measure having the form (A.8) we denote

$$\|\mu_{atom}\|_p = \left\{ \sum_{n=1}^{\infty} |\lambda_n|^p \right\}^{\frac{1}{p}}.$$

Clearly, $\|\mu_{atom}\|_p \leq \|\mu_{atom}\|_{\mathcal{M}}$, for any $p \in (1, \infty)$.

The estimate (3.16) implies in particular that, for any $\omega_0 \in L^1(\mathbb{R}^2)$,

$$\limsup_{t \to 0} t^{1-\frac{1}{p}} \|G_\nu(\cdot, t) * \omega_0\|_p = 0, \quad p \in (1, \infty].$$

The following claim is a generalization of this fact.

Claim A.9. *Let $\mu \in \mathcal{M}$ have the decomposition* (A.9). *Then*

$$\begin{aligned} \limsup_{t \to 0} t^{1-\frac{1}{p}} \|G_\nu(\cdot, t) * \mu\|_p &= \gamma(p, \nu) \|\mu_{atom}\|_p \\ &\leq \gamma(p, \nu) \|\mu_{atom}\|_{\mathcal{M}}, \quad p \in (1, \infty]. \end{aligned}$$

The exact value of $\gamma(p, \nu)$ is known: $\gamma(p, \nu) = (4\pi\nu)^{-1+\frac{1}{p}} p^{-\frac{1}{p}}$ [108, Lemma 1.2].

Since $G_\nu(\cdot, t) \in L^1(\mathbb{R}^2)$ for $t > 0$, it can be identified with a measure $\mu_{G_\nu(\cdot,t)}$. Using the terminology of weak* convergence, the basic property of the heat kernel is

$$\mu_{G_\nu(\cdot,t)} \xrightarrow[t \to 0]{weak^*} \delta_{\mathbf{x}=0}, \tag{A.13}$$

which is often simplified to

$$G_\nu(\cdot, t) \xrightarrow[t \to 0]{weak^*} \delta_{\mathbf{x}=0}.$$

PART II

Approximate Solutions

There are two things to be considered with regard to any scheme. In the first place, 'Is it good in itself?'. In the second, 'Can it be easily put into practice?'

J.-J. Rousseau, *Émile, or On Education* (1762) (trans. B. Foxley (1911)).

Chapter 7

Introduction

The purpose of this part of the monograph is to present a finite difference approach to the numerical approximation of solutions to the Navier–Stokes system in planar domains. It is based on the *pure streamfunction formulation* of the equations. In particular, this formulation involves the fourth-order biharmonic operator, which plays a central role here.

While this part is the "numerical counterpart" to the first part of this monograph, the issues addressed in the two parts are not entirely identical, due to the nature of the difficulties inherent in the two treatments. They may be summarized as follows.

(a) The primary issue in the first part is the well-posedness of the Navier–Stokes system with measure ("rough") initial data. This has only been studied during the last two decades and all the results presented are therefore quite recent. As pointed out in the Introduction to the first part, it corresponds to the physical setting of "singular objects," such as point vortices or vortex lines. The energy (L^2 norm of the velocity) is infinite, meaning that this case lies beyond the scope of the classical Leray theory. The interplay of the two kernels (heat and Biot–Savart) is brought to the fore. These kernels are much simpler in the case of the whole plane (or, equivalently, periodic boundary conditions). In fact, so far there have been no well-posedness results for the problem involving such singular initial data in bounded domains (say, with the "no-slip" boundary condition).

(b) Numerical simulations clearly force the implementation of boundary conditions; the presence of boundary layers is an essential feature in real fluid dynamical problems. By resorting to periodic boundary conditions this feature is completely lost. In this second part we have therefore chosen to consider domains with boundaries. The use of the streamfunction enables us to deal with the boundary conditions as well as to avoid the need of the aforementioned kernels. These points will be further expounded below.

We have made an effort to make the exposition self-contained; no familiarity with the numerical analysis of differential equations is required. In this spirit, we start with a full account of the treatment of the second-order derivative on an interval and its finite difference and spectral approximations. In addition to being useful in later developments, it serves as a convenient platform for the introduction of the "numerical language" of grid functions, difference operators, accuracy, convergence and so on.

The discrete approximation we describe relies on the pure streamfunction formulation and is based on our experience during the last decade.

The Navier–Stokes equations have been central to numerous applications in science and technology, from aircraft design to meteorology to blood flow. It is therefore no wonder that during the last half century, in parallel with the progress of computing technology, the numerical simulation of Navier–Stokes solutions has been at the forefront of applied mathematics.

Given the huge volume of the accumulated literature on this topic, it seems almost hopeless to provide a full survey of numerical methods for two-dimensional Navier–Stokes equations. We will give only a brief overview of some methods, categorized by the *formulation*, namely, *the flow variables* being used. Refer to [86, 120, 142] for more exhaustive reviews.

It seems that historically the basic system (see (1.1) in Part I) of the so-called *primitive variables* (pressure and velocity) was the starting point of numerical simulations. Within this formulation we can distinguish three primary approaches:

- *Finite differences*, where the differential operators are replaced by difference operators, making use of functional values on a discrete mesh (in space and time). In this category we note the early Marker-and-Cell (MAC) method (see [136] for a review of this method). The use of the velocity field entails the need to "numerically force" the incompressibility (null-divergence) condition. In this regard, we mention the "projection methods" suggested by Chorin [39] and Temam [170].
- *Spectral methods*, where the differential operators act in spaces spanned by a basis of special functions. The approximate solutions are considered in these spaces. For the Navier–Stokes system, this approach has its roots in the seminal work of Leray [125]. He used the eigenfunctions of the Stokes operator in order to establish the existence of a solution using an ascending sequence of finite-dimensional subspaces and a variational ("weak") formulation of the solution. This was therefore a constructive approach that could be implemented numerically.

It is generally referred to as the "Galerkin approach." General surveys of spectral methods can be found in [23, 33, 85, 96].

- *Finite element methods.* This methodology has been the most widely used in research as well as industrial applications. It is close in spirit to the previous one, in that it uses finite-dimensional functional subspaces. However, its "basis elements" are not eigenfunctions of an operator, but locally confined "structure functions," relying on a geometric triangulation of the domain. The discrete approximation is based on a "variational formulation" of the problem. These methods have been extensively used for time-dependent or steady-state partial differential equations. The books [80, 87, 89, 171] deal specifically with Navier–Stokes equations.
  Typically, these methods were of low-order accuracy, due to the arbitrary geometric structure of the underlying mesh. In recent years, the search for improved accuracy has led to schemes that can be considered as "offsprings" of the classical finite element methods. We mention in particular the *Spectral Element Method* [53, 147] and the *Discontinuous Galerkin Method* [97, 129, 131, 164].

While the above discussion referred to two-dimensional as well as three-dimensional computations, we henceforth restrict ourselves to the two-dimensional case.

The three approaches (finite differences, spectral methods and finite elements) have also been employed in the framework of the *velocity–vorticity* formulation, namely, the system (1.9)–(1.10). As mentioned in the Introduction to Part I, this formulation, restricted to the two-dimensional case, possesses some features that make it particularly attractive, both from the mathematical and from the fluid dynamical points of view. Indeed, the mathematical treatment in Part I is based on this formulation, which allows singular vorticity fields such as point vortices or vortex filaments. From the fluid dynamical point of view, the "passive transport" of the vorticity in the absence of viscosity (Euler's equations) allows a qualitative understanding of the flow. In the computational sphere, this formulation is the basis of the so-called "vortex methods." These methods typically combine (say by fractional steps [133]) the vorticity flow along streamlines and the diffusion of vorticity due to viscosity. The reconstruction of the velocity from the vorticity secures the null-divergence constraint. However, the need for *vorticity boundary values* (or *vorticity generation on the boundary* in the fluid dynamical language) has led to a variety of techniques, such as the

"Prandtl model." Refer to the books [50, 133] and the papers [40–42] and references therein for comprehensive presentations.

As already mentioned in the Introduction to Part I, the fact that the velocity field $\mathbf{u}$ is divergence-free implies (in a simply connected domain) the existence of a scalar streamfunction ψ such that

$$\mathbf{u} = (u, v) = \nabla^{\perp}\psi = (-\partial_y\psi, \partial_x\psi). \tag{7.1}$$

The streamfunction is uniquely (up to a constant) determined and the vorticity ω satisfies $\omega = \Delta\psi$ (see Equation (1.6).

We therefore arrive at the *vorticity-streamfunction* formulation, which involves ω and ψ as the two unknown scalar functions. We note that this formulation has mostly been used in the framework of finite difference methods.

In terms of the streamfunction, the "natural" (say "no-slip": vanishing velocity) boundary conditions imposed on the velocity are readily translated into equally "natural" conditions for the streamfunction. For example, the "no-slip" condition means the vanishing of the gradient of the streamfunction (on the boundary). However, in this formulation the evolution of the flow is represented as the evolution of the vorticity, hence the need for suitable vorticity boundary values persists. In [55, 56] this issue is handled by implementing Briley's formula (relating vorticity to streamfunction boundary values). Refer also to [52, 152, 153, 172] for works using the vorticity-streamfunction formulation.

Replacing the velocity field in (1.10) by $\boldsymbol{\nabla}^{\perp}\psi$ and the vorticity ω by $\Delta\psi$, we arrive at a single scalar equation for the streamfunction

$$\frac{\partial\Delta\psi}{\partial t} + \boldsymbol{\nabla}^{\perp}\psi \cdot \boldsymbol{\nabla}\Delta\psi - \nu\Delta^2\psi = 0. \tag{7.2}$$

It is supplemented by an initial condition

$$\psi(x, 0) = \psi_0(x), \tag{7.3}$$

and suitable boundary conditions to be discussed further below.

Equation (7.2) is truly remarkable; it reduces the vector system (1.1) to a single evolution equation for a scalar unknown function. This is the *pure streamfunction formulation* mentioned at the start of this Introduction.

As already stated above, the discrete approximation to solutions of this equation and the investigation of its accuracy and convergence properties are the focus of this part of the monograph.

Before moving on to a discussion of these aspects, let us comment on the differences between this equation and the vorticity equation (1.10).

Theoretically, they are interchangeable, using the identifications $\mathbf{u} = \nabla^{\perp}\psi$ and $\omega = \Delta\psi$, with boundary conditions imposed either on ψ or on $\mathbf{u}$. However, when using a discrete time-stepping of the vorticity equation, we need to determine at each step the associated velocity field and to verify that its divergence, in a suitable discrete sense, vanishes. As was already noted above in the primitive variables formulation this condition is satisfied by implementing a "projection method." In contrast, it is automatically satisfied in the streamfunction formulation, provided that a discrete analog of $\nabla \cdot \nabla^{\perp} = 0$ is established. An alternative point of view is to impose the condition $\omega = \Delta\psi$ (again using a suitable discrete analog), where the required boundary conditions are imposed on ψ. If we take, for example, the aforementioned "no-slip" condition, the vorticity belongs to the subspace

$$\{\omega = \Delta\psi, \quad \nabla\psi = 0 \text{ on the boundary}\}.$$

It is interesting to note [152] that this subspace is the orthogonal complement (in the L^2 sense) to the subspace of all harmonic functions. Projecting (the discrete vorticity obtained by time-stepping) on this subspace is equivalent to the projection of the discrete velocity on the subspace of divergence-free vector fields. However, the use of a pure streamfunction approach in this context spares us the need to deal with this dilemma.

Turning to the implementation of boundary conditions, we have already noted (when discussing the vorticity-streamfunction formulation) that some typical boundary conditions, imposed on the velocity, are readily expressed in terms of the streamfunction. Assume that some tangential velocity $\mathbf{u}$ is imposed on the boundary, thus "driving" the fluid inside the domain. It translates to

$$\nabla^{\perp}\psi = \mathbf{u}, \quad \text{on the boundary.} \tag{7.4}$$

In particular, $\nabla\psi$ is perpendicular to $\mathbf{u}$ on the boundary, hence perpendicular to the boundary. Since we consider here only simply connected domains, the streamfunction ψ is only determined up to a constant, so that this condition is equivalent to (with $\mathbf{t}$ being a unit tangent vector)

$$\psi = 0, \quad \frac{\partial\psi}{\partial n} = \mathbf{u} \cdot \mathbf{t} \quad \text{on the boundary.} \tag{7.5}$$

Notice that the highest-order differential operator in (7.2) is the biharmonic operator Δ^2. The boundary conditions (7.5) are natural (Dirichlet) boundary conditions for this elliptic operator.

The presence of the fourth-order operator may account for the fact that the pure streamfunction formulation was not central to the numerical analysis of the Navier–Stokes system (in two dimensions) at its early stages. In terms of the three discretization strategies discussed above, it appears to be easier for implementation by finite difference methods. Indeed, a finite difference approach (in the pure streamfunction formulation) was employed in [83, 84] some twenty years ago. The objective of these works was the numerical investigation of the Hopf bifurcation in the driven cavity problem. This benchmark problem will play a prominent role in Chapter 14.

We have already observed two beneficial aspects of discrete approximations based on the pure streamfunction formulation:

- The time evolution involves only one scalar unknown function – the streamfunction.
- The boundary conditions are naturally expressed in terms of the streamfunction; they conform to the general mathematical framework for boundary conditions in the presence of an elliptic fourth-order (in space) differential operator.

As can be expected, such nice benefits are not gained for free. A price to be paid is the need to deal with the biharmonic operator Δ^2. This operator plays a crucial role in Equation (7.2). However, it has proved to be a significant operator in a wide panorama of mathematical and physical studies. The reader is referred to the interesting historical review [137] of the two-dimensional biharmonic operator, along with its more than 750 references(!). In recent years there has been a renewed interest in the spectral theory of this operator, addressing some basic questions such as the eigenvalue distribution or features of the eigenfunctions [21, 28, 144]. The absence of suitable rigorous mathematical results stimulates the development of reliable numerical tools, like those described here.

Recall that solutions to Equation (7.2) do not share some of the outstanding properties of solutions to the convection–diffusion equation (Part I, Section 2.1), such as the Maximum principle or L^1 contraction.

The discrete approximation of Δ^2 is the pinnacle of the methodology expounded in this monograph. It should be emphasized that any method using the pure streamfunction formulation, namely, trying to discretize Equation (7.2), is confronted with the need to tackle the biharmonic operator. Here, however, we insist on a delicate balance: provide an accurate approximation to the biharmonic operator while retaining the natural treatment of the boundary conditions. Specifically, it means that the algorithm used in order to approximate the value of $\Delta^2\psi$ at a grid point should not rely on "distant points;" otherwise, near the boundary, it would require "ghost points" (outside the domain) or some manipulation of the given boundary conditions, so as to obtain approximate values of higher derivatives on the boundary.

Fortunately, there is a way to achieve this goal: implement a *compact finite difference scheme.* During the last three decades the idea of a "compact scheme" has gained considerable popularity among numerical analysts. It will be amply evident in the following chapters why this popularity is fully justified.

The methodology pursued here is based on two basic (and seemingly unrelated) building blocks:

- The *Hermitian derivative*, which allows "discrete differentiation" with fourth-order accuracy.
- A compact (nearest neighbors) approximation to the biharmonic operator.

These are introduced in full detail in Chapter 10, using the elementary one-dimensional setting. The treatment in this chapter goes beyond the "algorithmic" construction; in approximating fourth-order differentiation by a suitable discrete and compact analog, we seek, beyond the accuracy of the approximation, to establish discrete analogs of the coercivity and stability properties of the continuous (elliptic) operator. Since we insist on treating *boundary value problems*, the discretization at the boundary requires special care. Even though the accuracy at the boundary is naturally of lower order, the overall convergence is of fourth order, as shown in Section 10.7.

The approximation of the biharmonic operator in the two-dimensional setting is developed in Chapter 11. The rectangle is obviously the domain to be considered first, as it allows the use of a regular grid. The starting point is the discrete approximation proposed by Stephenson [168]; let f be a smooth function defined in the square $[-h, h] \times [-h, h]$ and consider the array consisting of the 3×3 points $\{(ih, jh),\ -1 \le i, j \le 1\}$. He used 13 monomials as in Equation (11.14), in order to construct a fourth-order polynomial $Q(x, y)$ interpolating nine functional values $\{f(ih, jh),\ -1 \le i, j \le 1\}$ and the four derivatives $\{f_x(\pm h, 0), f_y(0, \pm h)\}$. The approximation to $\Delta^2 f(0,0)$ is then *defined* to be $\Delta^2 Q(0,0)$. This yields only a *second-order* accurate approximation.

In order to obtain an optimal *fourth-order* approximation, we need to use polynomials of higher order. In the second part of Chapter 11 this is accomplished (in the rectangular case), with polynomials of order six, interpolating 19 data on the same compact (3×3) grid. In fact, the method is capable of providing a discrete approximation to the biharmonic operator in *irregular domains.* However, the geometric irregularity at the boundary reduces the accuracy in this case to second order.

We now return to the pure streamfunction equation (7.2). Taking a rectangular regular grid, and having at our disposal discrete (compact) approximations of all spatial differential operators, the equation can be discretized in space (see Section 12.2). We thus obtain a system of coupled ordinary differential equations for the approximating grid values (continuous in time) of the streamfunction. Section 12.3 is devoted to the proof of the convergence of the resulting grid function (as the grid is refined) to the exact solution ψ. It is here that all the previous work is rewarded, in particular the high order of accuracy and the coercivity of the discrete biharmonic operator. The reader may rightly regard this section as one of the highlights in this part of the monograph.

In actual numerical simulations, we use either the second-order (spatial) scheme described in Chapter 12 or the fourth-order scheme introduced in Chapter 13. In addition, we need to replace the time derivative with suitable discrete time-stepping algorithms. In Chapter 13 we review two time-stepping schemes; the first is a second-order scheme while the second is a higher-order one. These are illustrated by several test problems having analytical solutions.

Finally in Chapter 14 we display numerical results for the driven cavity problem. Convergence to a steady-state solution is observed for moderate Reynolds numbers. However, for high Reynolds numbers, a time-periodic solution emerges. This problem has been extensively studied in the literature, producing "reference results" (on very fine grids). The results displayed here compare very favorably with the latter.

Chapter 8

Notation

The notation for dealing with numerical aspects seems to be considerably more intricate than that employed in the continuous case. This is of course due to discretization, which adds to the functions in the continuous setting, discrete functions defined on a discrete set of grid points.

Even within the category of discrete functions we distinguish between *grid functions*, which retain the flavor of the continuous structure (for example, with a scalar product that accounts for the length of mesh intervals) and *finite-dimensional vectors*. The latter consist of the same components as the former, but are treated as vectors in the standard Euclidean space. Thus, *finite difference operators* acting on grid functions become *matrices* acting on their vector counterparts.

Having such issues in mind, we adopt notation that differs from that used in Part I of this monograph. In particular, all functions defined in the continuous setting are designated by lower case Latin letters and we use other types of letters for the variety of discrete functions, as detailed below.

8.1 One-dimensional discrete setting

Throughout this part, we deal with discrete schemes and therefore the notation of various discrete functions plays a significant role.

We will consider a fixed interval $I = [a, b]$, $a < b$, with length $L = b - a$. Small Latin letters are used for functions of the continuous variable $x \in I$, such as $u(x)$ and $v(x)$. In general our functions will be real valued, so that the scalar product in $L^2[a, b] = L^2[I]$ is given by

$$(u, v) = \int_a^b u(x)v(x)dx. \tag{8.1}$$

The $L^2[I]$ norm is defined by

$$\|u\|_{L^2[I]} = (u,u)^{1/2} = \left(\int_a^b u^2(x)dx \right)^{1/2}. \tag{8.2}$$

The $L^\infty[I]$ norm is denoted by

$$\|u\|_{L^\infty[I]} = \sup_{a \le x \le b} |u(x)|. \tag{8.3}$$

When the interval is understood from the context, we simplify to $\|u\|_{L^2}$ and $\|u\|_{L^\infty}$.

We lay out a uniform grid

$$a = x_0 < x_1 < \ldots < x_N = b, \tag{8.4}$$

where

$$x_i = a + ih, \quad i = 0, 1, \ldots, N, \quad h = L/N. \tag{8.5}$$

There are $N+1$ grid points numbered $i = 0, 1, \ldots, N-1, N$. The *internal grid points* are the ones indexed by $1, \ldots, N-1$. The two *boundary points* are x_0 and x_N and the *near-boundary points* are x_1 and x_{N-1} (see Figure 8.1).

The basic idea of discretization is to approximate a given function $u(x)$ by suitable values defined at the grid points. We now want to distinguish three categories of discrete function.

(i) A *grid function* is a function defined on the discrete grid $\{x_i\}_{i=0}^N$. We denote grid functions with Fraktur letters such as $\mathfrak{u}$, $\mathfrak{v}$. We have

$$\mathfrak{u} = (\mathfrak{u}(x_0), \mathfrak{u}(x_1), \ldots, \mathfrak{u}(x_{N-1}), \mathfrak{u}(x_N)). \tag{8.6}$$

We denote by l_h^2 the functional space of grid functions. This space is equipped with a scalar product and an associated norm

$$(\mathfrak{u}, \mathfrak{v})_h = h \sum_{i=0}^{N} \mathfrak{u}(x_i)\mathfrak{v}(x_i), \quad |\mathfrak{u}|_h = (\mathfrak{u}, \mathfrak{u})_h^{1/2}. \tag{8.7}$$

The subspace of grid functions, having zero boundary conditions at $x_0 = a$ and $x_N = b$, is denoted by $l_{h,0}^2$. For grid functions $\mathfrak{u}, \mathfrak{v} \in l_{h,0}^2$, we have

$$(\mathfrak{u}, \mathfrak{v})_h = h \sum_{i=1}^{N-1} \mathfrak{u}(x_i)\mathfrak{v}(x_i). \tag{8.8}$$

We also define the sup-norm for a grid function $\mathfrak{u}$

$$|\mathfrak{u}|_\infty = \max_{0 \le i \le N} |\mathfrak{u}(x_i)|. \tag{8.9}$$

(ii) Capital Latin letters denote vectors and matrices in real vector spaces. For example, the vector $U \in \mathbb{R}^{N+1}$ of the components of the grid function $\mathfrak{u} \in l^2_{h,0}$ is denoted by

$$U = [\mathfrak{u}_0, \mathfrak{u}_1, \ldots, \mathfrak{u}_N]^T = [\mathfrak{u}(x_0), \mathfrak{u}(x_1), \ldots, \mathfrak{u}(x_N)]^T. \tag{8.10}$$

The Euclidean norm in $\mathbb{R}^{N+1}$ is denoted by $|U|$. The induced norm for a matrix A of size $(N+1) \times (N+1)$ is

$$|A|_2 = \sup_{U \neq 0} \frac{|AU|}{|U|}. \tag{8.11}$$

If A is symmetric then $|A|_2 = \rho(A)$ where $\rho(A)$ is the spectral radius of A, namely the largest eigenvalue (in absolute value).
The sup-norm of a vector $U \in \mathbb{R}^{N+1}$ is defined as

$$|U|_\infty = \max_{0 \leq i \leq N+1} |U_i|. \tag{8.12}$$

The induced infinity norm of a matrix A of size $(N+1) \times (N+1)$ is

$$|A|_\infty = \sup_{U \neq 0} \frac{|AU|_\infty}{|U|_\infty} = \max_{1 \leq i \leq N+1} \sum_{j=1}^{N+1} |A_{i,j}|. \tag{8.13}$$

(iii) Given a function $u(x)$, we define the related grid function u^* by

$$u_i^* = u^*(x_i) = u(x_i), \quad i = 0, \ldots, N. \tag{8.14}$$

We distinguish between l^2_h, the space of grid functions, and $\mathbb{R}^{N+1}$ (and similarly between $l^2_{h,0}$ and $\mathbb{R}^{N-1}$), even though they are isomorphic. This distinction has several advantages. First it allows us to make a clear distinction between the two classes of operators:

- Finite difference operators, which act naturally on grid functions.
- Matrices, which act naturally on vectors.

Second, grid functions contain all the scalings needed in the physical context, whereas vectors are dimensionless. Third, reasoning at the discrete level is easier with grid functions and finite difference operators, whereas the actual coding has to be done within the linear algebraic framework.

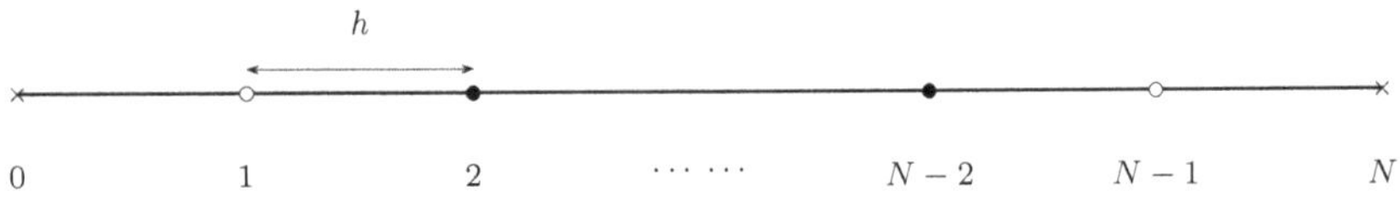

Fig. 8.1 One-dimensional grid. The two boundary points are $i = 0$ and $i = N$ (shown as "x" at the end of the line). The two near-boundary points are $i = 1$ and $N-1$ (shown as "o"). The points $i = 2, \ldots, N-2$ are internal points.

8.2 Two-dimensional discrete setting

The one-dimensional setting extends to the two-dimensional case in a natural way. For simplicity, we restrict the geometry to the square $\Omega = [a, b]^2$ with $L = b - a$.

The scalar product in $L^2(\Omega)$ is given by

$$(u, v) = \int_\Omega u(x, y)v(x, y)dxdy. \tag{8.15}$$

The $L^2(\Omega)$ norm is defined by

$$\|u\|_{L^2(\Omega)} = (u, u)^{1/2} = \left(\int_\Omega u^2(x, y)dxdy \right)^{1/2}. \tag{8.16}$$

The $L^\infty(\Omega)$ norm is denoted by

$$\|u\|_{L^\infty(\Omega)} = \sup_{(x,y)\in\Omega} |u(x, y)|. \tag{8.17}$$

When the domain Ω is understood from the context, we simplify to $\|u\|_{L^2}$ and $\|u\|_{L^\infty}$.

Partial derivatives are defined by $\partial_x u = \frac{\partial u}{\partial x}$ and $\partial_y u = \frac{\partial u}{\partial y}$.

We lay out a uniform grid (x_i, y_j) with $a = x_0 < \ldots < x_N = b$ and $a = y_0 < \ldots < y_N = b$, so that $x_{i+1} - x_i = y_{j+1} - y_j = h$. The internal points correspond to indices $1 \leq i, j \leq N - 1$. The boundary points are located on the four sides of the square. They correspond to

- Left
$$i = 0, \quad j = 1, \ldots, N - 1. \tag{8.18}$$
- Right
$$i = N, \quad j = 1, \ldots, N - 1. \tag{8.19}$$

- Bottom

$$i = 1, \ldots, N-1, \quad j = 0. \tag{8.20}$$

- Top

$$i = 1, \ldots, N-1, \quad j = N. \tag{8.21}$$

The *near-boundary* points are the internal points distant by at most h from a boundary point. In addition there are the four *corner points* $(i, j) = (0,0), (N,0), (0,N)$, and (N,N) (see Figure 8.2).

As in the one-dimensional setting there are different types of discrete function attached to the grid.

(i) A *grid function* is a function defined on the discrete grid (x_i, y_j). We denote grid functions with bold Greek letters such as $\boldsymbol{\psi}$ and $\boldsymbol{\phi}$. We have

$$\boldsymbol{\psi} = (\boldsymbol{\psi}(x_i, y_j))_{0 \le i,j \le N}. \tag{8.22}$$

We denote by L_h^2 the functional space of grid functions. This space is equipped with a scalar product and the associate norm

$$(\boldsymbol{\psi}, \boldsymbol{\phi})_h = h^2 \sum_{i,j=0}^{N} \boldsymbol{\psi}(x_i, y_j)\boldsymbol{\phi}(x_i, y_j), \quad |\boldsymbol{\psi}|_h = (\boldsymbol{\psi}, \boldsymbol{\psi})_h^{1/2}. \tag{8.23}$$

We also define the sup-norm for a grid function $\boldsymbol{\psi}$

$$|\boldsymbol{\psi}|_\infty = \max_{0 \le i,j \le N} |\boldsymbol{\psi}(x_i, y_j)|. \tag{8.24}$$

The subspace of grid functions, which have homogeneous boundary conditions at boundary and corner points is denoted by $L_{h,0}^2$. For grid functions $\boldsymbol{\psi}, \boldsymbol{\phi} \in L_{h,0}^2$, we have

$$(\boldsymbol{\psi}, \boldsymbol{\phi})_h = h^2 \sum_{i,j=1}^{N-1} \boldsymbol{\psi}(x_i, y_j)\boldsymbol{\phi}(x_i, y_j). \tag{8.25}$$

(ii) Given a function $\psi(x, y)$, we define the related grid function ψ^* by

$$\psi_{i,j}^* = \psi^*(x_i, y_j) = \psi(x_i, y_j), \quad i = 0, \ldots, N, \quad j = 0, \ldots, N. \tag{8.26}$$

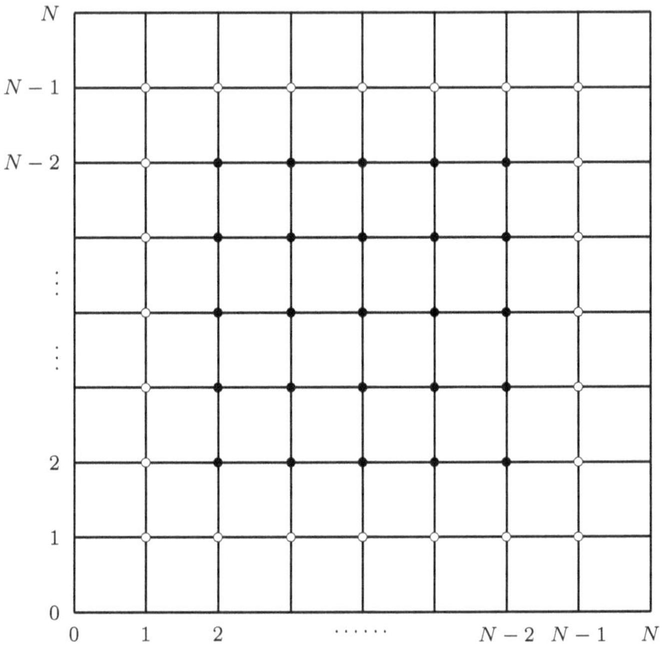

Fig. 8.2 Two-dimensional grid. The boundary points are $i = \{0, N\}$ or $j = \{0, N\}$. Near-boundary points are $i = \{1, N-1\}$ or $j = \{1, N-1\}$. The points $2 \le i, j \le N-2$ are the internal points.

Chapter 9

Finite Difference Approximation to Second-Order Boundary-Value Problems

In this chapter, we first introduce the concept of a finite difference operator and the principle of a finite difference scheme. These concepts are illustrated using the three-point discrete (one-dimensional) Laplacian and the associated scheme for the Sturm–Liouville second-order boundary-value problem.

9.1 The principle of finite difference schemes

Consider the second-order boundary-value problem

$$(9.1)\qquad \begin{cases} -u''(x) = f(x), & a < x < b, \\ u(a) = 0, \quad u(b) = 0. \end{cases}$$

Recall the discrete notation of Chapter 8. In particular, we have a uniform grid

$$(9.2)\qquad a = x_0 < x_1 < \ldots < x_N = b,$$

where

$$(9.3)\qquad x_i = a + ih, \quad i = 0, 1, ..., N, \quad h = L/N, \quad L = b - a.$$

Observe that at a point $x \in [a+h, b-h]$, if u is a smooth function, we can approximate $u''(x)$ within $O(h^2)$ by the quotient

$$(9.4)\qquad \frac{u(x+h) - 2u(x) + u(x-h)}{h^2}.$$

It is therefore plausible to define an approximate (discrete) solution to (9.1) by looking for a grid function $\mathfrak{u} = (\mathfrak{u}_0, \mathfrak{u}_1, \ldots, \mathfrak{u}_{N-1}, \mathfrak{u}_N)$ such that

- At each interior point, $1 \leq i \leq N-1$,

$$(9.5)\qquad -\frac{\mathfrak{u}_{i+1} - 2\mathfrak{u}_i + \mathfrak{u}_{i-1}}{h^2} = f^*(x_i) = f(x_i), \quad 1 \leq i \leq N-1.$$

- At the two endpoints

$$\text{(9.6)} \qquad \mathfrak{u}_0 = 0, \ \ \mathfrak{u}_N = 0.$$

Equations (9.5)–(9.6) constitute a system of $N+1$ linear equations for the $N+1$ unknowns. Natural questions are whether the linear system has a unique solution and if this solution is defined in a stable way in terms of the grid function f^*. The main objective is of course to find out whether or not the solution $\mathfrak{u}$ effectively approximates the grid function $u^* = (u(x_0), u(x_1), \ldots, u(x_{N-1}), u(x_N))$.

9.2 The three-point Laplacian

9.2.1 *General setting*

We consider the discrete boundary-value problem (9.5)–(9.6) and introduce the finite difference operator $\mathfrak{u} \mapsto \delta_x^2 \mathfrak{u}$, where δ_x^2 is defined by

$$\text{(9.7)} \qquad (\delta_x^2 \mathfrak{u})_i = \delta_x^2 \mathfrak{u}_i = \frac{\mathfrak{u}_{i+1} - 2\mathfrak{u}_i + \mathfrak{u}_{i-1}}{h^2}, \ \ 1 \le i \le N-1.$$

Then, the system (9.5)–(9.6) is equivalent to the problem:

Find $\mathfrak{u} \in l^2_{h,0}$ that solves

$$\text{(9.8)} \qquad \begin{cases} -\delta_x^2 \mathfrak{u}_i = f^*(x_i), & 1 \le i \le N-1, \\ \mathfrak{u}_0 = \mathfrak{u}_N = 0. \end{cases}$$

The grid function $u^* = (u(x_0), \ldots, u(x_N))$, where u is the exact solution of (9.1), does not satisfy Equation (9.8). In fact, the *truncation error* $\mathfrak{t}(x_i)$ at point x_i, associated with the discretization (9.8), measures the difference between $-\delta_x^2 u^*$ and f^*,

$$\text{(9.9)} \qquad \mathfrak{t}(x_i) = -\delta_x^2 u^*(x_i) - f^*(x_i), \quad i = 1, \ldots, N-1.$$

We estimate $\mathfrak{t}(x_i)$ in terms of a Taylor expansion as follows. Expanding $u(x_i + h)$ and $u(x_i - h)$, $1 \le i \le N-1$, we have

$$\text{(9.10)} \quad u(x_i + h) = u(x_i) + hu'(x_i) + \frac{h^2}{2!}u''(x_i) + \frac{h^3}{3!}u^3(x_i) + \frac{h^4}{4!}u^{(4)}(\eta_i)$$

and

$$\text{(9.11)} \quad u(x_i - h) = u(x_i) - hu'(x_i) + \frac{h^2}{2!}u''(x_i) - \frac{h^3}{3!}u^{(3)}(x_i) + \frac{h^4}{4!}u^{(4)}(\zeta_i).$$

Here, η_i and ζ_i are points in the intervals $(x_i, x_i + h)$ and $(x_i - h, x_i)$, respectively. Therefore,

$$
\begin{aligned}
\mathfrak{t}(x_i) &= -\delta_x^2 u^*(x_i) + (u'')^*(x_i) \\
&= -\frac{u(x_i+h) - 2u(x_i) + u(x_i-h)}{h^2} + u''(x_i) = -\frac{h^2}{12}u^{(4)}(\xi_i),
\end{aligned} \tag{9.12}
$$

where ξ_i is a point in the interval $(x_i - h, x_i + h)$. We refer to (9.12) as the *consistency* of the difference operator δ_x^2.

9.2.2 *Maximum principle analysis*

A natural way to analyze the discrete problem (9.8) is based on the discrete maximum principle.

Lemma 9.1. *Let* $\mathfrak{u} \in l_h^2$. *Suppose that* $-\delta_x^2 \mathfrak{u}_i \leq 0, \;\; 1 \leq i \leq N-1$. *Then the grid function* $\mathfrak{u}$ *satisfies*

$$
\max_{1 \leq i \leq N-1} \mathfrak{u}_i \leq \max(\mathfrak{u}_0, \mathfrak{u}_N), \quad \textit{(Maximum principle)}. \tag{9.13}
$$

That is $\mathfrak{u}$ *attains its maximum value at the boundary points* $i = 0$ *or* $i = N$. *Similarly, if* $-\delta_x^2 \mathfrak{u}_i \geq 0, \;\; 1 \leq i \leq N-1$, *then* $\mathfrak{u}$ *satisfies*

$$
\min_{1 \leq i \leq N-1} \mathfrak{u}_i \geq \min(\mathfrak{u}_0, \mathfrak{u}_N), \quad \textit{(Minimum principle)}. \tag{9.14}
$$

Proof. Consider the linear grid function $\mathfrak{v} \in l_h^2$ defined by

$$
\mathfrak{v}_i = \frac{(N-i)\mathfrak{u}_0 + i\mathfrak{u}_N}{N}. \tag{9.15}
$$

We have $\delta_x^2 \mathfrak{v}_i = 0$. Therefore, denoting $\mathfrak{w}_i = \mathfrak{u}_i - \mathfrak{v}_i$, we have by hypothesis

$$
-\delta_x^2 \mathfrak{w}_i = -\frac{\mathfrak{w}_{i+1} - 2\mathfrak{w}_i + \mathfrak{w}_{i-1}}{h^2} \leq 0, \;\; 1 \leq i \leq N-1. \tag{9.16}
$$

Therefore,

$$
\mathfrak{w}_i \leq \frac{1}{2}(\mathfrak{w}_{i+1} + \mathfrak{w}_{i-1}), \;\; 1 \leq i \leq N-1. \tag{9.17}
$$

This ensures that the maximum value of $\mathfrak{w}$ is attained at either x_0 or x_N, which means

$$
\mathfrak{w}_i \leq \max(\mathfrak{w}_0, \mathfrak{w}_N) = 0. \tag{9.18}
$$

It follows that $\mathfrak{u}_i \leq \mathfrak{v}_i$. In particular,

$$
\max_{0 \leq i \leq N} \mathfrak{u}_i \leq \max(\mathfrak{u}_0, \mathfrak{u}_N). \tag{9.19}
$$

In the same way, we get that if $-\delta_x^2\mathfrak{u}_i \geq 0$, then $\mathfrak{u}_i \geq \mathfrak{v}_i$, which leads to the Minimum principle (9.14). ■

Using this lemma, we establish stability and convergence in the following theorem.

Theorem 9.2. *(i) For any grid function $\mathfrak{v} \in l_{h,0}^2$ defined in the interval $[a,b]$, we have*

$$|\mathfrak{v}|_\infty \leq \frac{L^2}{8}|\delta_x^2\mathfrak{v}|_\infty, \tag{9.20}$$

where $L = b - a$.
(ii) The discrete system (9.8) has a unique solution $\mathfrak{u} \in l_{h,0}^2$.
(iii) There exists a constant C independent of h, such that the discrete solution $\mathfrak{u}$ satisfies the error estimate

$$\max_{1\leq i\leq N-1} |u^*(x_i) - \mathfrak{u}_i| \leq Ch^2\|u^{(4)}(x)\|_{L^\infty[a,b]}. \tag{9.21}$$

where $u(x)$ is the solution to (9.1) and $u^{(4)}(x)$ is its fourth-order derivative.

Proof. (i) To simplify the notation, we assume that $a = 0$ and $b = L$. Let $\mathfrak{g}$ be the grid function defined by

$$\mathfrak{g}_i := \delta_x^2\mathfrak{v}_i, \quad i = 1,\ldots,N-1, \tag{9.22}$$

and let $k = \max_i |\mathfrak{g}_i|$ so that

$$-k \leq \delta_x^2\mathfrak{v}_i \leq k, \quad 1 \leq i \leq N-1. \tag{9.23}$$

Define $\mathfrak{w}$ as

$$\mathfrak{w}_i = \frac{1}{2}\Big(x_i - \frac{L}{2}\Big)^2, \quad 0 \leq i \leq N. \tag{9.24}$$

Then, for $1 \leq i \leq N-1$, we have $\delta_x^2\mathfrak{w}_i = 1$ and

$$\begin{cases} \delta_x^2(\mathfrak{v}_i - \mathfrak{w}_i k) = \delta_x^2\mathfrak{v}_i - k \leq 0 \\ \delta_x^2(\mathfrak{v}_i + \mathfrak{w}_i k) = \delta_x^2\mathfrak{v}_i + k \geq 0. \end{cases} \tag{9.25}$$

Using the (discrete) Minimum principle, we have

$$-\frac{L^2}{8}k \leq \mathfrak{v}_i - \mathfrak{w}_i k \leq \mathfrak{v}_i, \quad i = 0,\ldots,N. \tag{9.26}$$

Similarly, from the (discrete) Maximum principle, we have

$$\mathfrak{v}_i \leq \mathfrak{v}_i + \mathfrak{w}_i k \leq \frac{L^2}{8}k, \quad i = 0,\ldots,N. \tag{9.27}$$

Thus,

$$-\frac{L^2}{8}k \le \mathfrak{v}_i \le \frac{L^2}{8}k, \quad i = 0, \ldots, N. \tag{9.28}$$

Finally, this gives

$$\max_{1 \le i \le N-1} |\mathfrak{v}_i| \le \frac{L^2}{8}k = \frac{L^2}{8} \max_{1 \le i \le N-1} |\delta_x^2 \mathfrak{v}_i|. \tag{9.29}$$

(ii) The uniqueness of the solution to (9.8) results from the preceding observation. If $\mathfrak{u}_1$ and $\mathfrak{u}_2$ are two solutions, then $\mathfrak{u} := \mathfrak{u}_1 - \mathfrak{u}_2$ satisfies $\delta_x^2 \mathfrak{u} = 0$. Therefore, $\mathfrak{u} = 0$ by (i). Since the problem is linear and finite-dimensional, the existence is proved as well.
(iii) In view of (9.12), we have for $1 \le i \le N-1$,

$$-Ch^2 \|u^{(4)}(x)\|_{L^\infty[a,b]} \le \delta_x^2(\mathfrak{u}_i - u_i^*) \le Ch^2 \|u^{(4)}(x)\|_{L^\infty[a,b]}. \tag{9.30}$$

Using (i) and (9.30) and denoting $\mathfrak{e}_i = \mathfrak{u}_i - u_i^*$, we obtain

$$\begin{aligned} \max_{1 \le i \le N-1} |\mathfrak{e}_i| = \max_{1 \le i \le N-1} |\mathfrak{u}_i - u_i^*| &\le \frac{L^2}{8} \max_{1 \le i \le N-1} |\delta_x^2(\mathfrak{u}_i - u_i^*)| \\ &\le Ch^2 \|u^{(4)}(x)\|_{L^\infty[a,b]}. \end{aligned} \tag{9.31}$$

■

9.2.3 *Coercivity and energy estimate*

We now can derive a *coercivity* estimate as follows.

Let $\mathfrak{u}, \mathfrak{v} \in l_{h,0}^2$. Using the definition (9.7) of $\delta_x^2 \mathfrak{u}_i$, we have

$$-\sum_{i=1}^{N-1} \delta_x^2 \mathfrak{u}_i \, \mathfrak{v}_i = \sum_{i=0}^{N-1} \frac{\mathfrak{u}_{i+1} - \mathfrak{u}_i}{h} \frac{\mathfrak{v}_{i+1} - \mathfrak{v}_i}{h}. \tag{9.32}$$

Taking $\mathfrak{v} = \mathfrak{u}$ we obtain the *coercivity* equality

$$-(\delta_x^2 \mathfrak{u}, \mathfrak{u})_h = h \sum_{i=0}^{N-1} \left| \frac{\mathfrak{u}_{i+1} - \mathfrak{u}_i}{h} \right|^2. \tag{9.33}$$

This is the discrete analog of the equality $-(u'', u) = \|u\|_{L^2}^2$, which is valid for every $u \in L^2[a, b]$ vanishing at the endpoints.

Definition 9.3 (Coercivity property). *The equality (9.33) expresses the* ***coercivity property*** *of the operator* $-\delta_x^2$.

This means that $-(\delta_x^2\mathfrak{u}, \mathfrak{v})_h$ can be used a scalar product.

Choosing $\mathfrak{v} = \mathfrak{u}$ in (9.32) and applying the Cauchy–Schwarz inequality, we get

$$\sum_{i=0}^{N-1}\left(\frac{\mathfrak{u}_{i+1}-\mathfrak{u}_i}{h}\right)^2 \le \left(\sum_{i=1}^{N-1}(\delta_x^2\mathfrak{u}_i)^2\right)^{1/2}\left(\sum_{i=1}^{N-1}\mathfrak{u}_i^2\right)^{1/2}. \tag{9.34}$$

The right-hand side can be further estimated by using

$$\begin{aligned}\mathfrak{u}_i &= (\mathfrak{u}_i-\mathfrak{u}_{i-1})+(\mathfrak{u}_{i-1}-\mathfrak{u}_{i-2})+\cdots+(\mathfrak{u}_1-\mathfrak{u}_0)\\ &= h\sum_{k=0}^{i-1}\frac{\mathfrak{u}_{k+1}-\mathfrak{u}_k}{h}.\end{aligned}$$

So that, since $i \le N = Lh^{-1}$,

$$\begin{aligned}|\mathfrak{u}_i| &\le h\sum_{k=0}^{i-1}\frac{|\mathfrak{u}_{k+1}-\mathfrak{u}_k|}{h} \le \left(\sum_{k=0}^{i-1}\left(\frac{\mathfrak{u}_{k+1}-\mathfrak{u}_k}{h}\right)^2\right)^{1/2} i^{1/2}h\\ &\le L^{1/2}\left(\sum_{k=0}^{i-1}\left(\frac{\mathfrak{u}_{k+1}-\mathfrak{u}_k}{h}\right)^2\right)^{1/2} h^{1/2}.\end{aligned}$$

Summation over $i = 1$ to $N-1$ of $|\mathfrak{u}_i|^2$ yields the following

$$\sum_{i=1}^{N-1}|\mathfrak{u}_i|^2 \le L^2\sum_{k=0}^{N-1}\left(\frac{\mathfrak{u}_{k+1}-\mathfrak{u}_k}{h}\right)^2 \quad \textit{(Discrete Poincaré inequality)}. \tag{9.35}$$

Plugging the last estimate into the right-hand side of (9.34), we obtain

$$\sum_{i=0}^{N-1}\left(\frac{\mathfrak{u}_{i+1}-\mathfrak{u}_i}{h}\right)^2 \le L^2\sum_{i=1}^{N-1}(\delta_x^2\mathfrak{u}_i)^2, \tag{9.36}$$

which expresses also a discrete Poincaré inequality, now between the first- and second-order discrete derivatives. Combined with the discrete Poincaré inequality (9.35) above, we finally get

$$\sum_{i=1}^{N-1}\mathfrak{u}_i^2 \le L^4\sum_{i=1}^{N-1}(\delta_x^2\mathfrak{u}_i)^2. \tag{9.37}$$

This is called a *stability* estimate. It implies the convergence in the l_h^2 norm of the discrete solution to the continuous one without resorting to the maximum principle, as follows. In view of (9.37) and (9.30), we obtain

$$\begin{aligned}|u^*-\mathfrak{u}|_h &\le L^2|\delta_x^2(u^*-\mathfrak{u})|_h = L^2|\delta_x^2u^*+f^*|_h\\ &= L^2|\delta_x^2u^*-(u'')^*|_h \le CL^2h^2\|u^{(4)}\|_{L^\infty[a,b]}.\end{aligned} \tag{9.38}$$

Observe that this estimate may be also derived from (9.31).

The method employed in the derivation of (9.38) is known as the *energy method*. It will prove to be crucial in later chapters.

9.3 Matrix representation of the three-point Laplacian

9.3.1 *Continuous and discrete eigenfunctions*

The Sturm–Liouville problem for (9.1) leads to the classical eigenfunction problem:

Find the values $\mu \in \mathbb{R}$ for which there exists a non-zero function $z(x)$ satisfying

$$\begin{cases} (a) & -z''(x) = \mu z(x), \qquad 0 \leq x \leq L, \\ (b) & z(0) = 0, \quad z(L) = 0. \end{cases} \tag{9.39}$$

The general non-zero solution of this equation is

$$z(x) = A\cos(\sqrt{\mu}x) + B\sin(\sqrt{\mu}x), \tag{9.40}$$

where $\mu \geq 0$ and A, B are two complex constants, not both zero. The boundary conditions $z(0) = 0$ and $z(L) = 0$ yield

$$\begin{cases} z(0) = A = 0, \\ z(L) = B\sin(\sqrt{\mu}L) = 0. \end{cases} \tag{9.41}$$

In order to have a non-zero solution, we infer from the second equation that $B \neq 0$. Therefore, μ must solve $\sin(\sqrt{\mu}L) = 0$. This yields $\sqrt{\mu}L = k\pi$ for $k = 1, 2, \ldots$. Thus, all eigenfunctions are proportional to $\sin(\frac{k\pi x}{L})$. Two eigenfunctions z^k, z^l with $\mu_k \neq \mu_l$ are orthogonal with respect to the L^2 scalar product, because

$$\mu_k(z^k, z^l) = ((z^k)', (z^l)') = \mu_l(z^k, z^l), \tag{9.42}$$

which gives $(z^k, z^l) = 0$.

Proposition 9.4. *The normalized solutions of (9.39) form a sequence* $\{\mu_k, z^k(x)\}_{k=1}^{\infty}$, *with*

$$\begin{cases} \mu_k = \left(\dfrac{k\pi}{L}\right)^2, \quad k \geq 1 \\ z^k(x) = \left(\dfrac{2}{L}\right)^{1/2} \sin\left(\dfrac{k\pi x}{L}\right), \quad k \geq 1. \end{cases} \tag{9.43}$$

The functions $\{z^k(x)\}_{k=1}^{\infty}$ *form an orthonormal basis of* $L^2[0, L]$.

Refer to [25, 150, 178] for a full treatment of Fourier series on an interval.

Let us now turn to the discrete setting and consider the action of the operator $-\delta_x^2$ on grid functions. The eigenvalue problem (9.39) is transformed into the following: Find $\mathfrak{z} \in l_{h,0}^2$ such that

$$\begin{cases} -\delta_x^2 \mathfrak{z} = \tilde{\lambda}\mathfrak{z}, \\ \mathfrak{z}_0 = 0, \quad \mathfrak{z}_N = 0. \end{cases} \tag{9.44}$$

Consider the grid function defined by the evaluation of $z^k(x)$ at grid points $\{x_i\}_{i=0}^N$ [see (9.2)],

$$(9.45)\qquad \mathfrak{z}_i^k = (z^k)^*(x_i) = \left(\frac{2}{L}\right)^{1/2} \sin\left(\frac{k\pi x_i}{L}\right) = \left(\frac{2}{L}\right)^{1/2} \sin\left(\frac{ki\pi}{N}\right).$$

Using the trigonometric identity

$$\begin{aligned}\sin(i+1)\theta - 2\sin i\theta + \sin(i-1)\theta &= 2\sin\left(\frac{\theta}{2}\right)\left(\cos\left(i+\frac{1}{2}\right)\theta - \cos\left(i-\frac{1}{2}\right)\theta\right)\\ &= -4\sin^2\left(\frac{\theta}{2}\right)\sin i\theta\end{aligned}$$

and the definition (9.45) of $\mathfrak{z}_i^k$, we find that

$$(9.46)\quad -(\delta_x^2\mathfrak{z}^k)_i = \frac{4}{h^2}\sin^2\left(\frac{\pi k}{2N}\right)(\mathfrak{z}^k)_i, \quad 1 \le k \le N-1, \quad 1 \le i \le N-1,$$

so that $\{\mathfrak{z}^k\}_{k=1}^{N-1}$ are eigenfunctions of $-\delta_x^2$. The corresponding eigenvalues are the $\tilde{\lambda}_k$ given by

$$(9.47)\qquad \tilde{\lambda}_k = \frac{4}{h^2}\sin^2\left(\frac{\pi k}{2N}\right), \quad 1 \le k \le N-1.$$

By simple trigonometric identities, it is seen that the grid functions $\mathfrak{z}^k$, $1 \le k \le N-1$, satisfy $|\mathfrak{z}^k|_h = 1$. This set of eigenfunctions is linearly independent, since eigenfunctions associated with different eigenvalues are linearly independent. Thus any grid function $\mathfrak{u} \in l_{h,0}^2$ may be represented in terms of the basis $\{\mathfrak{z}^k\}_{k=1}^{N-1}$ as

$$(9.48)\qquad \mathfrak{u} = \sum_{k=1}^{N-1} \hat{\mathfrak{u}}_k \mathfrak{z}^k,$$

where the component $\hat{\mathfrak{u}}^k$ is given by

$$(9.49)\qquad \hat{\mathfrak{u}}_k = (\mathfrak{u}, \mathfrak{z}^k)_h = h\left(\frac{2}{L}\right)^{1/2} \sum_{i=1}^{N-1} \mathfrak{u}_i \sin\left(\frac{ki\pi}{N}\right), \quad 1 \le k \le N-1.$$

Now we turn to the vector setting, which translates the above considerations on grid functions to the linear algebraic setting. It is this setting that is used in actual implementation and simulation. Let $U \in \mathbb{R}^{N-1}$ be the vector corresponding to $\mathfrak{u} \in l_{h,0}^2$. We have, with $\mathfrak{u}_i = \mathfrak{u}(x_i)$,

$$(9.50)\qquad U = [\mathfrak{u}_1, \ldots, \mathfrak{u}_{N-1}]^T.$$

Let T be the $(N-1)\times(N-1)$ tridiagonal positive definite matrix T

$$T = \begin{bmatrix} 2 & -1 & 0 & \dots & 0 \\ -1 & 2 & -1 & \dots & 0 \\ \vdots & \vdots & \vdots & \dots & \vdots \\ 0 & \dots & -1 & 2 & -1 \\ 0 & \dots & 0 & -1 & 2 \end{bmatrix}. \tag{9.51}$$

It is easy to verify that $TU \in \mathbb{R}^{N-1}$ corresponds to the grid function $-h^2\delta_x^2\mathfrak{u}$ (recall that $\mathfrak{u} \in l_{h,0}^2$). Therefore, the eigenvalue problem (9.44) is equivalent to

$$TU = \lambda U, \tag{9.52}$$

where

$$U = [\mathfrak{u}_1, \dots, \mathfrak{u}_{N-1}]^T, \quad \lambda = h^2\tilde{\lambda}. \tag{9.53}$$

By the equivalence between (9.44) and (9.52), we conclude that the vector $Z^k = h^{1/2}[\mathfrak{z}_1^k, \dots, \mathfrak{z}_{N-1}^k]^T$ is an eigenvector of T corresponding to the eigenvalue λ_k where

$$\lambda_k = 4\sin^2\left(\frac{k\pi}{2N}\right), \quad 1 \le k \le N-1. \tag{9.54}$$

Furthermore, $Z^k \in \mathbb{R}^{N-1}$ is a unit vector with respect to the Euclidean norm. It is given by

$$Z_i^k = \left(\frac{2}{N}\right)^{1/2} \sin\left(\frac{ki\pi}{N}\right), \quad 1 \le k, i \le N-1. \tag{9.55}$$

Let Z be the $(N-1)\times(N-1)$ matrix whose kth column is Z^k. Observe that by (9.55) Z is symmetric, so that

$$Z^2 = ZZ^T = I_{N-1} \tag{9.56}$$

and

$$T = Z\Lambda Z^T, \quad \Lambda = \operatorname{diag}(\lambda_1, \dots, \lambda_{N-1}). \tag{9.57}$$

9.3.2 *Convergence analysis*

In Subsection 9.2.3 we derived an error estimate in the context of grid functions. Here we use the spectral structure of the matrix T in order to reformulate this energy estimate. Recall that $\mathfrak{u} \in l_{h,0}^2$ is the solution to the discrete problem (9.5) and that u^* is the grid function corresponding to the exact solution of (9.1). Let us denote by $U, U^* \in \mathbb{R}^{N-1}$ the vectors corresponding to $\mathfrak{u}$ and u^*.

Let $\mathfrak{e} = \mathfrak{u} - u^*$ be the error grid function. It follows from (9.9) that

$$-\delta_x^2 \mathfrak{e}_i = -\delta_x^2 (\mathfrak{u} - u^*) = -\mathfrak{t}_i, \quad i = 1, \ldots, N-1. \tag{9.58}$$

In addition, it results from (9.12) that

$$\mathfrak{t}_i = -\frac{h^2}{4} (u^{(4)})^* (\xi_i), \quad i = 1, \ldots, N-1. \tag{9.59}$$

Let $E, Q \in \mathbb{R}^{N-1}$ be the vectors corresponding to $\mathfrak{e}, \mathfrak{t}$, respectively. Then, Equation (9.58) can be recast as

$$TE = -h^2 Q. \tag{9.60}$$

By (9.57) $T^{-1} = Z\Lambda^{-1} Z^T$. Using

$$\sin x \geq (2/\pi) x, \quad 0 \leq x \leq \pi/2, \tag{9.61}$$

we have that

$$|T^{-1}|_2 = \max_{i=1,\ldots,N-1} |\lambda_i^{-1}| \leq Ch^{-2}, \tag{9.62}$$

so that

$$|E| \leq h^2 |T^{-1}|_2 |Q| \leq C|Q| \leq C_1 h^{3/2}. \tag{9.63}$$

Therefore,

$$|\mathfrak{e}|_h \leq C_1 h^2. \tag{9.64}$$

■

9.4 Notes for Chapter 9

- A classical source on finite difference methods for partial difference equations is [156]. Recent references include [94, 169].
- The treatment of the discrete maximum principle can be found in [169].
- A general introduction to numerical methods for differential equations is [101].
- The algebraic treatment either by matrices or by the Fast Fourier Transform (FFT) of finite difference operators is addressed in [94,175].

Chapter 10

From Hermitian Derivative to the Compact Discrete Biharmonic Operator

In this chapter we introduce the Hermitian derivative and the three-point biharmonic operator, which approximates the one-dimensional biharmonic operator. The consistency and the convergence of the three-point biharmonic scheme are proved.

10.1 The Hermitian derivative operator

In the preceding chapter, we discussed various aspects of the difference operator δ_x^2. It was introduced as the discrete counterpart of the second-order derivative in the continuous context.

In this section, we consider a more basic operator, namely the first-order differentiation operator and its discrete counterpart. Using the terminology of grid functions, we have an obvious candidate for the discrete derivative given by

$$\delta_x \mathfrak{u}_i = \frac{\mathfrak{u}_{i+1} - \mathfrak{u}_{i-1}}{2h}, \quad i = 1, \ldots, N-1. \tag{10.1}$$

Using (9.10) and (9.11), we have

$$\delta_x u^*(x_i) = u'(x_i) + \frac{h^2}{6} u'''(\xi_i), \tag{10.2}$$

with $\xi_i \in (x_{i-1}, x_{i+1})$. Let $U \in \mathbb{R}^{N-1}$ be the vector corresponding to the grid function $\mathfrak{u} \in l_{h,0}^2$,

$$U = [\mathfrak{u}_1, \ldots, \mathfrak{u}_{N-1}]^T. \tag{10.3}$$

The vector corresponding to the grid function $\delta_x \mathfrak{u}$ is

$$\frac{1}{2h} KU, \tag{10.4}$$

where $K = (K_{i,m})_{1 \le i,m \le N-1}$ is the skew-symmetric matrix

$$(10.5) \qquad K = \begin{bmatrix} 0 & 1 & 0 & \dots & 0 \\ -1 & 0 & 1 & \dots & 0 \\ \vdots & \vdots & \vdots & \dots & \vdots \\ 0 & \dots & -1 & 0 & 1 \\ 0 & \dots & 0 & -1 & 0 \end{bmatrix}.$$

We observe from (10.2) that the approximation $\delta_x \mathfrak{u}_i$ is only second-order accurate. For reasons that will be clear as we proceed towards the discretization of the full Navier–Stokes system, we shall need a *fourth-order accurate* approximation of the gradient. Such an approximation is provided by the *Hermitian operator*, which we describe next.

Consider a smooth function $u(x)$ defined on the interval $I = [a, b]$. Let $v(x) = u'(x)$. We take a uniform grid $x_i = a+ih$, $0 \le i \le N$, $h = (b-a)/N$. Integrating $v(x)$ over $[x_{i-1}, x_{i+1}]$, we obtain

$$(10.6) \qquad \frac{1}{2h}\int_{x_{i-1}}^{x_{i+1}} u'(x)dx = \frac{u(x_{i+1}) - u(x_{i-1})}{2h}.$$

Approximating the integral on the left-hand side by the Simpson rule and using a well-known formula for the remainder [48, p. 308], we obtain

$$(10.7) \quad \frac{1}{6}v(x_{i-1}) + \frac{2}{3}v(x_i) + \frac{1}{6}v(x_{i+1}) = \frac{u(x_{i+1}) - u(x_{i-1})}{2h} + \frac{1}{180}h^4 u^{(5)}(\xi_i),$$

where $\xi_i \in [x_{i-1}, x_{i+1}]$. This expression motivates our definition of the Hermitian derivative. Let $\mathfrak{u} \in l_h^2$ be a given grid function. We define the grid function $\mathfrak{v} \in l_h^2$ by

$$(10.8) \qquad \frac{1}{6}\mathfrak{v}_{i-1} + \frac{2}{3}\mathfrak{v}_i + \frac{1}{6}\mathfrak{v}_{i+1} = \delta_x \mathfrak{u}_i, \quad 1 \le i \le N-1,$$

where the boundary values $\mathfrak{v}_0$, $\mathfrak{v}_N$ are known.

It is obvious that the system (10.8) has a unique solution for the vector $[\mathfrak{v}_1, \dots, \mathfrak{v}_{N-1}]^T$. We now define the **Hermitian derivative** of $\mathfrak{u}$ as $\mathfrak{u}_x$ by

$$(10.9) \qquad (\mathfrak{u}_x)_i = \mathfrak{v}_i, \quad 1 \le i \le N-1.$$

If we introduce the three-point operator σ_x on grid functions by

$$(10.10) \qquad \sigma_x \mathfrak{u}_i = \frac{1}{6}\mathfrak{u}_{i-1} + \frac{2}{3}\mathfrak{u}_i + \frac{1}{6}\mathfrak{u}_{i+1}, \quad 1 \le i \le N-1,$$

we can rewrite (10.8) and (10.9) as

$$(10.11) \qquad \sigma_x(\mathfrak{u}_x)_i = \delta_x \mathfrak{u}_i, \quad 1 \le i \le N-1.$$

Recall that $(\mathfrak{u}_x)_0$, $(\mathfrak{u}_x)_N$ need to be known in order to solve (10.11). Observe that we have the following operator equality (in the space of grid functions)

$$\sigma_x = I + \frac{h^2}{6}\delta_x^2. \tag{10.12}$$

We refer to σ_x as the *Simpson operator.*

In order to estimate the error incurred in approximating a derivative of a smooth function by the Hermitian derivative, we first define P to be the positive definite $(N-1) \times (N-1)$ matrix

$$P = \begin{bmatrix} 4 & 1 & 0 \dots & 0 \\ 1 & 4 & 1 \dots & 0 \\ \vdots & \vdots & \vdots \dots & \vdots \\ 0 \dots & 1 & 4 & 1 \\ 0 \dots & 0 & 1 & 4 \end{bmatrix}. \tag{10.13}$$

The matrix P is related to T by [see (9.51)]

$$P = 6I - T. \tag{10.14}$$

The matrix corresponding to the operator σ_x (restricted to $l^2_{h,0}$) is $P/6 = I - T/6$. Thus, Equation (10.11) can be written as

$$\frac{1}{6}PU_x = \frac{1}{2h}KU, \tag{10.15}$$

where U and U_x are the vectors corresponding to $\mathfrak{u}$ and $\mathfrak{u}_x$, respectively. In order to give a complete description of the Hermitian derivative operator σ_x, we must compute the truncation error.

Recall that in the space of grid functions, with v^* being the grid function corresponding to u', Equation (10.7) gives

$$|\sigma_x v_i^* - \delta_x u_i^*| \le Ch^4 \|u^{(5)}\|_{L^\infty}, \quad i = 1, \dots, N-1, \tag{10.16}$$

or equivalently

$$|\sigma_x \left(v^* - \sigma_x^{-1}(\delta_x u^*)\right)|_\infty \le Ch^4 \|u^{(5)}\|_{L^\infty}. \tag{10.17}$$

Using the definition of $\mathfrak{u}_x$ in (10.11), we have for $\mathfrak{u} = u^*$,

$$\sigma_x \left(\mathfrak{u}_x - \sigma_x^{-1}(\delta_x \mathfrak{u})\right) = 0. \tag{10.18}$$

Subtracting (10.18) from (10.17) we find

$$|\sigma_x(v^* - \mathfrak{u}_x)|_\infty \le Ch^4 \|u^{(5)}\|_{L^\infty}. \tag{10.19}$$

In order to estimate $v^* - \mathfrak{u}_x$, it is more convenient to work with the matrix equivalent of Equation (10.19). We denote by V the vector corresponding to v^* (note that V and v^* are identical but considered in different spaces, namely $V \in \mathbb{R}^{N-1}$ and $v^* \in l^2_{h,0}$). Using the fact that

$$(\mathfrak{u}_x)_0 = (u')^*(x_0), \quad (\mathfrak{u}_x)_N = (u')^*(x_N), \tag{10.20}$$

we can therefore write (10.19) as

$$\left|\frac{1}{6}P(V - U_x)\right|_\infty \le Ch^4 \|u^{(5)}\|_{L^\infty}. \tag{10.21}$$

Thus,

$$|V - U_x|_\infty \le C|(P/6)^{-1}|_\infty h^4 \|u^{(5)}\|_{L^\infty}. \tag{10.22}$$

Since $P/6 = I - T/6$ then

$$(P/6)^{-1} = \sum_{k=0}^{\infty} \left(\frac{T}{6}\right)^k. \tag{10.23}$$

Therefore, using (8.13), we have

$$|(P/6)^{-1}|_\infty \le \sum_{k=0}^{\infty} \left(\frac{|T|_\infty}{6}\right)^k \le \sum_{k=0}^{\infty} \left(\frac{4}{6}\right)^k = 3. \tag{10.24}$$

From (10.22) and (10.24), we conclude that

$$|V - U_x|_\infty \le Ch^4 \|u^{(5)}\|_{L^\infty}, \tag{10.25}$$

or equivalently

$$|(u')^* - u_x|_\infty \le Ch^4 \|u^{(5)}\|_{L^\infty}. \tag{10.26}$$

We summarize the above in the following lemma.

Lemma 10.1. *Suppose that $u(x)$ is a smooth function on $[0,1]$ and let $\mathfrak{u} = u^*$. Then, the Hermitian derivative $\mathfrak{u}_x$, as obtained from the values $u(x_i)$, $0 \le i \le N$, by*

$$(\sigma_x \mathfrak{u}_x)_i = (\delta_x u^*)_i, \quad 1 \le i \le N-1 \tag{10.27}$$

and

$$(\mathfrak{u}_x)_0 = (u')^*(x_0), \quad (\mathfrak{u}_x)_N = (u')^*(x_N), \tag{10.28}$$

has a truncation error $\mathfrak{u}_x - (u')^*$ of order $O(h^4)$. More precisely

$$|\mathfrak{u}_x - (u')^*|_\infty \le Ch^4 \|u^{(5)}\|_{L^\infty}. \tag{10.29}$$

Remark 10.2. Note that local truncation errors cannot be used with the Hermitian derivative since its approximation is global in nature.

The following example illustrates how the definitions of the Hermitian derivative and the Simpson operator are implemented for a particular class of functions. This computation will be needed in Appendix B.

Example 10.3. Consider the function depending on the six parameters $A, B, C, D, a, b \in \mathbb{R}$,

$$v(x) = A\cos(ax) + B\sin(bx) + Cx + D, \quad x \in [0, 1]. \tag{10.30}$$

Let $\mathfrak{v} = v^*$ be the corresponding grid function $\mathfrak{v}(x_i) = v(x_i)$. The computation of the Hermitian derivative of v^* proceeds as follows. We obtain readily [see Equation (9.46)] that for $1 \le i \le N-1$

$$\begin{aligned} \delta_x \mathfrak{v}(x_i) &= \big[-A\sin(ax_i)\sin(ah) + B\cos(bx_i)\sin(bh)\big]h^{-1} + C, \\ \delta_x^2 \mathfrak{v}(x_i) &= -4\left[A\cos(ax_i)\sin^2\left(\frac{ah}{2}\right) + B\sin(bx_i)\sin^2\left(\frac{bh}{2}\right)\right]h^{-2}. \end{aligned} \tag{10.31}$$

For the Simpson operator [see Equation (10.12)] we therefore have for $1 \le i \le N-1$

$$\sigma_x \mathfrak{v}(x_i) = \mathfrak{v}(x_i) - \frac{2}{3}\left[A\cos(ax_i)\sin^2\left(\frac{ah}{2}\right) + B\sin(bx_i)\sin^2\left(\frac{bh}{2}\right)\right].$$

We now assume that the coefficients A, B, C, D, a and b are chosen so that the function $v(x)$ and its derivative $v'(x)$ vanish at both endpoints $x = 0, 1$. We impose

$$\mathfrak{v}(x_0) = \mathfrak{v}_x(x_0) = \mathfrak{v}(x_N) = \mathfrak{v}_x(x_N) = 0.$$

Assume that

$$\mathfrak{v}_x(x_i) = [A'\sin(ax_i) + B'\cos(bx_i)]h^{-1} + C', \quad 1 \le i \le N-1, \tag{10.32}$$

then the relation $\sigma_x \mathfrak{v}_x = \delta_x \mathfrak{v}$ [see Equation (10.11)] yields, at all interior points $1 \le i \le N-1$,

$$\begin{aligned} \mathfrak{v}_x(x_i) - \frac{2}{3}&\left[B'\cos(bx_i)\sin^2\left(\frac{bh}{2}\right) + A'\sin(ax_i)\sin^2\left(\frac{ah}{2}\right)\right]h^{-1} \\ &= \big[-A\sin(ax_i)\sin(ah) + B\cos(bx_i)\sin(bh)\big]h^{-1} + C. \end{aligned}$$

After simplification, we find that the Hermitian derivative of v^* is effectively given by (10.32) with the constants A', B' and C' given by

$$\begin{cases} A' = -\dfrac{\sin(ah)}{1 - \frac{2}{3}\sin^2(\frac{ah}{2})}A \\[2ex] B' = \dfrac{\sin(bh)}{1 - \frac{2}{3}\sin^2(\frac{bh}{2})}B \\[2ex] C' = C. \end{cases} \tag{10.33}$$

10.2 A finite element approach to the Hermitian derivative

We can construct, using finite element theory, piecewise-linear continuous functions corresponding to $\mathfrak{u}$ and $\mathfrak{u}_x$.

We define the space $P^1_{c,0}$ of such functions by

$$P^1_{c,0} = \Big\{ v_h(x) \;\; | \;\; v_h(x) \text{ is continuous, linear in each } [x_i, x_{i+1}], \quad 0 \le i \le N-1, \; v_h(x_0) = v_h(x_N) = 0 \Big\}. \tag{10.34}$$

For each grid function $\mathfrak{v} \in l^2_{h,0}$, we match the function $v_h(x) \in P^1_{c,0}$ by taking

$$v_h(x_i) = \mathfrak{v}_i, \quad i = 0, \dots, N. \tag{10.35}$$

Clearly, this is a linear isomorphism from $l^2_{h,0}$ onto $P^1_{c,0}$. In addition, starting with $\mathfrak{v} \in l^2_{h,0}$, we introduce the two piecewise constant functions $\bar{v}_h$ and $v_{h,x}$ defined in each interval $K_{i+1/2} = [x_i, x_{i+1}]$, $i = 0, \dots, N-1$, by

$$\bar{v}_h(x) = \frac{\mathfrak{v}_i + \mathfrak{v}_{i+1}}{2}, \quad v_{h,x}(x) = \frac{\mathfrak{v}_{i+1} - \mathfrak{v}_i}{h}, \quad x \in K_{i+1/2}, \tag{10.36}$$

so that for $i = 0, \dots, N-1$

$$v_h(x) = \bar{v}_h(x_{i+1/2}) + v_{h,x}(x_{i+1/2})(x - x_{i+1/2}), \quad x \in K_{i+1/2}, \tag{10.37}$$

where $x_{i+1/2} = \frac{1}{2}(x_i + x_{i+1})$.

It is clear from the definitions above that the functions $\bar{v}_h(x)$ and $v_{h,x}(x_{i+1/2})(x - x_{i+1/2})$ are orthogonal in $K_{i+1/2}$, so that we have, in terms of the L^2 scalar product,

$$(u_h, v_h) = (\bar{u}_h, \bar{v}_h) + \frac{h^2}{12}(u_{h,x}, v_{h,x}) \tag{10.38}$$

for all functions u_h, v_h corresponding to $\mathfrak{u}, \mathfrak{v} \in l^2_{h,0}$.

An important aspect of using $P^1_{c,0}$ in the study of finite difference schemes is that it allows us to streamline analytic operations such as integration by parts or averaged quantities over the intervals $K_{i+1/2}$.

The scalar products in $l^2_{h,0}$ and in L^2 are related as follows.

Lemma 10.4. *For any $\mathfrak{u}, \mathfrak{v} \in l^2_{h,0}$, let $u_h(x), v_h(x) \in P^1_{c,0}$ be the corresponding functions as defined above. Then,*
(i)

$$(\mathfrak{u}, \mathfrak{v})_h = (u_h, v_h) + \frac{h^2}{6}(u_{h,x}, v_{h,x}), \tag{10.39}$$

$$(\delta_x \mathfrak{u}, \mathfrak{v})_h = (u_{h,x}, v_h), \tag{10.40}$$

$$(\delta_x^2 \mathfrak{u}, \mathfrak{v})_h = -(u_{h,x}, v_{h,x}). \tag{10.41}$$

(ii) Let $\mathfrak{u}_x$ be the Hermitian derivative of $\mathfrak{u}$. Assume that $\mathfrak{u}_x \in l^2_{h,0}$ and let $p_h \in P^1_{c,0}$ be its corresponding function. Then $p_h(x)$ is the orthogonal projection of the piecewise constant function $u_{h,x}$ into $P^1_{c,0}$. In other words, it is the unique function $p_h \in P^1_{c,0}$ solving

$$(p_h, q_h) = (u_{h,x}, q_h), \quad \forall q_h \in P^1_{c,0}. \tag{10.42}$$

Proof. (i) Using the linearity of the correspondence between $l^2_{h,0}$ and $P^1_{c,0}$, it suffices to check that (10.39), (10.40) and (10.41) hold for the basis functions $\mathfrak{u}(x_i) = \delta_{ik}$ and $\mathfrak{v}(x_i) = \delta_{il}$, $i = 1, \ldots, N-1$, where $k, l \in \{1, \ldots, N-1\}$ are fixed. In this case the equations follow readily by an easy computation.

(ii) Taking the scalar product of (10.11) with any function $\mathfrak{q} \in l^2_{h,0}$ yields

$$(\mathfrak{u}_x, \mathfrak{q})_h + \frac{1}{6} h^2 (\delta_x^2 \mathfrak{u}_x, \mathfrak{q})_h = (\delta_x \mathfrak{u}, \mathfrak{q})_h, \qquad \forall \mathfrak{q} \in l^2_{h,0}. \tag{10.43}$$

We can rewrite (10.43) as

$$\begin{aligned}
(u_{h,x}, q_h) &\overset{(10.40)}{=} (\delta_x \mathfrak{u}, \mathfrak{q})_h \overset{(10.43)}{=} (\mathfrak{u}_x, \mathfrak{q})_h + \frac{h^2}{6} (\delta_x^2 \mathfrak{u}_x, \mathfrak{q})_h \\
&\overset{(10.39),(10.41)}{=} (p_h, q_h) + \frac{h^2}{6} (p_{h,x}, q_{h,x}) - \frac{h^2}{6} (p_{h,x}, q_{h,x}) \\
&= (p_h, q_h),
\end{aligned}$$

which gives (10.42). ■

10.3 The three-point biharmonic operator

In Section 10.1 we developed the concept of the Hermitian derivative in order to approximate $u'(x)$ to fourth-order accuracy at all grid points. We shall now use the Hermitian derivative in order to approximate the fourth-order derivative $u^{(4)}(x)$ with fourth-order accuracy.

Consider the one-dimensional biharmonic equation

$$\begin{cases} (a) & u^{(4)}(x) = f(x), \quad a < x < b, \\ (b) & u(a) = g_a, \; u(b) = g_b, \\ (c) & u'(a) = g'_a, \; u'(b) = g'_b. \end{cases} \tag{10.44}$$

By analogy with the treatment of the second-order equation in Chapter 9, we would like to derive an approximation for $u^{(4)}(x)$. Assuming that $u(x)$ is a smooth function in $[a,b]$ and using the notation introduced in Chapter 8, we let $u^*(x_i) = u(x_i)$ be the corresponding grid function, where $x_i = a + ih, i = 0,\dots,N, h = (b-a)/L$. An obvious approximation for $u^{(4)}(x_i)$ at a point x_i, $2 \le i \le N-2$, is

(10.45)
$$\delta_x^2\delta_x^2 u_i^* = \frac{u^*(x_{i+2}) - 4u^*(x_{i+1}) + 6u^*(x_i) - 4u^*(x_{i-1}) + u^*(x_{i-2})}{h^4}.$$

A Taylor expansion shows that for $i = 2,\dots,N-2$

$$(\delta_x^2)^2 u^*(x_i) = (u^{(4)})^*(x_i) + \frac{1}{6}h^2(u^{(6)})^*(x_i) + O(h^4). \tag{10.46}$$

Therefore, the approximation $\delta_x^2\delta_x^2 u_i^*$ of $(u^{(4)})^*(x_i)$ is second-order accurate. In addition, the evaluation of $\delta_x^2\delta_x^2$ can be carried out only at the interior points $x_2,\dots,x_{N-2}$. Thus, it does not provide any approximation at near-boundary points, x_1, x_{N-1}. Hence, the near-boundary points, x_1, x_{N-1}, necessitate a special treatment.

Our purpose is to avoid the difficulties mentioned above, that is to obtain a higher-order approximation for $u^{(4)}$, which is valid also at near-boundary points x_1, x_{N-1}.

Assume that the values $u^*(x_{i-1})$, $u^*(x_i)$ and $u^*(x_{i+1})$ are given for $i = 1,\dots,N-1$. In addition, we are given the values $u_x^*(x_{i-1}), u_x^*(x_i)$ and $u_x^*(x_{i+1})$, which approximate $(u')^*(x_{i-1}), (u')^*(x_i)$ and $(u')^*(x_{i+1})$, respectively. Let $q(x)$ be a fourth-order polynomial given by

$$q(x) = a_0 + a_1(x-x_i) + a_2(x-x_i)^2 + a_3(x-x_i)^3 + a_4(x-x_i)^4. \tag{10.47}$$

We require that $q(x)$ interpolate the five data

$$u^*(x_{i-1}),\ u^*(x_i),\ u^*(x_{i+1}),\ u_x^*(x_{i-1}),\ u_x^*(x_{i+1}) \tag{10.48}$$

as follows.

$$\begin{aligned} &q(x_{i-1}) = u^*(x_{i-1}),\ q(x_i) = u^*(x_i),\ q(x_{i+1}) = u^*(x_{i+1}),\\ &q'(x_{i-1}) = u_x^*(x_{i-1}),\ q'(x_{i+1}) = u_x^*(x_{i+1}). \end{aligned} \tag{10.49}$$

The coefficients a_0, a_1, a_2, a_3 and a_4 of the polynomial $q(x)$ may easily be computed, and we get for $1 \le i \le N-1$

$$\begin{cases} (a) & a_0 = u^*(x_i),\\ (b) & a_1 = \dfrac{3}{2}\delta_x u^*(x_i) - \dfrac{1}{4}\Big(u_x^*(x_{i+1}) + u_x^*(x_{i-1})\Big),\\ (c) & a_2 = \delta_x^2 u^*(x_i) - \dfrac{1}{2}\delta_x u_x^*(x_i),\\ (d) & a_3 = \dfrac{1}{4h^2}\Big[(u_x^*(x_{i+1}) + u_x^*(x_{i-1})) - 2\delta_x u^*(x_i)\Big],\\ (e) & a_4 = \dfrac{1}{2h^2}\big(\delta_x u_x^*(x_i) - \delta_x^2 u^*(x_i)\big). \end{cases} \tag{10.50}$$

Therefore, a natural candidate for the approximation of $(u^{(4)})^*(x_i)$ is $q^{(4)}(x_i) = 24a_4$, that is

$$\frac{12}{h^2}\left(\delta_x u_x^*(x_i) - \delta_x^2 u^*(x_i)\right). \tag{10.51}$$

Note, that this approximation includes the near-boundary points x_1, x_{N-1}.

The expression in (10.50)(e) for the fourth-order derivative $(u^{(4)})^*(x_i)$ requires the knowledge of u_x^* at x_{i-1} and x_{i+1}.

In general we are given *a priori* only the values of u^* at all grid points but we are not given the exact values of $(u')^*$. Thus, we need to provide approximate values for the first-order derivative. These values were denoted above by u_x^*. A natural candidate for $u_x^*(x_i)$ is $q'(x_i)$. From (10.47) we have that $q'(x_i) = a_1$. Thus, by (10.50)(b)

$$u_x^*(x_i) = \frac{3}{2}\delta_x u^*(x_i) - \frac{1}{4}\Big(u_x^*(x_{i+1}) + u_x^*(x_{i-1})\Big). \tag{10.52}$$

Equation (10.52) may be recast as

$$\frac{1}{6}u_x^*(x_{i-1}) + \frac{2}{3}u_x^*(x_i) + \frac{1}{6}u_x^*(x_{i+1}) = \delta_x u^*(x_i). \tag{10.53}$$

Observe that this relation is identical to the relation (10.11) for the Hermitian derivative of u^*. *Thus, we conclude that u_x^* should be taken as the Hermitian derivative $\mathfrak{u}_x$ of $\mathfrak{u} = u^*$.*

The two boundary conditions $(\mathfrak{u}_x)_0 = u'(a)$ and $(\mathfrak{u}_x)_N = u'(b)$ [see (10.44)(c)] are needed in order to obtain $\mathfrak{u}_x$ in terms of $\mathfrak{u}$ at all interior points $x_1, \ldots, x_{N-1}$.

We summarize the preceding derivation in the following definition.

Definition 10.5 (Three-point biharmonic operator). *Let $\mathfrak{u} \in l_h^2$ be a given grid function. The* ***three-point biharmonic operator*** *is defined by*

$$\delta_x^4 \mathfrak{u}_i = \frac{12}{h^2}(\delta_x \mathfrak{u}_{x,i} - \delta_x^2 \mathfrak{u}_i), \quad 1 \le i \le N-1. \tag{10.54}$$

Here $\mathfrak{u}_x$ is the Hermitian derivative of $\mathfrak{u}$ satisfying (10.11) with given boundary values $\mathfrak{u}_{x,0}$ and $\mathfrak{u}_{x,N}$.

Expressing the operator $\delta_x^4 \mathfrak{u}$ in more detail, we obtain, for $1 \le i \le N-1$,

$$\delta_x^4 \mathfrak{u}_i = \frac{12}{h^2}\left(\frac{\mathfrak{u}_{x,i+1} - \mathfrak{u}_{x,i-1}}{2h} - \frac{\mathfrak{u}_{i+1} - 2\mathfrak{u}_i + \mathfrak{u}_{i-1}}{h^2}\right). \tag{10.55}$$

In the next section [see Equation (10.64)] we shall derive a detailed estimate for the truncation error $\delta_x^4 u^*(x_i) - (u^{(4)})^*(x_i)$, showing fourth-order accuracy (in a suitable sense) at all interior points.

Using(10.55) we obtain an approximate solution to the biharmonic problem (10.44) as follows.

Definition 10.6 (One-dimensional Stephenson scheme). *The approximation to the biharmonic problem (10.44) given by*

$$\text{(10.56)} \qquad \begin{cases} (a) & \delta_x^4 \mathfrak{u}_i = f^*(x_i), \qquad 1 \le i \le N-1, \\ (b) & \mathfrak{u}_0 = g_a,\ \mathfrak{u}_N = g_b,\ \mathfrak{u}_{x,0} = g_a',\ \mathfrak{u}_{x,N} = g_b', \end{cases}$$

is called the ***one-dimensional Stephenson scheme*** *to the problem (see Figure 10.1).*

This scheme can be more explicitly written as

$$\text{(10.57)} \qquad \begin{aligned} &(a)\ \frac{12}{h^2}\left(\frac{\mathfrak{u}_{x,i+1}-\mathfrak{u}_{x,i-1}}{2h} - \frac{\mathfrak{u}_{i+1}+\mathfrak{u}_{i-1}-2\mathfrak{u}_i}{h^2}\right) = f^*(x_i), \\ &\hspace{20em} 1 \le i \le N-1, \\ &(b)\ \frac{1}{6}\mathfrak{u}_{x,i-1} + \frac{2}{3}\mathfrak{u}_{x,i} + \frac{1}{6}\mathfrak{u}_{x,i+1} = \delta_x \mathfrak{u}_i, \quad 1 \le i \le N-1, \\ &(c)\ \mathfrak{u}_0 = g_a,\ \mathfrak{u}_N = g_b,\ \mathfrak{u}_{x,0} = g_a',\ \mathfrak{u}_{x,N} = g_b'. \end{aligned}$$

The terminology *Stephenson scheme* is justified by the fact that the scheme (10.57) is the one-dimensional restriction of the scheme proposed by Stephenson [168] for the two-dimensional biharmonic problem. In Chapter 11 we will study the scheme in the two-dimensional case.

Example 10.7. Continuing Example 10.3, let $\mathfrak{v}$ be the grid function corresponding to the function $v(x)$ in (10.30). Using Equation (10.54), we compute $\delta_x^4\mathfrak{v}$. From Equations (10.31) and (10.32) and $\mathfrak{v}(x_0) = \mathfrak{v}_x(x_0) = \mathfrak{v}(x_N) = \mathfrak{v}_x(x_N) = 0$, we get

$$\text{(10.58)} \qquad \begin{aligned} \delta_x^4\mathfrak{v}(x_i) &= \frac{12}{h^4}\Bigg[(-B'\sin(bx_i)\sin(bh) + A'\cos(ax_i)\sin(ah)) \\ &\qquad + 4\left(A\cos(ax_i)\sin^2\left(\frac{ah}{2}\right) + B\sin(bx_i)\sin^2\left(\frac{bh}{2}\right)\right)\Bigg] \\ &= \frac{12}{h^4}\left(A''\cos(ax_i) + B''\sin(bx_i)\right), \quad 1 \le i \le N-1, \end{aligned}$$

where

$$\text{(10.59)} \qquad \begin{cases} A'' = \dfrac{4A\sin^4(\frac{ah}{2})}{3-2\sin^2(\frac{ah}{2})}, \\[2ex] B'' = \dfrac{4B\sin^4(\frac{bh}{2})}{3-2\sin^2(\frac{bh}{2})} \end{cases}$$

[we have used the expressions for A' and B' in (10.33)].

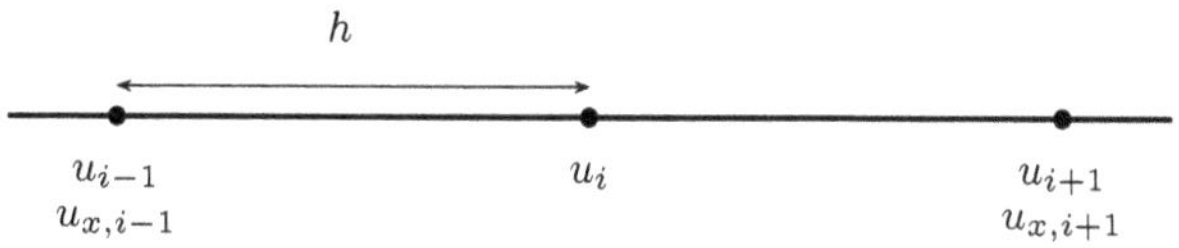

Fig. 10.1 Stephenson's scheme for $u^{(4)} = f$. The finite difference operator $\delta_x^4 u_i$ at point i is $q^{(4)}(x_i)$ where $q(x) \in P^4[x_{i-1}, x_{i+1}]$ is defined by the five collocated values for u_{i-1}, u_i, u_{i+1}, $u_{x,i-1}$ and $u_{x,i+1}$.

10.4 Accuracy of the three-point biharmonic operator

In Section 9.2 we studied the accuracy of the three-point Laplacian operator δ_x^2 [see (9.12)] which for any smooth function u is the difference $\delta_x^2 u^*(x_i) - u''(x_i)$. In this section we carry out an analogous treatment for the one-dimensional three-point biharmonic operator δ_x^4 defined in (10.54).

Let $u(x)$ be a smooth function on $[a, b]$ such that $u(a) = u(b) = 0$ and $u'(a) = u'(b) = 0$. According to our general notational convention (see Chapter 8), we denote by u^* the related grid function.

We will first study the action of operators on a grid function $\mathrm{u} \in l_{h,0}^2$. We begin by considering the action of $\sigma_x \delta_x^4$ at the interior point x_i.

$$\sigma_x \delta_x^4 \mathrm{u}_i = \frac{1}{6}\delta_x^4 \mathrm{u}_{i-1} + \frac{2}{3}\delta_x^4 \mathrm{u}_i + \frac{1}{6}\delta_x^4 \mathrm{u}_{i+1}, \quad 2 \le i \le N-2, \tag{10.60}$$

where σ_x is the Simpson operator defined in (10.10). Using the definition of δ_x^4, Equation (10.60) can be rewritten as

$$\frac{1}{6}\delta_x^4 \mathrm{u}_{i-1} + \frac{2}{3}\delta_x^4 \mathrm{u}_i + \frac{1}{6}\delta_x^4 \mathrm{u}_{i+1} = \frac{12}{h^2}\left[\left(\frac{1}{6}\delta_x \mathrm{u}_{x,i-1} + \frac{2}{3}\delta_x \mathrm{u}_{x,i} + \frac{1}{6}\delta_x \mathrm{u}_{x,i+1}\right) - \left(\frac{1}{6}\delta_x^2 \mathrm{u}_{i-1} + \frac{2}{3}\delta_x^2 \mathrm{u}_i + \frac{1}{6}\delta_x^2 \mathrm{u}_{i+1}\right)\right].$$

Using the definition of u_x, the first term on the right-hand side is

$$\begin{aligned}&\frac{1}{6}\delta_x \mathrm{u}_{x,i-1} + \frac{2}{3}\delta_x \mathrm{u}_{x,i} + \frac{1}{6}\delta_x \mathrm{u}_{x,i+1}\\ &= \sigma_x \delta_x \mathrm{u}_{x,i} = \delta_x \sigma_x \mathrm{u}_{x,i} = \delta_x \delta_x \mathrm{u}_i\\ &= \frac{1}{4h^2}(\mathrm{u}_{i+2} - 2\mathrm{u}_i + \mathrm{u}_{i-2}), \quad 2 \le i \le N-2.\end{aligned} \tag{10.61}$$

The second term may be written as

$$\begin{aligned}&\frac{1}{6}\delta_x^2 \mathrm{u}_{i-1} + \frac{2}{3}\delta_x^2 \mathrm{u}_i + \frac{1}{6}\delta_x^2 \mathrm{u}_{i+1}\\ &= \frac{1}{6h^2}(\mathrm{u}_{i-2} + 2\mathrm{u}_{i-1} - 6\mathrm{u}_i + 2\mathrm{u}_{i+1} + \mathrm{u}_{i+2}), \quad 2 \le i \le N-2.\end{aligned} \tag{10.62}$$

Therefore, inserting (10.61) and (10.62) in (10.60), we have for $2 \le i \le N-2$,

$$\sigma_x \delta_x^4 \mathfrak{u}_i = \frac{1}{h^4}(\mathfrak{u}_{i-2} - 4\mathfrak{u}_{i-1} + 6\mathfrak{u}_i - 4\mathfrak{u}_{i+1} + \mathfrak{u}_{i+2}) = \delta_x^2 \delta_x^2 \mathfrak{u}_i. \tag{10.63}$$

Thus, in the absence of boundaries, $\sigma_x \delta_x^4 \mathfrak{u} = (\delta_x^2)^2$. Explicit estimates for $\sigma_x \delta_x^4 u_i$ at near-boundary points x_1, x_{N-1} are given below [see (10.65)]. From (10.63) it results that $\sigma_x \delta_x^4$ actually coincides with the operator $(\delta_x^2)^2$ at points x_i, $2 \le i \le N-2$. Only at near-boundary points, $i = 1, N-1$, do we have a "numerical boundary layer" effect. We will now investigate the accuracy of the three-point biharmonic operator.

Proposition 10.8. *Suppose that $u(x)$ is a smooth function on $[a,b]$. Assume, in addition, that $u(a) = u(b) = 0$ and $u'(a) = u'(b) = 0$. Let $u_i^* = u(x_i)$ and $(u^{(4)})^*(x_i) = u^{(4)}(x_i)$ be the grid functions corresponding, respectively, to u and $u^{(4)}$. Then the three-point biharmonic operator δ_x^4 satisfies the following accuracy properties:*
- *At interior points $2 \le i \le N-2$ we have*

$$|\sigma_x \delta_x^4 u_i^* - \sigma_x (u^{(4)})^*(x_i)| \le C h^4 \|u^{(8)}\|_{L^\infty}. \tag{10.64}$$

- *At near-boundary points $i = 1$ and $i = N-1$, the fourth-order accuracy of (10.64) drops to first order,*

$$|\sigma_x \delta_x^4 u_1^* - \sigma_x (u^{(4)})^*(x_1)| \le C h \|u^{(5)}\|_{L^\infty}, \tag{10.65}$$

with a similar estimate for $i = N-1$. Inequality (10.65) holds also for interior points $i = 2, \ldots, N-2$ as a suboptimal estimate.
- *The error in the energy norm is given by*

$$|\delta_x^4 u^* - (u^{(4)})^*|_h \le C h^{3/2} (\|u^{(5)}\|_{L^\infty} + \|u^{(8)}\|_{L^\infty}). \tag{10.66}$$

In the above estimates C is a constant that is independent of u and h.

Proof. According to (10.63), we have

$$\sigma_x \delta_x^4 u_i^* = (\delta_x^2)^2 u_i^*, \qquad i = 2, ..., N-2. \tag{10.67}$$

We now expand $(\delta_x^2)^2$ as a Taylor series. As in the computation leading to (9.12), we have

$$\delta_x^2 u_i = u''(x_i) + \frac{h^2}{12} u^{(4)}(x_i) + \frac{h^4}{360} u^{(6)}(\xi_i), \quad 1 \le i \le N-1, \tag{10.68}$$

where $\xi_i \in (x_i - h, x_i + h)$.

We note that for any smooth function $v(x)$ defined in $[a,b]$, if $[x-h, x+h] \subseteq [a,b]$ then

$$\frac{v(x+h) - 2v(x) + v(x-h)}{h^2} = v''(\eta), \tag{10.69}$$

where $\eta \in (x-h, x+h)$.

Applying δ_x^2 to Equation (10.68) at the interior points x_i, $2 \le i \le N-2$, we obtain

$$\sigma_x \delta_x^4 u_i^* = (\delta_x^2)^2 u_i^* = u^{(4)}(x_i) + \frac{h^2}{6} u^{(6)}(x_i) + p_i, \tag{10.70}$$

where

$$|p_i| \le C_1 h^4 \|u^{(8)}\|_{L^\infty}, \quad 2 \le i \le N-2. \tag{10.71}$$

On the other hand, $\sigma_x (u^{(4)})^*(x_i)$ may be expanded around x_i, $2 \le i \le N-2$, as follows.

$$\begin{aligned} \sigma_x (u^{(4)})^*(x_i) &= \left(I + \frac{h^2}{6}\delta_x^2\right)(u^{(4)})^*(x_i) \\ &= u^{(4)}(x_i) + \frac{h^2}{6} u^{(6)}(x_i) + q_i, \end{aligned}$$

where

$$|q_i| \le C_2 h^4 \|u^{(8)}\|_{L^\infty}, \quad 2 \le i \le N-2. \tag{10.72}$$

Therefore, subtracting this equation from (10.70), we obtain the estimate (10.64). Consider now the near-boundary point x_1. We set $\delta_x^4 u_0^* = (u^{(4)})^*(x_0)$ and then, using the definition of $\sigma_x \delta_x^4$, we have

$$\begin{aligned} &\sigma_x \delta_x^4 u_1^* - \sigma_x (u^{(4)})^*(x_1) \\ &= \left(\tfrac{2}{3}\delta_x^4 u_1^* + \tfrac{1}{6}\delta_x^4 u_2^*\right) - \left(\tfrac{2}{3}(u^{(4)})^*(x_1) + \tfrac{1}{6}(u^{(4)})^*(x_2)\right) \\ &= \tfrac{2}{3}\left(\delta_x^4 u_1^* - (u^{(4)})^*(x_1)\right) + \tfrac{1}{6}\left(\delta_x^4 u_2^* - (u^{(4)})^*(x_2)\right). \end{aligned} \tag{10.73}$$

First, we consider the terms evaluated at x_1. Recall that

$$\delta_x^4 u_1^* = \frac{12}{h^2}\left((\delta_x \mathfrak{u}_x)_1 - \delta_x^2 u_1^*\right), \tag{10.74}$$

where $\mathfrak{u}_x$ is the Hermitian derivative of u^*. Using the boundary values $u_0^* = \mathfrak{u}_{x,0} = 0$, we have, from (10.29),

$$(\delta_x \mathfrak{u}_x)_1 = \frac{\mathfrak{u}_{x,2}}{2h} = u''(x_1) + \frac{h^2}{6} u^{(4)}(x_1) + r_1, \quad |r_1| \le C h^3 \|u^{(5)}\|_{L^\infty}, \tag{10.75}$$

and

$$\delta_x^2 u_1^* = u''(x_1) + \frac{h^2}{12} u^{(4)}(x_1) + r_2, \quad |r_2| \le Ch^3 \|u^{(5)}\|_{L^\infty}. \tag{10.76}$$

Inserting the estimates (10.75) and (10.76) into Equation (10.74), we obtain

$$\delta_x^4 u_1^* = u^{(4)}(x_1) + r_3, \quad |r_3| \le Ch \|u^{(5)}\|_{L^\infty}. \tag{10.77}$$

Next, for x_2 we have

$$\delta_x^4 u_2^* = \frac{12}{h^2}\left((\delta_x \mathfrak{u}_x)_2 - \delta_x^2 u_2^*\right). \tag{10.78}$$

Expanding the term $(\delta_x \mathfrak{u}_x)_2$ and again using (10.29), we have

$$(\delta_x \mathfrak{u}_x)_2 = \frac{\mathfrak{u}_{x,3} - \mathfrak{u}_{x,1}}{2h} = u''(x_2) + \frac{h^2}{6} u^{(4)}(x_2) + s_1, \tag{10.79}$$

where

$$|s_1| \le Ch^3 \|u^{(5)}\|_{L^\infty}. \tag{10.80}$$

For the second term $\delta_x^2 u_2^*$ we have, as in (10.76),

$$\delta_x^2 u_2^* = u''(x_2) + \frac{h^2}{12} u^{(4)}(x_2) + s_2, \quad |s_2| \le Ch^3 \|u^{(5)}\|_{L^\infty}. \tag{10.81}$$

Inserting the estimates (10.79) and (10.81) into Equation (10.78), we obtain, as for (10.77),

$$\delta_x^4 u_2^* = u^{(4)}(x_2) + s_3, \quad |s_3| \le Ch \|u^{(5)}\|_{L^\infty}. \tag{10.82}$$

Combining the estimates for r_3 and s_3 and inserting them into (10.73), we obtain

$$|\sigma_x \delta_x^4 u_1^* - \sigma_x (u^{(4)})^*(x_1))| \le Ch \|u^{(5)}\|_{L^\infty}, \tag{10.83}$$

which proves (10.65).

Let $\mathfrak{t}_i = \delta_x^4 u^*(x_i) - (u^{(4)})^*(x_i)$ be the truncation error for the approximation of the fourth-order derivative. We have

$$\sigma_x \mathfrak{t} = \mathfrak{v}, \tag{10.84}$$

where $\mathfrak{v} \in l_{h,0}^2$ satisfies the estimates established in the previous parts of the lemma

$$\begin{cases} |\mathfrak{v}_1|, |\mathfrak{v}_{N-1}| \le Ch \|u^{(5)}\|_{L^\infty}, \\ |\mathfrak{v}_i| \le Ch^4 \|u^{(8)}\|_{L^\infty}, \quad 2 \le i \le N-2. \end{cases} \tag{10.85}$$

The representative matrix of σ_x restricted to $l^2_{h,0}$ is $P/6 = I - T/6$. The eigenvalues of $P/6$ are [see (9.54)]

$$1 - \frac{2}{3}\lambda_k = 1 - \frac{2}{3}\sin^2\left(\frac{k\pi}{2N}\right), \quad 1 \le k \le N-1. \tag{10.86}$$

The L^2 matrix norm of its inverse is [see (8.11)]

$$|(P/6)^{-1}|_2 = \max_{k=1,\dots,N-1}\left|\frac{1}{1 - \frac{2}{3}\sin^2\left(\frac{k\pi}{2N}\right)}\right| \le 3. \tag{10.87}$$

From (10.84), (10.85), and (10.87) we obtain

$$|\mathfrak{t}|_h \le C|\mathfrak{v}|_h. \tag{10.88}$$

Finally, since

$$\begin{aligned}|\mathfrak{v}|_h^2 &\le Ch\left(2h^2 + \sum_{i=2}^{N-2} h^8\right)\left(\|u^{(5)}\|^2_{L^\infty} + \|u^{(8)}\|^2_{L^\infty}\right)\\ &\le Ch^3\left(\|u^{(5)}\|^2_{L^\infty} + \|u^{(8)}\|^2_{L^\infty}\right),\end{aligned}$$

we get (10.66). ■

10.5 Coercivity and stability properties of the three-point biharmonic operator

In Subsection 9.2.3 we established the coercivity and stability properties of the discrete Laplacian. In particular, we found the stability estimate (9.37)

$$|\mathfrak{u}|_h^2 \le L^4|\delta_x^2\mathfrak{u}|_h^2, \quad \mathfrak{u} \in l^2_{h,0}.$$

As noted there, this stability is a fundamental tool in the proof of convergence of the discrete solution to the exact one (in a suitable Hilbert-space setting).

Our goal is to derive analogous estimates for the discrete biharmonic operator (10.54). More precisely, we wish to prove a discrete counterpart to the elliptic identity

$$(u^{(4)}, v) = (u, v^{(4)}) = (u'', v''), \quad \forall v, u \in H^4(a,b) \cap H_0^2(a,b), \tag{10.89}$$

where (u, v) is the $L^2[a,b]$ scalar product. This gives the coercivity of the biharmonic operator $(d/dx)^4$, and it can be supplemented by the Poincaré inequalities

$$\|u\|_{L^2[a,b]} \le C_1\|u'\|_{L^2[a,b]} \le C_2\|u''\|_{L^2[a,b]}. \tag{10.90}$$

Therefore for all $u \in H^4(a,b) \cap H_0^2(a,b)$,

$$(u^{(4)}, u) = \|u''\|^2_{L^2[a,b]} \geq C_2^{-2}\|u\|^2_{L^2[a,b]}. \tag{10.91}$$

Here, we consider the problem (10.44) with homogeneous boundary Dirichlet conditions imposed on u and u'.

The following two lemmas discuss the two basic properties, self-adjointness and coercivity, of the discrete biharmonic operator.

In what follows we let $\mathfrak{u}, \mathfrak{v} \in l^2_{h,0}$ and assume also that the Hermitian derivatives $\mathfrak{u}_x, \mathfrak{v}_x \in l^2_{h,0}$. Let $u_h(x)$, $v_h(x)$, $p_h(x)$, $q_h(x) \in P^1_{c,0}$ be the functions corresponding, respectively, to $\mathfrak{u}, \mathfrak{v}, \mathfrak{u}_x, \mathfrak{v}_x$, as defined in Section 10.2. As with Lemma 10.4, we rely on the correlation between the discrete and continuous scalar products (connecting grid functions and their associated $P^1_{c,0}$ elements).

Lemma 10.9. *The three-point discrete biharmonic operator δ_x^4 is self-adjoint for $\mathfrak{u}, \mathfrak{v}, \mathfrak{u}_x, \mathfrak{v}_x \in l^2_{h,0}$, that is*

$$(\delta_x^4 \mathfrak{u}, \mathfrak{v})_h = (\mathfrak{u}, \delta_x^4 \mathfrak{v})_h. \tag{10.92}$$

Proof. The three-point biharmonic operator is

$$\delta_x^4 \mathfrak{u}_i = \frac{12}{h^2}\Big((\delta_x \mathfrak{u}_x)_i - \delta_x^2 \mathfrak{u}_i \Big). \tag{10.93}$$

Using (10.40) and (10.41) yields

$$\begin{aligned}(\delta_x^4 \mathfrak{u}, \mathfrak{v})_h &= \frac{12}{h^2}\Big((\delta_x \mathfrak{u}_x, \mathfrak{v})_h - (\delta_x^2 \mathfrak{u}, \mathfrak{v})_h \Big) \\ &= \frac{12}{h^2}\Big((p_{h,x}, v_h) + (u_{h,x}, v_{h,x}) \Big).\end{aligned}$$

Since $(p_{h,x}, v_h) = -(p_h, v_{h,x})$, we obtain

$$(\delta_x^4 \mathfrak{u}, \mathfrak{v})_h = \frac{12}{h^2}\Big(v_{h,x}, u_{h,x} - p_h \Big). \tag{10.94}$$

Subtracting $0 = (q_h, u_{h,x} - p_h)$ [see (10.42)] from (10.94) yields

$$\langle \mathfrak{u}, \mathfrak{v} \rangle_h = (\delta_x^4 \mathfrak{u}, \mathfrak{v})_h = \frac{12}{h^2}\Big(u_{h,x} - p_h, v_{h,x} - q_h \Big). \tag{10.95}$$

The symmetry of this expression proves Equation (10.92). ∎

Assuming that $\mathfrak{u}, \mathfrak{v} \in l^2_{h,0}$ and $\mathfrak{u}_x, \mathfrak{v}_x \in l^2_{h,0}$, we define the inner product $\langle \mathfrak{u}, \mathfrak{v} \rangle_h$ by

$$\langle \mathfrak{u}, \mathfrak{v} \rangle_h = (\delta_x^4 \mathfrak{u}, \mathfrak{v})_h = (\mathfrak{u}, \delta_x^4 \mathfrak{v})_h. \tag{10.96}$$

Lemma 10.10. *We have the identity*

(10.97)

$$\begin{aligned}\langle \mathfrak{u}, \mathfrak{v} \rangle_h &= h \sum_{i=0}^{N-1} \frac{\mathfrak{u}_{x,i+1} - \mathfrak{u}_{x,i}}{h} \frac{\mathfrak{v}_{x,i+1} - \mathfrak{v}_{x,i}}{h} \\ &+ \frac{12}{h^2} h \sum_{i=0}^{N-1} \left(\frac{\mathfrak{u}_{i+1} - \mathfrak{u}_i}{h} - \frac{1}{2}(\mathfrak{u}_{x,i} + \mathfrak{u}_{x,i+1}) \right) \left(\frac{\mathfrak{v}_{i+1} - \mathfrak{v}_i}{h} - \frac{1}{2}(\mathfrak{v}_{x,i} + \mathfrak{v}_{x,i+1}) \right) \\ &= (p_{h,x}, q_{h,x}) + \tfrac{12}{h^2} \left(u_{h,x} - \bar{p}_h, v_{h,x} - \bar{q}_h \right).\end{aligned}$$

Proof. Using (10.37), we can replace p_h and q_h in (10.95) with

$$\begin{aligned} p_h(x) &= \bar{p}_h + p_{h,x}(x - x_{i+1/2}), \\ q_h(x) &= \bar{q}_h + q_{h,x}(x - x_{i+1/2}), \end{aligned} \tag{10.98}$$

respectively. Thus,

$$\langle \mathfrak{u}, \mathfrak{v} \rangle_h = \frac{12}{h^2} \big(u_{h,x} - \bar{p}_h - p_{h,x}(x - x_{i+1/2}), v_{h,x} - \bar{q}_h - q_{h,x}(x - x_{i+1/2}) \big).$$

This yields

$$\begin{aligned}\langle \mathfrak{u}, \mathfrak{v} \rangle_h &= \frac{12}{h^2} \big(u_{h,x} - \bar{p}_h, v_{h,x} - \bar{q}_h \big) \\ &+ \frac{12}{h^2} \big(p_{h,x}(x - x_{i+1/2}), q_{h,x}(x - x_{i+1/2}) \big),\end{aligned}$$

which results in

$$\langle \mathfrak{u}, \mathfrak{v} \rangle_h = \frac{12}{h^2} \big(u_{h,x} - \bar{p}_h, v_{h,x} - \bar{q}_h \big) + (p_{h,x}, q_{h,x}). \tag{10.99}$$

This completes the proof. ∎

In the following proposition, we highlight three very important inequalities that follow directly from this lemma.

Proposition 10.11. *There exists a constant C, independent of h, such that if $\mathfrak{u}, \mathfrak{u}_x \in l^2_{h,0}$, then*

$$\langle \mathfrak{u}, \mathfrak{u} \rangle_h \geq h \sum_{i=0}^{N-1} \left(\frac{\mathfrak{u}_{x,i+1} - \mathfrak{u}_{x,i}}{h} \right)^2, \tag{10.100}$$

$$\langle \mathfrak{u}, \mathfrak{u} \rangle_h \geq Ch \sum_{i=0}^{N-1} \left(\frac{\mathfrak{u}_{i+1} - \mathfrak{u}_i}{h} \right)^2, \tag{10.101}$$

$$\langle \mathfrak{u}, \mathfrak{u} \rangle_h \geq Ch \sum_{i=1}^{N-1} (\mathfrak{u}_i)^2. \tag{10.102}$$

Proof. Inequality (10.100) follows directly from (10.97), since $p_{h,x}$ is a piecewise constant function, which coincides in the interval $K_{i+\frac{1}{2}} = (x_i, x_{i+1})$ with $(\mathfrak{u}_{x,i+1} - \mathfrak{u}_{x,i})/h$. Note that this inequality is a sharp Gårding inequality for the (fourth-order) discrete biharmonic operator.
For (10.101) we denote as before by u_h, p_h the $P^1_{c,0}$ functions associated with $\mathfrak{u}, \mathfrak{u}_x$ respectively. In view of (10.42) and (10.95) we have

$$
\begin{aligned}
h \sum_{i=0}^{N-1} \left(\frac{\mathfrak{u}_{i+1} - \mathfrak{u}_i}{h} \right)^2 &= \|u_{h,x}\|^2_{L^2[a,b]} \\
&\overset{(10.42)}{=} (u_{h,x} - p_h, u_{h,x} - p_h) + (p_h, p_h) \\
&\overset{(10.95)}{=} \frac{h^2}{12} \langle \mathfrak{u}, \mathfrak{u} \rangle_h + \|p_h\|^2_{L^2[a,b]}.
\end{aligned}
\tag{10.103}
$$

Applying the *continuous* Poincaré inequality to the function p_h, we have

$$
\|p_h\|^2_{L^2[a,b]} \le Ch \sum_{i=0}^{N-1} \left(\frac{\mathfrak{u}_{x,i+1} - \mathfrak{u}_{x,i}}{h} \right)^2 \le C \langle \mathfrak{u}, \mathfrak{u} \rangle_h, \tag{10.104}
$$

where in the above inequality we used (10.100). Inserting this inequality into (10.103), we obtain (10.101). Inequality (10.102) follows from (10.101) and the discrete Poincaré inequality (9.35). ■

Definition 10.12. *We refer to the inequality (10.100) as the (discrete)* ***coercivity property*** *of the operator δ_x^4.*

The right-hand side of the coercivity inequality (10.100) is (almost) the square of the $l^2_{h,0}$ norm of the grid function

$$
\widetilde{\mathfrak{w}}_i = \frac{1}{h}(\mathfrak{u}_{x,i+1} - \mathfrak{u}_{x,i}), \quad 1 \le i \le N-1,
$$

except for the additional term $h(\frac{\mathfrak{u}_{x,1}}{h})^2$ in the sum.

The grid function $\widetilde{\mathfrak{w}}$ is a discrete version of the second derivative of $\mathfrak{u}$, analogous to the more natural $\delta_x^2 \mathfrak{u}$. The reader is therefore justified in wondering why we did not use the latter in that inequality. We shall now establish an analog to the coercivity result (10.100), stated in terms of $\delta_x^2 \mathfrak{u}$. However, the proof is somewhat more complicated than that of (10.100) [which was a straightforward consequence of Equation (10.97)].

Proposition 10.13. *There exists a constant $C > 0$, independent of h, such that*

$$
\langle \mathfrak{u}, \mathfrak{u} \rangle_h \ \ge \ Ch \sum_{i=1}^{N-1} (\delta_x^2 \mathfrak{u}_i)^2 = C|\delta_x^2 \mathfrak{u}|_h^2. \tag{10.105}
$$

Proof. In view of Equation (10.97) we have

$$\frac{12}{h^2}h\sum_{i=0}^{N-1}\Big(\frac{\mathfrak{u}_{i+1}-\mathfrak{u}_i}{h}-\frac{1}{2}(\mathfrak{u}_{x,i+1}+\mathfrak{u}_{x,i})\Big)^2 \le \langle \mathfrak{u},\mathfrak{u}\rangle_h.$$

We now subtract from it the same inequality with i replaced by $i-1$, thus obtaining

$$\frac{12}{h^2}h\sum_{i=0}^{N-1}\Big(\frac{\mathfrak{u}_{i+1}-2\mathfrak{u}_i+\mathfrak{u}_{i-1}}{h}-\frac{1}{2}(\mathfrak{u}_{x,i+1}-\mathfrak{u}_{x,i-1})\Big)^2 \le 4\langle \mathfrak{u},\mathfrak{u}\rangle_h.$$

Observe that we take above $\mathfrak{u}_{-1}=\mathfrak{u}_{x,-1}=0$.

Similarly, from the inequality (10.100) we obtain (replacing i by $i-1$ and adding)

$$\frac{12}{h^2}h\sum_{i=0}^{N-1}\Big(\frac{\mathfrak{u}_{x,i+1}-\mathfrak{u}_{x,i-1}}{2}\Big)^2 \le 12\langle \mathfrak{u},\mathfrak{u}\rangle_h.$$

Adding these two estimates we get

$$\frac{12}{h^2}h\sum_{i=0}^{N-1}\Big(\frac{\mathfrak{u}_{i+1}-2\mathfrak{u}_i+\mathfrak{u}_{i-1}}{h}\Big)^2 \le 32\langle \mathfrak{u},\mathfrak{u}\rangle_h,$$

which is the estimate (10.105). ∎

We can now formulate the analog of Theorem 9.2 for the biharmonic operator.

Theorem 10.14. *Let u^* be the grid function corresponding to the exact solution $u(x)$ of Equation (10.44) and let $\mathfrak{u}$ be the discrete solution of (10.56). Then the following error estimate holds.*

$$\langle u^*-\mathfrak{u},u^*-\mathfrak{u}\rangle_h^{1/2} \le Ch^{3/2}(\|f'\|_{L^\infty}+\|f^{(4)}\|_{L^\infty}), \tag{10.106}$$

where the constant C is independent of h.

Proof. In view of the common boundary conditions for u^* and $\mathfrak{u}$, we conclude that $u^*-\mathfrak{u}$ and $(u')^*-\mathfrak{u}_x$ vanish at the endpoints, that is, they belong to $l^2_{h,0}$. This allows us to use the coercivity estimate of Proposition 10.11. Refer to Section 9.2 for the principle of the proof. The error is estimated by combining the truncation error with the coercivity property. Using the definition of the norm associated with the scalar product $\langle \mathfrak{u},\mathfrak{v}\rangle_h$, we have

$$\langle u^*-\mathfrak{u},u^*-\mathfrak{u}\rangle_h^{1/2} = \sup_{\mathfrak{v}\neq 0}\frac{\langle u^*-\mathfrak{u},\mathfrak{v}\rangle_h}{\langle \mathfrak{v},\mathfrak{v}\rangle_h^{1/2}}. \tag{10.107}$$

We have
(10.108)

$$\langle u^* - \mathfrak{u}, \mathfrak{v}\rangle_h = (\delta_x^4(u^* - \mathfrak{u}), \mathfrak{v})_h = (\delta_x^4 u^* - f^*, \mathfrak{v})_h = h\sum_{i=1}^{N-1}(\delta_x^4 u_i^* - f_i)\mathfrak{v}_i,$$

so that, from the inequality (10.66),

$$\begin{aligned}|\langle u^* - \mathfrak{u}, \mathfrak{v}\rangle_h| &\le |\delta_x^4 u^* - f^*|_h|\mathfrak{v}|_h \\ &= |\delta_x^4 u^* - (u^{(4)})^*|_h|\mathfrak{v}|_h \\ &\le Ch^{3/2}(\|f'\|_{L^\infty} + \|f^{(4)}\|_{L^\infty})|\mathfrak{v}|_h.\end{aligned}$$

Using $|\mathfrak{v}|_h \le C\langle \mathfrak{v}, \mathfrak{v}\rangle_h^{1/2}$ [see (10.102)], we obtain

$$|\langle u^* - \mathfrak{u}, \mathfrak{v}\rangle_h| \le Ch^{3/2}\langle \mathfrak{v}, \mathfrak{v}\rangle_h^{1/2}(\|f'\|_{L^\infty} + \|f^{(4)}\|_{L^\infty}), \tag{10.109}$$

which proves (10.106). ■

10.6 Matrix representation of the three-point biharmonic operator

In Section 9.3 we introduced the matrix representation of the three-point Laplacian (see Chapter 8 for the correspondence between grid functions and their vector representations). Our aim in this section is to derive an analogous representation for the three-point biharmonic operator δ_x^4. We consider this operator as a positive definite operator in $l_{h,0}^2$ (both for the functions and their derivatives), as in the previous section.

Let $\mathfrak{u}, \mathfrak{u}_x \in l_{h,0}^2$ and let $U, U_x \in \mathbb{R}^{N-1}$ be the related vectors

$$U = [\mathfrak{u}_1, \ldots, \mathfrak{u}_{N-1}]^T, \quad U_x = [\mathfrak{u}_{x,1}, \ldots, \mathfrak{u}_{x,N-1}]^T. \tag{10.110}$$

Let the $(N-1)\times(N-1)$ matrix S be the matrix related to δ_x^4. Thus, SU corresponds to $\delta_x^4\mathfrak{u}$. Note that U_x is related to U via the Hermitian derivative. The matrix forms of the operators $\mathfrak{u}\mapsto\delta_x\mathfrak{u}$, $\mathfrak{u}\mapsto\delta_x^2\mathfrak{u}$ and $\mathfrak{u}\mapsto\mathfrak{u}_x$ were introduced in (10.4), (9.51) and (10.15), respectively. Using the definition (10.54) of δ_x^4, we can express the corresponding matrix S as

$$S = \frac{12}{h^2}\left(\frac{3}{2h^2}KP^{-1}K + \frac{1}{h^2}T\right) = \frac{6}{h^4}\left(3KP^{-1}K + 2T\right). \tag{10.111}$$

Let $\{e_i\}_{i=1}^{N-1}$ be the standard basis[1] of $\mathbb{R}^{N-1}$. We regard $e_i = [0,\ldots,0,1,0,\ldots,0]^T$ as a column vector and define the $(N-1)\times(N-1)$

[1]As an exception to the convention of Chapter 8, we denote here these vectors by small Latin letters throughout this section

matrix E_i, $1 \le i \le N-1$, by

$$(10.112) \qquad E_i = (row\ i \to) \begin{bmatrix} 0 & 0 & 0 \dots & 0 \\ 0 & 0 & 0 \dots & 0 \\ \vdots & \vdots & \vdots \dots & \vdots \\ 0 & \vdots & 1 & \vdots & 0 \\ \vdots & \vdots & \vdots \dots & \vdots \\ 0 \dots & 0 & 0 & 0 \\ 0 \dots & 0 & 0 & 0 \end{bmatrix} = e_i e_i^T .$$

Note that all the entries of E_i are zero, except for the element (i,i) of the matrix, which is equal to one.

While the matrices P and K do not commute, their commutator is limited to values near the boundary. This fact will allow us to obtain a more amenable form for S. The precise form of the commutator is given in the following lemma. Its straightforward proof is left to the reader.

Lemma 10.15. *(i) The commutator of the $(N-1)\times(N-1)$ matrices P and K is*

$$(10.113) \qquad [P,K] = PK - KP = 2\left(E_{N-1} - E_1\right).$$

(ii) The commutator of P^{-1} and K is

$$(10.114) \qquad [P^{-1},K] = P^{-1}K - KP^{-1} = -2P^{-1}(E_{N-1} - E_1)P^{-1}.$$

(iii) The symmetric matrix K^2 is related to T by

$$(10.115) \qquad K^2 = T^2 - 4T + 2(E_1 + E_{N-1}).$$

Recall that (see Section 10.1 of this chapter) the matrix corresponding to the operator σ_x is $P/6 = I - T/6$. In the following propositions we use this fact (as well as the commutator relations of Lemma 10.15) in our simplification of (10.111).

Proposition 10.16. *The matrix $\frac{1}{6}PS$ can be expressed as*

$$(10.116) \qquad \frac{1}{6}PS = \frac{1}{h^4}T^2 + \frac{6}{h^4}\left[e_1(e_1 + KP^{-1}e_1)^T + e_{N-1}(e_{N-1} - KP^{-1}e_{N-1})^T\right].$$

Note that $\frac{1}{6}PS$ corresponds to the operator $\sigma_x\delta_x^4$, where the image of δ_x^4 is projected onto $l_{h,0}^2$.

Proof. If $U, V \in \mathbb{R}^{N-1}$ are column vectors, then $U^T, V^T \in \mathbb{R}^{N-1}$ are row vectors. If A is an $(N-1) \times (N-1)$ matrix, then

$$U(V^T A) = (UV^T)A = U(A^T V)^T. \tag{10.117}$$

Multiplying Equation (10.111) on the left by P gives, using (10.113),

$$\begin{aligned} PS &= \frac{6}{h^4}(3PKP^{-1}K + 2PT) \\ &= \frac{6}{h^4}(3[P,K]P^{-1}K + 3KPP^{-1}K + 2PT) \\ &= \frac{6}{h^4}(6(E_{N-1} - E_1)P^{-1}K + 3K^2 + 2(6I - T)T). \end{aligned}$$

Expressing K^2 in terms of T [see (10.115)] and using $(P^{-1})^T = P^{-1}$ and $K^T = -K$, we get

$$\begin{aligned} PS = \frac{6}{h^4}\Big[&6e_{N-1}e_{N-1}^T P^{-1}K - 6e_1 e_1^T P^{-1}K \\ &+ 3(T^2 - 4T + 2e_1e_1^T + 2e_{N-1}e_{N-1}^T) + 12T - 2T^2\Big] \\ = \frac{6}{h^4}\Big[&- 6e_{N-1}e_{N-1}^T (P^{-1})^T K^T + 6e_1e_1^T (P^{-1})^T K^T + T^2 \\ &+ 6e_1e_1^T + 6e_{N-1}e_{N-1}^T\Big] \\ = \frac{6}{h^4}T^2 &+ \frac{36}{h^4}\Big[e_1(e_1 + KP^{-1}e_1)^T + e_{N-1}(e_{N-1} - KP^{-1}e_{N-1})^T\Big], \end{aligned}$$

which proves (10.116). ∎

We can now proceed to derive a simplified expression for S. The proof of the following proposition is slightly more technical and may be skipped by the reader on a first reading.

Proposition 10.17. *The expression (10.111) for S may be simplified as*

$$S = \frac{6}{h^4}P^{-1}T^2 + \frac{36}{h^4}\left(V_1V_1^T + V_2V_2^T\right), \tag{10.118}$$

where the (column) vectors V_1 and V_2 are

$$\begin{cases} V_1 = (\alpha - \beta)^{1/2} P^{-1}\left(\dfrac{\sqrt{2}}{2}e_1 - \dfrac{\sqrt{2}}{2}e_{N-1}\right), \\ V_2 = (\alpha + \beta)^{1/2} P^{-1}\left(\dfrac{\sqrt{2}}{2}e_1 + \dfrac{\sqrt{2}}{2}e_{N-1}\right), \end{cases} \tag{10.119}$$

and the constants α *and* β *are*

$$(10.120) \qquad \begin{cases} \alpha = 2(2 - e_1^T P^{-1} e_1) = 2(2 - e_{N-1}^T P^{-1} e_{N-1}), \\ \\ \beta = 2 e_{N-1}^T P^{-1} e_1. \end{cases}$$

Remark 10.18. In view of Equation (10.24) and the positivity of P^{-1}, we have $0 \le e_1^T P^{-1} e_1 \le 1/2$ and $|e_{N-1}^T P^{-1} e_1| \le 1/2$, so that $3 \le \alpha \le 4$ and $|\beta| \le 1$. Thus, $(\alpha \pm \beta)^{1/2}$ are well defined.

Proof. Multiplying (10.116) by P^{-1} on the left, we obtain

$$\begin{aligned} S &= \frac{6}{h^4} P^{-1} T^2 + \frac{36}{h^4} \Big[P^{-1} e_1 (e_1 + K P^{-1} e_1)^T \\ &\qquad + P^{-1} e_{N-1} (e_{N-1} - K P^{-1} e_{N-1})^T \Big] \\ &= \frac{6}{h^4} P^{-1} T^2 + \frac{36}{h^4} G_1. \end{aligned}$$

The matrix G_1 can be expanded as

$$\begin{aligned} G_1 &= P^{-1} e_1 e_1^T - P^{-1} e_1 e_1^T P^{-1} K + P^{-1} e_{N-1} e_{N-1}^T + P^{-1} e_{N-1} e_{N-1}^T P^{-1} K \\ &= P^{-1} e_1 e_1^T - P^{-1} e_1 e_1^T K P^{-1} + P^{-1} e_1 e_1^T [K, P^{-1}] \\ &\quad + P^{-1} e_{N-1} e_{N-1}^T + P^{-1} e_{N-1} e_{N-1}^T K P^{-1} + P^{-1} e_{N-1} e_{N-1}^T [P^{-1}, K]. \end{aligned}$$

We now use the commutator relations (10.113) and (10.114) and find that $G_1 = P^{-1} G_2 P^{-1}$, where G_2 is

$$\begin{aligned} G_2 &= e_1 e_1^T P - e_1 e_1^T K - 2 e_1 e_1^T P^{-1} e_1 e_1^T + 2 e_1 e_1^T P^{-1} e_{N-1} e_{N-1}^T \\ &\quad + e_{N-1} e_{N-1}^T P + e_{N-1} e_{N-1}^T K + 2 e_{N-1} e_{N-1}^T P^{-1} e_1 e_1^T \\ &\quad - 2 e_{N-1} e_{N-1}^T P^{-1} e_{N-1} e_{N-1}^T \\ &= \Big(- 2 e_1 e_1^T P^{-1} e_1 e_1^T - 2 e_{N-1} e_{N-1}^T P^{-1} e_{N-1} e_{N-1}^T \\ &\quad + 2 (e_1 e_1^T P^{-1} e_{N-1} e_{N-1}^T + e_{N-1} e_{N-1}^T P^{-1} e_1 e_1^T) \Big) \\ &\quad + \Big(e_1 \big((P + K) e_1 \big)^T + e_{N-1} \big((P - K) e_{N-1} \big)^T \Big). \end{aligned}$$

It is readily seen that

$$(10.121) \qquad \begin{cases} (P + K) e_1 = 4 e_1 \\ (P - K) e_{N-1} = 4 e_{N-1}, \end{cases}$$

therefore, we can rewrite G_1 as

$$G_1 = P^{-1}\Big(-2e_1e_1^TP^{-1}e_1e_1^T - 2e_{N-1}e_{N-1}^TP^{-1}e_{N-1}e_{N-1}^T + 2e_1e_1^TP^{-1}e_{N-1}e_{N-1}^T + 2e_{N-1}e_{N-1}^TP^{-1}e_1e_1^T + 4e_1e_1^T + 4e_{N-1}e_{N-1}^T\Big)P^{-1}.$$

This expression can be further simplified to

$$G_1 = P^{-1}\begin{bmatrix} e_1, e_{N-1}\end{bmatrix}\begin{bmatrix}\alpha & \beta \\ \beta & \alpha\end{bmatrix}\begin{bmatrix} e_1^T \\ e_{N-1}^T\end{bmatrix}P^{-1}, \tag{10.122}$$

where α and β are as in (10.120). Since we have

$$\begin{bmatrix}\alpha & \beta \\ \beta & \alpha\end{bmatrix} = \begin{bmatrix}\frac{\sqrt{2}}{2} & \frac{\sqrt{2}}{2} \\ -\frac{\sqrt{2}}{2} & \frac{\sqrt{2}}{2}\end{bmatrix}\begin{bmatrix}\alpha-\beta & 0 \\ 0 & \alpha+\beta\end{bmatrix}\begin{bmatrix}\frac{\sqrt{2}}{2} & -\frac{\sqrt{2}}{2} \\ \frac{\sqrt{2}}{2} & \frac{\sqrt{2}}{2}\end{bmatrix}, \tag{10.123}$$

we can finally write G_1 as

$$G_1 = \begin{bmatrix} V_1, V_2\end{bmatrix}\begin{bmatrix} V_1^T \\ V_2^T\end{bmatrix}, \tag{10.124}$$

where the vectors V_1 and V_2 are given in (10.119). This completes the proof of the proposition. ■

10.7 Convergence analysis using the matrix representation

In the proof of Theorem 10.14 (Section 10.5) we used energy methods to obtain an error estimate for the one-dimensional biharmonic equation. In this section we enhance this result by using the matrix representation of δ_x^4 given in the preceding section.

Consider again the biharmonic equation (10.44) and its approximation by the Stephenson scheme (10.56). Let $\mathfrak{u}$ be the solution of the latter. For notational simplicity we take the unit interval $a = 0,\ b = 1$, so that $h = 1/N$.

Let u^* be the grid function corresponding to u. It satisfies

$$\delta_x^4 u_i^* = f^*(x_i) + \mathfrak{r}_i, \qquad 1 \le i \le N-1, \tag{10.125}$$

where $\mathfrak{r}$ is by definition the truncation error. We later refer to Proposition 10.8 for estimates of $\mathfrak{r}$.

Denote by $\mathfrak{e}$ the error $\mathfrak{e} = \mathfrak{u} - u^*$. It satisfies

$$\begin{aligned} &\delta_x^4 \mathfrak{e}_i = -\mathfrak{r}_i, \qquad 1 \le i \le N-1, \\ &\mathfrak{e}_0 = 0,\ \mathfrak{e}_N = 0,\ \mathfrak{e}_{x,0} = 0,\ \mathfrak{e}_{x,N} = 0. \end{aligned} \tag{10.126}$$

We prove the following error estimate.

Theorem 10.19. *Let u be the exact solution of (10.44) and assume that u has continuous derivatives up to order eight on $[0,1]$. Let $\mathfrak{u}$ be the approximation to u given by (10.56). Let u^* be the grid function corresponding to u. Then, the error $\mathfrak{e} = \mathfrak{u} - u^*$ satisfies*

$$|\mathfrak{e}|_h \le Ch^4, \tag{10.127}$$

where C depends only on f.

Proof. Let $U, U^* \in \mathbb{R}^{N-1}$ be the vectors corresponding to $\mathfrak{u}$ and u^*, respectively, and let F be the vector corresponding to f^*. We denote by $E = U - U^*$ and R the vectors corresponding to $\mathfrak{e} = \mathfrak{u} - u^*$ and $\mathfrak{r}$, respectively.

Using the matrix representation (10.118), we can write Equations (10.56) and (10.125) in the form

$$SU = F, \tag{10.128}$$

and

$$SU^* = F + R. \tag{10.129}$$

We therefore have

$$SE = -R. \tag{10.130}$$

In view of (10.118) and (10.119) we have

$$PSP = \frac{6}{h^4} T^2 P + \frac{36}{h^4} JJ^T, \tag{10.131}$$

where

$$J = \frac{\sqrt{2}}{2} \left[(\alpha - \beta)^{1/2} (e_1 - e_{N-1}), (\alpha + \beta)^{1/2} (e_1 + e_{N-1}) \right]. \tag{10.132}$$

Inverting PSP and multiplying by PR, we have

$$-P^{-1}E = P^{-1}S^{-1}R = (PSP)^{-1}PR. \tag{10.133}$$

Our goal is to bound the elements of $P^{-1}E$ by Ch^4. Note that by Proposition 10.8 we have

$$\begin{aligned} &|(PR)_1|, \quad |(PR)_{N-1}| \le Ch, \\ &|(PR)_j| \le Ch^4, \quad 2 \le j \le N-2. \end{aligned} \tag{10.134}$$

Thus, we need to estimate $(PSP)^{-1}PR$. We decompose PSP as follows

$$PSP = GH^{-1}, \tag{10.135}$$

where

$$G = I + 6JJ^TP^{-1}T^{-2}, \quad H = \frac{h^4}{6}P^{-1}T^{-2}, \tag{10.136}$$

so that

$$(PSP)^{-1} = HG^{-1}. \tag{10.137}$$

Note that with $L = (6/h^4)H$ and $Q = 6JJ^T$, we have

$$G = I + QL. \tag{10.138}$$

We first estimate the elements of the matrix H.

Estimate of the elements of H

In the following we use C for various constants that do not depend on h. As in (9.57) we can diagonalize H by

$$H = Z\Lambda' Z^T,$$

where the jth column of the matrix Z is Z^j as defined in (9.55). Recall that $P = 6I - T$ [see (10.14)] and that the eigenvalues λ_j of T are given by (9.54). Therefore, the eigenvalues κ_j, $1 \le j \le N-1$ of P are given by

$$\kappa_j = 6 - \lambda_j = 6 - 4\sin^2\left(\frac{j\pi}{2N}\right), \quad 1 \le j \le N-1. \tag{10.139}$$

The diagonal matrix Λ' contains the eigenvalues of H, which can be written as

$$\theta_j = \frac{h^4}{6}\lambda_j^{-2}\kappa_j^{-1} = \frac{h^4}{96}\frac{1}{\sin^4(\frac{j\pi}{2N})\left(6 - 4\sin^2(\frac{j\pi}{2N})\right)}, \quad j = 1, \cdots, N-1.$$

The element $H_{i,k}$ of the matrix H is

$$H_{i,k} = \sum_{j=1}^{N-1} Z_{i,j}\theta_j Z_{j,k},$$

where we have used the notation $Z_{i,j}$ instead of Z_j^i in (9.55). An explicit expression for $H_{i,k}$ is

$$H_{i,k} = \sum_{j=1}^{N-1} \frac{h^4}{96}\frac{2}{N}\sin\left(\frac{ij\pi}{N}\right)\frac{\sin(\frac{jk\pi}{N})}{\sin^4\left(\frac{j\pi}{2N}\right)\left(6 - 4\sin^2(\frac{j\pi}{2N})\right)}. \tag{10.140}$$

We can now estimate the order of magnitude of the elements of H as functions of h. In fact, we shall separately inspect the first and last columns of H and the rest ($k = 2, \ldots, N-2$). The reason is that from writing

$$(HG^{-1}PR)_i = \sum_{k=1}^{N-1} H_{i,k}(G^{-1}PR)_k, \tag{10.141}$$

we shall see that $(G^{-1}PR)_1$ and $(G^{-1}PR)_{N-1}$ can only be estimated by Ch^2 [see (10.162) below] so that the additional accuracy should come from $H_{i,1}$ and $H_{i,N-1}$.

Consider first the elements (i,k) of H for $k = 1, N-1$. It suffices to consider $k = 1$.

$$H_{i,1} = \sum_{j=1}^{N-1} \frac{h^4}{96} \frac{2}{N} \sin\left(\frac{ij\pi}{N}\right) \frac{1}{\sin^4(\frac{j\pi}{2N})\left(6 - 4\sin^2(\frac{j\pi}{2N})\right)} \sin\left(\frac{j\pi}{N}\right). \tag{10.142}$$

Recall the elementary inequalities

$$\sin x \geq \frac{2}{\pi} x, \qquad 0 \leq x \leq \frac{\pi}{2}, \tag{10.143}$$

$$|\sin x| \leq |x|, \quad 2 \leq 6 - 4\sin^2\left(\frac{j\pi}{2N}\right) \leq 6. \tag{10.144}$$

Noting that $h = 1/N$ and using the estimate $|\sin(\frac{ij\pi}{N})| \leq 1$, we obtain

$$|H_{i,1}| = |H_{1,i}| \leq C \sum_{j=1}^{N-1} h^5 \frac{1}{(jh)^4}(jh) \leq Ch^2, \quad i = 2, \ldots, N-2. \tag{10.145}$$

Similarly, we have

$$C_1 h^3 \leq H_{1,1} \leq C \sum_{j=1}^{N-1} h^5 \frac{1}{(jh)^4}(jh)^2 \leq C_2 h^3. \tag{10.146}$$

This estimate holds equally for $H_{N-1,N-1}$. For the other corner elements of H we have

$$|H_{1,N-1}| = |H_{N-1,1}| \leq C_2 h^3. \tag{10.147}$$

For $i, k = 2, \ldots, N-2$ we have

$$|H_{i,k}| \leq C \sum_{j=1}^{N-1} h^5 \frac{1}{(jh)^4} \leq Ch. \tag{10.148}$$

Estimate of the elements of G^{-1}

We will show that G is invertible and we will estimate its elements. First note that the elements of L are the elements of H multiplied by $6/h^4$. In the following we will use the fact that L is symmetric not only with respect to the diagonal $i = j$, but also with respect to the secondary diagonal $i + j = N$. This follows from the fact that both P and T have these properties. Explicitly, this means that $L_{i,j} = L_{j,i}$ and $L_{i,j} = L_{N-j,N-i}$.

The estimates of the elements of L are deduced from (10.145)–(10.148), as follows.

- $L_{1,1}$, $L_{N-1,N-1}$, $L_{1,N-1}$, and $L_{N-1,N}$ are bounded by $Ch^3/h^4 = C/h$.
- $L_{1,k}$, $L_{k,1}$, $L_{N-1,k}$, and $L_{k,N-1}$ are bounded by $Ch^2/h^4 = C/h^2$, for $k = 2, \ldots, N-2$.
- All other elements of L are bounded by $Ch/h^4 = C/h^3$.

Therefore, the orders of magnitude of the elements of L can be displayed as

$$\begin{bmatrix} C/h & C/h^2 & \ldots & C/h^2 & C/h \\ C/h^2 & C/h^3 & \ldots & C/h^3 & C/h^2 \\ \vdots & \vdots & \ldots\ \ldots & & \vdots \\ C/h^2 & C/h^3 & \ldots & C/h^3 & C/h^2 \\ C/h & C/h^2 & \ldots & C/h^2 & C/h \end{bmatrix}. \tag{10.149}$$

The matrix Q is $(N-1) \times (N-1)$, but it has only four non-zero components at the corner positions,

$$Q_{1,1} = Q_{N-1,N-1} = 6\alpha, \quad Q_{1,N-1} = Q_{N-1,1} = 6\beta. \tag{10.150}$$

Therefore, QL has only two non-zero rows – the first and the last. The first row is given by

$$(QL)_{1,j} = 6(\alpha L_{1,j} + \beta L_{N-1,j}), \quad j = 1, \ldots, N-1$$

and the last row is

$$(QL)_{N-1,j} = 6(\beta L_{1,j} + \alpha L_{N-1,j}), \quad j = 1, \ldots, N-1.$$

Thus,

$$\begin{cases} G_{1,1} = 1 + 6(\alpha L_{1,1} + \beta L_{N-1,1}) =: a_1, \\ G_{1,N-1} = 6(\alpha L_{1,N-1} + \beta L_{N-1,N-1}) =: a_{N-1}, \\ G_{1,j} = 6(\alpha L_{1,j} + \beta L_{N-1,j}) =: b_j, \quad j = 2, \ldots, N-2 \end{cases} \tag{10.151}$$

and

$$\begin{cases} G_{N-1,1} = 6(\beta L_{1,1} + \alpha L_{N-1,1}) = G_{1,N-1} = a_{N-1}, \\ G_{N-1,N-1} = 1 + 6(\beta L_{1,N-1} + \alpha L_{N-1,N-1}) = G_{1,1} = a_1, \\ G_{N-1,j} = 6(\beta L_{1,j} + \alpha L_{N-1,j}) = b_{N-j}, \quad j = 2, \ldots, N-2, \end{cases} \tag{10.152}$$

where the symmetries of L have been used.

In rows $2, 3, \ldots, N-2$ the elements of matrix G are 1 on the diagonal and otherwise they are zero.

The orders of magnitude of a_1, a_{N-1} and b_j $(2 \le j \le N-2)$ follow from those of the elements of L. Namely, $|a_1|$, $|a_{N-1}| \le C/h$ and $|b_j| \le C/h^2$ for $j = 2, \ldots, N-2$.

In the following we shall need lower bounds for a_1 and $a_1^2 - a_{N-1}^2$. From their definitions above it can be seen that we need to inspect the terms $L_{1,1} = (6/h^4)H_{1,1}$ and $L_{1,N-1} = (6/h^4)H_{1,N-1}$. Note that all the terms in the sum for $H_{1,1}$ are positive. On the other hand the terms for $H_{N-1,1}$ [obtained by setting $i = N-1$ in (10.142)] have the same absolute value but alternate in sign. Therefore,

$$
\begin{aligned}
&L_{1,1} > |L_{1,N-1}|,\\
&L_{1,1} \mp L_{1,N-1} = \tfrac{h}{4} \sum_{\substack{j=1 \\ j \text{ even or odd}}}^{N-1} \frac{\sin^2 \frac{j\pi}{N}}{\sin^4 \frac{j\pi}{2N}(6 - 4\sin^2 \frac{j\pi}{2N})} \\
&\qquad \geq Ch \sum_{\substack{j=1 \\ j \text{ even or odd}}}^{N-1} \frac{(jh)^2}{(jh)^4} = \frac{C}{h}.
\end{aligned}
\tag{10.153}
$$

In the above, take $j =$ *even* for "$-$" and $j =$ *odd* for "$+$."

We first provide a lower bound for a_1. Using (10.151) and the bounds $3 \leq \alpha \leq 4$, $|\beta| \leq 1$ (Remark 10.18), we get from (10.153) and (10.146)

$$
|a_1| \geq 6(3L_{1,1} - |L_{N-1,1}|) - 1 \geq 12L_{1,1} - 1 \geq \frac{C}{h}. \tag{10.154}
$$

Next, we consider the difference $a_1^2 - a_{N-1}^2$. Since we have the upper bound $|a_1^2 - a_{N-1}^2| \leq C_1/h^2$, we need only a lower bound. That is, we want to show that

$$
|a_1^2 - a_{N-1}^2| \geq C_2/h^2. \tag{10.155}
$$

We write the difference $a_1^2 - a_{N-1}^2$ as

$$
a_1^2 - a_{N-1}^2 = \Big(1 + 6(\alpha + \beta)(L_{1,1} + L_{1,N-1})\Big) \cdot \Big(1 + 6(\alpha - \beta)(L_{1,1} - L_{1,N-1})\Big),
$$

using the symmetries of L. From (10.153) and $\alpha \geq 3$, $|\beta| \leq 1$, we obtain the lower bound (10.155).

To compute the inverse of G, we apply Gaussian elimination using the following method. We perform operations on rows of G and apply the same operations to the identity matrix I. When G is transformed to the identity matrix, I is transformed to G^{-1}.

We divide the first and last rows of G by a_1 and annihilate the terms $j = 2, \ldots, N-2$ of both rows by subtracting suitable multiples of rows

$2, \ldots, N-2$. The result is G_1, where

$$G_1 = \begin{bmatrix} 1 & 0\,0 \ldots & 0 & 0 & \frac{a_{N-1}}{a_1} \\ 0 & 1\,0 \ldots & 0 & 0 & 0 \\ 0 & 0\,1 \ldots & 0 & 0 & 0 \\ \vdots & \vdots\,\vdots \ldots & \ldots & \ldots & \vdots \\ 0 & 0\,0 \ldots & 0 & 1 & 0 \\ \frac{a_{N-1}}{a_1} & 0\,0 \ldots & 0 & 0 & 1 \end{bmatrix}. \tag{10.156}$$

The same operations on the identity matrix give

$$I_1 = \begin{bmatrix} \frac{1}{a_1} & \frac{-b_2}{a_1} & \frac{-b_3}{a_1} & \ldots & \frac{-b_{N-3}}{a_1} & \frac{-b_{N-2}}{a_1} & 0 \\ 0 & 1 & -0 & \ldots & 0 & 0 & 0 \\ 0 & 0 & 1 & \ldots & 0 & 0 & 0 \\ \vdots & \vdots & \vdots & \ldots & \ldots & \ldots & \vdots \\ 0 & 0 & 0 & \ldots & 0 & 1 & 0 \\ 0 & \frac{-b_{N-2}}{a_1} & \frac{-b_{N-3}}{a_1} & \ldots & \frac{-b_3}{a_1} & \frac{-b_2}{a_1} & \frac{1}{a_1} \end{bmatrix}. \tag{10.157}$$

In order to eliminate the non-zero element of G_1 in position $(N-1, 1)$, we subtract a suitable multiple of the first row and add the result to the last row, thus getting the transformed matrix G_2

$$G_2 = \begin{bmatrix} 1\,0\,0 \ldots & 0 & 0 & \frac{a_{N-1}}{a_1} \\ 0\,1\,0 \ldots & 0 & 0 & 0 \\ 0\,0\,1 \ldots & 0 & 0 & 0 \\ \vdots\,\vdots\,\vdots \ldots & \ldots & \ldots & \vdots \\ 0\,0\,0 \ldots & 0 & 1 & 0 \\ 0\,0\,0 \ldots & 0 & 0 & \frac{a_1^2 - a_{N-1}^2}{a_1^2} \end{bmatrix}. \tag{10.158}$$

The corresponding matrix I_2 (obtained similarly from I_1) is

$$I_2 = \begin{bmatrix} \frac{1}{a_1} & \frac{-b_2}{a_1} & \frac{-b_3}{a_1} & \ldots & \frac{-b_{N-3}}{a_1} & \frac{-b_{N-2}}{a_1} & 0 \\ 0 & 1 & 0 & \ldots & 0 & 0 & 0 \\ 0 & 0 & 1 & \ldots & 0 & 0 & 0 \\ \vdots & \vdots & \vdots & \ldots & \ldots & \ldots & \vdots \\ 0 & 0 & 0 & \ldots & 0 & 1 & 0 \\ \frac{-a_{N-1}}{a_1^2} & \frac{b_2 a_{N-1} - a_1 b_{N-2}}{a_1^2} & \frac{b_3 a_{N-1} - a_1 b_{N-3}}{a_1^2} & \ldots & \frac{b_{N-3} a_{N-1} - a_1 b_3}{a_1^2} & \frac{b_{N-2} a_{N-1} - a_1 b_2}{a_1^2} & \frac{1}{a_1} \end{bmatrix}.$$

Now we divide the last row of G_2 and of I_2 by $\frac{a_1^2-a_{N-1}^2}{a_1^2}$. We get

$$G_3 = \begin{bmatrix} 1\,0\,0 \ldots & 0 & 0 & \frac{a_{N-1}}{a_1} \\ 0\,1\,0 \ldots & 0 & 0 & 0 \\ 0\,0\,1 \ldots & 0 & 0 & 0 \\ \vdots\;\vdots\;\vdots \ldots & \ldots & \ldots & \vdots \\ 0\,0\,0 \ldots & 0 & 1 & 0 \\ 0\,0\,0 \ldots & 0 & 0 & 1 \end{bmatrix} \tag{10.159}$$

and I_3 is

$$I_3 = \begin{bmatrix} \frac{1}{a_1} & \frac{-b_2}{a_1} & \frac{-b_3}{a_1} & \cdots & \frac{-b_{N-3}}{a_1} & \frac{-b_{N-2}}{a_1} & 0 \\ 0 & 1 & 0 & \cdots & 0 & 0 & 0 \\ 0 & 0 & 1 & \cdots & 0 & 0 & 0 \\ \vdots & \vdots & \vdots & \cdots & \cdots & \cdots & \vdots \\ 0 & 0 & 0 & \cdots & 0 & 1 & 0 \\ \frac{-a_{N-1}}{a_1^2-a_{N-1}^2} & \frac{b_2 a_{N-1}-a_1 b_{N-2}}{a_1^2-a_{N-1}^2} & \frac{b_3 a_{N-1}-a_1 b_{N-3}}{a_1^2-a_{N-1}^2} & \cdots & \frac{b_{N-3} a_{N-1}-a_1 b_3}{a_1^2-a_{N-1}^2} & \frac{b_{N-2} a_{N-1}-a_1 b_2}{a_1^2-a_{N-1}^2} & \frac{a_1}{a_1^2-a_{N-1}^2} \end{bmatrix}.$$

Finally, we eliminate the $(1, N-1)$ element in G_3 by subtracting a multiple of the last row. The corresponding operation on I_3 yields the inverse G^{-1} as

$$G^{-1} = \begin{bmatrix} \frac{a_1}{a_1^2-a_{N-1}^2} & \frac{a_{N-1} b_{N-2}-a_1 b_2}{a_1^2-a_{N-1}^2} & \frac{a_{N-1} b_{N-3}-a_1 b_3}{a_1^2-a_{N-1}^2} & \cdots & \frac{a_{N-1} b_3-a_1 b_{N-3}}{a_1^2-a_{N-1}^2} & \frac{a_{N-1} b_2-a_1 b_{N-2}}{a_1^2-a_{N-1}^2} & \frac{-a_{N-1}}{a_1^2-a_{N-1}^2} \\ 0 & 1 & 0 & \cdots & 0 & 0 & 0 \\ 0 & 0 & 1 & \cdots & 0 & 0 & 0 \\ \vdots & \vdots & \vdots & \cdots & \cdots & \cdots & \vdots \\ 0 & 0 & 0 & \cdots & 0 & 1 & 0 \\ \frac{-a_{N-1}}{a_1^2-a_{N-1}^2} & \frac{a_{N-1} b_2-a_1 b_{N-2}}{a_1^2-a_{N-1}^2} & \frac{a_{N-1} b_3-a_1 b_{N-3}}{a_1^2-a_{N-1}^2} & \cdots & \frac{a_{N-1} b_{N-3}-a_1 b_3}{a_1^2-a_{N-1}^2} & \frac{a_{N-1} b_{N-2}-a_1 b_2}{a_1^2-a_{N-1}^2} & \frac{a_1}{a_1^2-a_{N-1}^2} \end{bmatrix}.$$

We can now give accurate estimates for the non-trivial elements of G^{-1}, those in the first and last rows. Using (10.154), (10.155), and the corresponding upper bounds, we can readily see that

$$|(G^{-1})_{1,1}|,\ |(G^{-1})_{N-1,N-1}|,\ |(G^{-1})_{N-1,1}|,\ |(G^{-1})_{1,N-1}| \le Ch. \tag{10.160}$$

Recalling also that $|b_j| \le C/h^2$, we get

$$|(G^{-1})_{1,j}|),\ |(G^{-1})_{N-1,j}| \le \frac{C}{h}, \quad j = 2, \ldots, N-2. \tag{10.161}$$

Turning back to (10.141), we obtain bounds for the elements of $G^{-1}PR$ using (10.160), (10.161), and (10.134),

$$
\begin{aligned}
|(G^{-1}PR)_1| &\le \sum_{k=1}^{N-1} |(G^{-1})_{1,k}| \cdot |(PR)_k| \\
&= |(G^{-1})_{1,1}| \cdot |(PR)_1| + \sum_{k=2}^{N-2} |(G^{-1})_{1,k}| \cdot |(PR)_k| \\
&\quad + |(G^{-1})_{1,N-1}| \cdot |(PR)_{N-1}| \\
&\le C_1 h \cdot h + C_2(N-3)(1/h) \cdot h^4 \le Ch^2.
\end{aligned} \tag{10.162}
$$

Similarly, we have $|(G^{-1}PR)_{N-1}| \le Ch^2$.

For $i = 2, \ldots, N-2$

$$|(G^{-1}PR)_i| = |(PR)_i| \le Ch^4. \tag{10.163}$$

Finally we consider the product $HG^{-1}PR$ [see (10.133) and (10.137)]

$$-P^{-1}E = HG^{-1}PR. \tag{10.164}$$

Combining the estimates (10.145)–(10.148) with (10.162)–(10.163), we obtain

$$
\begin{aligned}
|(HG^{-1}PR)_i| &\le \sum_{k=1}^{N-1} |H_{i,k}| \cdot |(G^{-1}PR)_k| \\
&= |H_{i,1}| \cdot |(G^{-1}PR)_1| + \sum_{k=2}^{N-2} |H_{i,k}| \cdot |(G^{-1}PR)_k| \\
&\quad + |H_{i,N-1}| \cdot |(G^{-1}PR)_{N-1}| \\
&\le C_1 h^2 h^2 + C_2(N-3)hh^4 \le Ch^4.
\end{aligned} \tag{10.165}
$$

Therefore,

$$|(P^{-1}E)_i| = |(HG^{-1}PR)_i| \le Ch^4, \qquad 1 \le i \le N-1. \tag{10.166}$$

Conclusion of the proof of Theorem 10.19.
Using (10.166) we find that the Euclidean norm of the vector $E = U - U^*$ satisfies the estimate

$$|E| = |PP^{-1}E| \le C|P^{-1}E| \overset{(10.166)}{\le} C\sqrt{\sum_{i=1}^{N-1} (h^4)^2} = Ch^{-1/2}h^4. \tag{10.167}$$

Thus, in view of the definition of the norm (8.8)

$$|\mathfrak{e}|_h \le Ch^4. \tag{10.168}$$

This proves the fourth-order error estimate result. ∎

10.8 Notes for Chapter 10

- The Hermitian derivatives (of all orders) were introduced in [46, Chapter III]. The term was used only in the English translation, inspired by the term "Hermitian interpolation operator" when derivative values are involved. The approach of Collatz [46] is indeed *interpolatory*. It consists of writing a sum (at the ith grid point)

$$\sum_{j=-k}^{k} (a_j y_{i+k} + b_k y'_{i+j}),$$

  and try (using a Taylor's expansion) to eliminate as many powers of h as possible, by choosing approximate values of the coefficients a_j and b_j. For $k = 1$ the result is identical to Equation (10.11), which we obtained using Simpson's rule. In [93, Section 4.3] these discrete derivatives are called compact "Padé type difference operators" and the order of accuracy is determined using Fourier symbols of difference operators. Unlike the above references, careful attention is given here to an accurate approximation even near the boundary, as expressed in Lemma 10.1.
- Sections 10.2–10.5 follow [12].
- Section 10.6 follows [13].
- Section 10.7 follows [66]. Note in particular that even though the accuracy deteriorates near the boundary (Proposition 10.8), the convergence of the one-dimensional Stephenson scheme to the exact solution is of fourth order (Theorem 10.19).

Chapter 11

Polynomial Approach to the Discrete Biharmonic Operator

11.1 The biharmonic problem in a rectangle

11.1.1 *General setting*

We begin our treatment of two-dimensional problems with the biharmonic equation in a rectangle. For simplicity we actually take the square $\Omega = [0,1]^2$, so that the problem is the following:

$$\text{(11.1)} \qquad \begin{cases} (a) & \Delta^2\psi(x,y) = f(x,y), \quad (x,y) \in \Omega, \\ (b) & \psi = g_1(x,y), \quad (x,y) \in \partial\Omega, \\ (c) & \frac{\partial \psi}{\partial n} = g_2(x,y), \quad (x,y) \in \partial\Omega, \end{cases}$$

where $\frac{\partial \psi}{\partial n}$ denotes the (outward) normal derivative. The biharmonic operator is

$$\text{(11.2)} \qquad \Delta^2\psi = \partial_x^4\psi + 2\partial_x^2\partial_y^2\psi + \partial_y^4\psi.$$

Equation (11.1) is encountered in various physical applications, such as elasticity theory and fluid dynamics (the Stokes equation in the streamfunction formulation). An extensive review [137] gives a detailed account of the applications of the biharmonic operator, as well as special solutions and historical notes. Refer to [21] for a standard finite difference approach to the biharmonic operator in the two-dimensional case and an FFT method for inverting the resulting set of linear equations. Refer also to [20, 45] for other finite difference operators involving splines.

In our treatment of the Navier–Stokes system, the biharmonic operator plays a fundamental role, since we focus on the streamfunction formulation. Fluid dynamical applications naturally require boundary conditions and it is for this reason that we concentrate on the *boundary-value problem* (11.1). It is well-known that the imposition of boundary conditions leads to *boundary layers*, which pose considerable difficulties, at both the theoretical and numerical levels. In fact, the numerical effect of the boundary layers has already been encountered in the one-dimensional case (see Proposition 10.8). When the boundary conditions are replaced by periodic ones, there are no boundary layers and therefore no special numerical treatment is needed at the boundary.

As in the one-dimensional case discussed in the preceding chapter, we focus on *high-order compact* schemes. Such schemes are more easily presented (and analyzed) in a rectangular domain. Note, however, that the standard elliptic theory pertaining to bounded domains requires regularity of the boundary. The presence of corners in the square $[0,1]^2$ poses additional theoretical difficulties.

Here is a summary of "design principles" for the scheme presented in this section:

- The discrete approximation of the pure fourth-order derivative is precisely the one obtained in the preceding chapter.
- The mixed term $\partial_x^2\partial_y^2$ is replaced by a second-order approximation based on the operators introduced in Chapter 9. See Section 13.1 for a full fourth-order approximation.
- The scheme uses data only from the eight neighboring grid points.
- In particular, the boundary conditions in the discrete version are the exact analogs of the continuous ones in Equation (11.1).
- There is no special treatment of "near-boundary" points; they are treated as regular internal points.
- The resulting scheme is *stable*, implying the convergence of the approximate solution to the exact one.

We begin by recalling the standard (second-order) finite difference approximation for the biharmonic operator. Following the notation introduced in Section 8.2, we consider a uniform grid $(x_i, y_j) = (ih, jh), 0 \le i, j \le N$ (see Figure 8.2 in Chapter 8). We use bold Greek letters to denote grid functions such as $\boldsymbol{\psi} = (\psi(x_i, y_j))_{0\le i,j\le N}$. *A second-order* approximation of ∂_x^4 is given in (10.45). For a second-order approximation of the mixed term $\partial_x^2\partial_y^2$, we invoke the difference operator $\delta_x^2\delta_y^2$. Thus, we arrive at the following finite difference approximation of Δ^2, designated as Δ_{13}^2,

$$\Delta_{13}^2\boldsymbol{\psi} = \delta_x^2\delta_x^2\boldsymbol{\psi} + 2\delta_x^2\delta_y^2\boldsymbol{\psi} + \delta_y^2\delta_y^2\boldsymbol{\psi}. \tag{11.3}$$

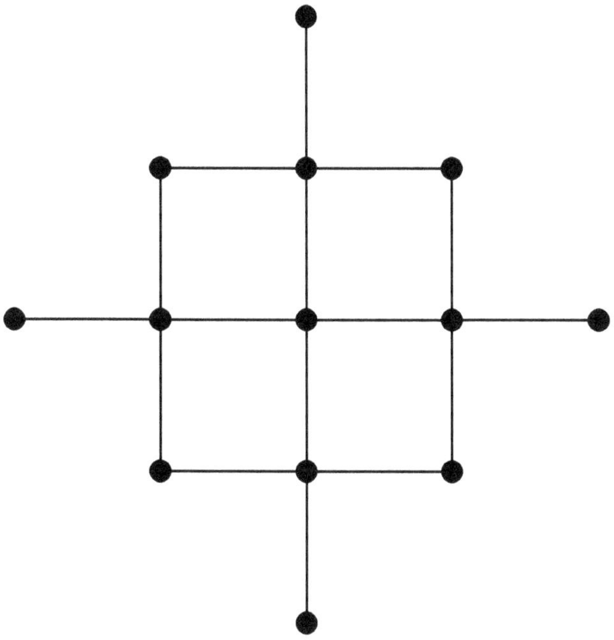

Fig. 11.1 Stencil of the 13-point biharmonic operator Δ_{13}^2.

An explicit representation of $\Delta_{13}^2\psi$ is

$$\begin{aligned}\Delta_{13}^2\psi_{i,j} = \frac{1}{h^4}\Big(& (\psi_{i+2,j}+\psi_{i-2,j}+\psi_{i,j+2}+\psi_{i,j-2}) \\ & +2\,(\psi_{i+1,j+1}+\psi_{i-1,j+1}+\psi_{i+1,j-1}+\psi_{i-1,j-1}) \\ & -8\,(\psi_{i,j+1}+\psi_{i,j-1}+\psi_{i+1,j}+\psi_{i-1,j}) + 20\psi_{i,j}\Big).\end{aligned} \tag{11.4}$$

Note that the values of ψ at the 13 points are required (hence the notation Δ_{13}^2); see Figure 11.1.

Let $\psi(x,y)$ be a given function and ψ^* the corresponding grid function. It is clear from Equations (10.46) and (9.12) that

$$|\Delta_{13}^2\psi^*(x_i,y_j) - (\Delta^2\psi)^*(x_i,y_j)| \le Ch^2, \quad 2 \le i,j \le N-2, \tag{11.5}$$

where $C > 0$ depends on various sixth-order derivatives of ψ.

Note that the near-boundary points $i = 1, N-1$ or $j = 1, N-1$ are not included in this argument, and they require special approximations and accuracy estimates.

11.1.2 *The nine-point compact biharmonic operator*

Recall that in Definition 10.5, we introduced the three-point biharmonic operator δ_x^4. Instead of the five-point data needed for the operator $\delta_x^2\delta_x^2$ [see Equation (10.45)], only three points are needed for δ_x^4 [see Equation (10.55)]. This is what characterizes it as a *compact scheme*. At the same time, the order of accuracy is improved from two to four. A major ingredient of this compact improvement is the inclusion of the *Hermitian derivative* in the operator. We follow a similar procedure in the two-dimensional setting, so as to obtain a **nine-point biharmonic operator**.

Let $\psi \in L_h^2$ be a grid function. Extending Equation (10.1) to the two-dimensional case, we first introduce the finite difference operators δ_x and δ_y,

$$(11.6)\qquad \begin{cases} \delta_x\psi_{i,j} = \dfrac{\psi_{i+1,j}-\psi_{i-1,j}}{2h}, & 1\le i\le N-1,\ 0\le j\le N, \\[2mm] \delta_y\psi_{i,j} = \dfrac{\psi_{i,j+1}-\psi_{i,j-1}}{2h}, & 0\le i\le N,\ 1\le j\le N-1. \end{cases}$$

The centered operators δ_x^2 and δ_y^2 are

$$(11.7)\qquad \begin{cases} \delta_x^2\psi_{i,j} = \dfrac{\psi_{i+1,j}+\psi_{i-1,j}-2\psi_{i,j}}{h^2}, & 1\le i\le N-1,\ 0\le j\le N, \\[2mm] \delta_y^2\psi_{i,j} = \dfrac{\psi_{i,j+1}+\psi_{i,j-1}-2\psi_{i,j}}{h^2}, & 0\le i\le N,\ 1\le j\le N-1. \end{cases}$$

The Simpson operators σ_x and σ_y are [see Equation (10.12)]

$$(11.8)\qquad \sigma_x = I + \frac{h^2}{6}\delta_x^2, \quad \sigma_y = I + \frac{h^2}{6}\delta_y^2.$$

We designate by ψ_x and ψ_y the Hermitian derivatives in the x and y directions, respectively. They are evaluated, as in the one-dimensional case, by

$$(11.9)\qquad \begin{cases} \sigma_x(\psi_x)_{i,j} = \delta_x\psi_{i,j}, & 1\le i,j\le N-1, \\ \sigma_y(\psi_y)_{i,j} = \delta_y\psi_{i,j}, & 1\le i,j\le N-1. \end{cases}$$

In the following we simplify notation by writing

$$(11.10)\qquad (\psi_x)_{i,j} = \psi_{x,i,j}, \quad (\psi_y)_{i,j} = \psi_{y,i,j}.$$

We can now introduce a discrete version of the biharmonic operator as follows.

Definition 11.1 (Nine-point biharmonic operator). *Let $\psi \in L_h^2$ be a given grid function. The* ***nine-point biharmonic operator*** *is defined by*

$$(11.11)\qquad \Delta_h^2\psi_{i,j} = \delta_x^4\psi_{i,j} + \delta_y^4\psi_{i,j} + 2\delta_x^2\delta_y^2\psi_{i,j}, \quad 1\le i,j\le N-1.$$

The operators δ_x^4 and δ_y^4 are explicitly given by

(11.12)
$$\begin{cases} \delta_x^4\psi_{i,j} = \dfrac{12}{h^2}\left(\delta_x\psi_{x,i,j} - \delta_x^2\psi_{i,j}\right), & 1 \le i,j \le N-1 \\ \delta_y^4\psi_{i,j} = \dfrac{12}{h^2}\left(\delta_y\psi_{y,i,j} - \delta_y^2\psi_{i,j}\right), & 1 \le i,j \le N-1. \end{cases}$$

Observe that the boundary data in (11.1)(b) and (11.1)(c) are used as boundary values of $\psi_{i,j}$, $\psi_{x,i,j}$ and $\psi_{y,i,j}$ for $i \in \{0,N\}, j \in \{0,N\}$.

In analogy with the one-dimensional case [see Equation (10.47)], Equation (11.11) can be derived using a two-dimensional interpolating polynomial approach. In fact, this approach was used in the original work of Stephenson [168]. Assume that the values $\psi^*_{i,j}$ are given at the grid points $0 \le i,j \le N$. Assume in addition that we are given $\psi^*_{x,i,j}$ and $\psi^*_{y,i,j}$, approximating $(\partial_x\psi)^*(x_i,y_j)$ and $(\partial_y\psi)^*(x_i,y_j)$, respectively. At each interior point (x_i,y_j), $1 \le i,j \le N-1$, let $Q(x,y)$ be a fourth-order polynomial $Q(x,y)$ of the following form

(11.13)
$$\begin{aligned} Q(x,y) = a_{0,0} &+ a_{1,0}(x-x_i) + a_{0,1}(y-y_j) \\ &+ a_{2,0}(x-x_i)^2 + a_{1,1}(x-x_i)(y-y_j) + a_{0,2}(y-y_j)^2 \\ &+ a_{3,0}(x-x_i)^3 + a_{2,1}(x-x_i)^2(y-y_j) \\ &+ a_{1,2}(x-x_i)(y-y_j)^2 + a_{0,3}(y-y_j)^3 \\ &+ a_{4,0}(x-x_i)^4 + a_{2,2}(x-x_i)^2(y-y_j)^2 + a_{0,4}(y-y_j)^4, \end{aligned}$$

which is in the span of the 13 monomials

(11.14)
$$\mathcal{V} = \{1, x, y, x^2, y^2, xy, x^3, x^2y, xy^2, y^3, x^4, x^2y^2, y^4\}$$

shifted to (x_i,y_j). We require that Q satisfies the 13 conditions (see Figure 11.2),

(11.15)
- $Q(x_l,y_m) = \psi^*(x_l,y_m), \quad |l-i| \le 1, |m-j| \le 1,$ 9 equations,
- $\partial_x Q(x_l,y_j) = \psi^*_{x,l,j}, \quad l = i \pm 1,$ 2 equations,
- $\partial_y Q(x_i,y_m) = \psi^*_{y,i,m}, \quad m = j \pm 1,$ 2 equations.

It can be verified that a polynomial of this form must vanish if all the 13 data vanish. Thus, there is a unique polynomial $Q(x,y)$ satisfying (11.15).

Its coefficients are given by

(11.16)

$$(a)\quad a_{0,0} = \psi^*_{i,j},$$

$$(b)\quad a_{1,0} = \frac{3}{2}\delta_x\psi^*_{i,j} - \frac{1}{4}(\psi^*_{x,i+1,j} + \psi^*_{x,i-1,j}),$$

$$a_{0,1} = \frac{3}{2}\delta_y\psi^*_{i,j} - \frac{1}{4}(\psi^*_{y,i,j+1} + \psi^*_{y,i,j-1}),$$

$$(c)\quad a_{2,0} = \delta_x^2\psi^*_{i,j} - \frac{1}{2}(\delta_x\psi^*_x)_{i,j},\quad a_{0,2} = \delta_y^2\psi^*_{i,j} - \frac{1}{2}(\delta_y\psi^*_y)_{i,j},$$

$$a_{1,1} = \delta_x\delta_y\psi^*_{i,j},$$

$$(d_1)\quad a_{3,0} = \frac{1}{6}(\delta_x^2\psi^*_x)_{i,j},\quad a_{0,3} = \frac{1}{6}(\delta_y^2\psi^*_y)_{i,j},$$

$$(d_2)\quad a_{2,1} = \frac{1}{2}(\delta_x^2\delta_y\psi^*)_{i,j},\quad a_{1,2} = \frac{1}{2}(\delta_y^2\delta_x\psi^*)_{i,j},$$

$$(e_1)\quad a_{4,0} = \frac{1}{2h^2}\left((\delta_x\psi^*_x)_{i,j} - \delta_x^2\psi^*_{i,j}\right),\quad a_{0,4} = \frac{1}{2h^2}\left((\delta_y\psi^*_y)_{i,j} - \delta_y^2\psi^*_{i,j}\right),$$

$$(e_2)\quad a_{2,2} = \frac{1}{4}(\delta_x^2\delta_y^2\psi^*)_{i,j}.$$

Note that in (d_1) the values of $\psi^*_{x,i,j}$ and $\psi^*_{y,i,j}$ are needed. They are expressed in terms of the neighboring values as in Equation (11.18) below. As in the one-dimensional case [see (10.52)], natural candidates for $\psi^*_{x,i,j}$ and $\psi^*_{y,i,j}$ are

$$(11.17)\qquad \begin{cases} \psi^*_{x,i,j} = \partial_x Q(x_i, y_j) = a_{1,0}, \\ \psi^*_{y,i,j} = \partial_y Q(x_i, y_j) = a_{0,1}. \end{cases}$$

Thus (11.16)(b) shows that ψ^*_x and ψ^*_y are connected to the data $\psi^*(x_i, y_j)$ by

$$(11.18)\qquad \begin{cases} \frac{1}{6}\psi^*_{x,i-1,j} + \frac{2}{3}\psi^*_{x,i,j} + \frac{1}{6}\psi^*_{x,i+1,j} = \delta_x\psi^*_{i,j}, \\ \frac{1}{6}\psi^*_{y,i,j-1} + \frac{2}{3}\psi^*_{y,i,j} + \frac{1}{6}\psi^*_{y,i,j+1} = \delta_y\psi^*_{i,j}. \end{cases}$$

According to (11.9), this means that ψ^*_x and ψ^*_y *should be taken as the Hermitian derivatives of* ψ^*. The discrete biharmonic operator is then defined by evaluating $\Delta^2 Q$ at (x_i, y_j):

$$(11.19)\qquad \Delta^2 Q(x_i, y_j) = 24a_{4,0} + 8a_{2,2} + 24a_{0,4}.$$

From Equation (11.16) we find that $24a_{4,0} = \delta_x^4\psi^*$, $24a_{0,4} = \delta_y^4\psi^*$, and $8a_{2,2} = 2\delta_x^2\delta_y^2\psi^*$, thus we observe that $\Delta^2 Q(x_i, y_j) = \Delta_h^2\psi^*_{i,j}$, as given in (11.11). Finally, Equation (11.11) can be explicitly written as

$$\Delta_h^2 \psi_{i,j} = \frac{1}{h^4}\Big(56\psi_{i,j} - 16\left(\psi_{i+1,j} + \psi_{i,j+1} + \psi_{i-1,j} + \psi_{i,j-1}\right) \tag{11.20}$$
$$+2\left(\psi_{i+1,j+1} + \psi_{i-1,j+1} + \psi_{i-1,j-1} + \psi_{i+1,j-1}\right)$$
$$+6h\Big((\psi_x)_{i+1,j} - (\psi_x)_{i-1,j} + (\psi_y)_{i,j+1} - (\psi_y)_{i,j-1}\Big)\Big).$$

Using the biharmonic equation (11.1), we now introduce the two-dimensional Stephenson scheme, which generalizes the one-dimensional scheme introduced in Definition 10.6.

Definition 11.2 (Two-dimensional Stephenson scheme). *The approximation to the biharmonic problem (11.1), given by*

$$\Delta_h^2 \psi_{i,j} = f^*(x_i, y_j), \quad 1 \le i,j \le N-1, \tag{11.21}$$

and subject to the boundary conditions

$$\begin{cases} \psi_{i,j} = g_1^*(x_i, y_j), \ \{i = 0, N, \quad 0 \le j \le N\} \\ \qquad\qquad\qquad \text{or } \{j = 0, N, \quad 0 \le i \le N\}, \\ \psi_{x,i,j} = -g_2^*(x_i, y_j), \quad i = 0, \quad 0 \le j \le N, \\ \psi_{x,i,j} = g_2^*(x_i, y_j), \quad i = N, \quad 0 \le j \le N, \\ \psi_{y,i,j} = -g_2^*(x_i, y_j), \quad j = 0, \quad 0 \le i \le N, \\ \psi_{y,i,j} = g_2^*(x_i, y_j), \quad j = N, \quad 0 \le i \le N, \end{cases} \tag{11.22}$$

is called the ***two-dimensional Stephenson scheme*** *for the problem (see Figure 11.2).*

Remark 11.3. Values of ψ_x (respectively ψ_y) on the horizontal (respectively vertical) sides are obtained from g_2^*.

Proposition 11.4 (Commutation of the x and y derivatives). *The mixed Hermitian derivatives satisfy the equality*

$$(\psi_y)_x = (\psi_x)_y. \tag{11.23}$$

Proof. We have $(\psi_y)_x = \sigma_x^{-1}\delta_x(\sigma_y)^{-1}\delta_y\psi$ and $(\psi_x)_y = \sigma_y^{-1}\delta_y(\sigma_x)^{-1}\delta_x\psi$. Since the operators $\sigma_x^{-1}\delta_x$ and $\sigma_y^{-1}\delta_y$ clearly commute, the result is proved. Here we assume that $\psi_y = \psi_x = 0$ as well as $(\psi_y)_x = (\psi_x)_y = 0$ on the boundaries. ∎

Note that this proposition will be needed later in obtaining the discrete incompressibility condition.

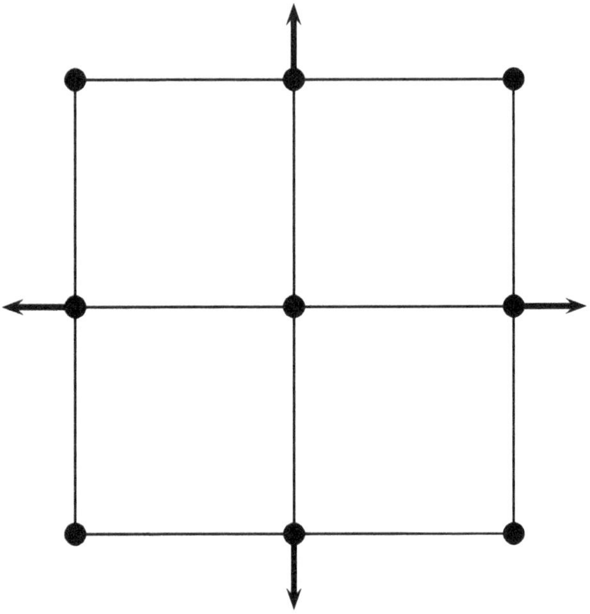

Fig. 11.2 Stencil of the 13-data compact biharmonic operator Δ_h^2.

11.1.3 *Accuracy of the discrete biharmonic operator*

In Chapter 10 (see Proposition 10.8) we saw that the one-dimensional three-point biharmonic operator is fourth-order accurate at interior points but only first-order accurate at near-boundary points. In analogy with this proposition, we have the following result.

Claim 11.5 (Accuracy of the nine-point biharmonic). *Let ψ be a smooth function in Ω and suppose that it vanishes along with its gradient on $\partial\Omega$. Then the nine-point biharmonic operator Δ_h^2 satisfies the following accuracy properties:*

- *At interior points $2 \leq i, j \leq N-2$, we have*

(11.24)
$$|\sigma_x\sigma_y\Delta_h^2\psi_{i,j}^* - \sigma_x\sigma_y(\Delta^2\psi)_{i,j}^*| \leq Ch^4\left(\|\partial_x^8\psi\|_{L^\infty} + \|\partial_y^8\psi\|_{L^\infty}\right)$$
$$+ Ch^2(\|\partial_x^2\partial_y^4\psi\|_{L^\infty} + \|\partial_y^2\partial_x^4\psi\|_{L^\infty}$$
$$+ \|\partial_x^6\psi\|_{L^\infty} + \|\partial_y^6\psi\|_{L^\infty}).$$

- *At near-boundary points $i = 1, N-1,\ \ 1 \leq j \leq N-1$, or $j = 1, N-1$,*

$1 \le i \le N-1$, *we have*

(11.25)
$$|\sigma_x\sigma_y\Delta_h^2\psi_{i,j}^* - \sigma_x\sigma_y(\Delta^2\psi)_{i,j}^*|$$
$$\le Ch\left(\|\partial_x^5\psi\|_{L^\infty} + \|\partial_y^5\psi\|_{L^\infty}\right)$$
$$+ Ch^2\left(\|\partial_x^2\partial_y^4\psi\|_{L^\infty} + \|\partial_y^2\partial_x^4\psi\|_{L^\infty} + \|\partial_x^6\psi\|_{L^\infty} + \|\partial_y^6\psi\|_{L^\infty}\right).$$

In the above estimates $C > 0$ is a constant independent of h and ψ.

Proof. In view of (10.64), we have for $2 \le i \le N-2$, $1 \le j \le N-1$,

(11.26) $$|\sigma_x\delta_x^4\psi_{i,j}^* - \sigma_x(\partial_x^4\psi)_{i,j}^*| \le Ch^4\|\partial_x^8\psi\|_{L^\infty},$$

and similarly for the y direction.

Expanding the left-hand side of (11.24), we have for $2 \le i,j \le N-2$

(11.27) $$\sigma_x\sigma_y(\Delta_h^2\psi_{i,j}^* - (\Delta^2\psi)_{i,j}^*) = \sigma_y(\sigma_x\delta_x^4\psi_{i,j}^* - \sigma_x(\partial_x^4\psi)_{i,j}^*)$$
$$+ \sigma_x(\sigma_y\delta_y^4\psi_{i,j}^* - \sigma_y(\partial_y^4\psi)_{i,j}^*)$$
$$+ 2\sigma_x\sigma_y(\delta_x^2\delta_y^2\psi_{i,j}^* - (\partial_x^2\partial_y^2\psi)_{i,j}^*).$$

Let us first consider $\sigma_y(\sigma_x\delta_x^4\psi_{i,j}^* - \sigma_x(\partial_x^4\psi)_{i,j}^*)$. Using Equation (11.8) and applying (11.26) to j and $j \pm 1$, we get

(11.28) $$|\sigma_y(\sigma_x\delta_x^4\psi_{i,j}^* - \sigma_x(\partial_x^4\psi)_{i,j}^*)| \le Ch^4\|\partial_x^8\psi\|_{L^\infty}, \quad 2 \le i,j \le N-2.$$

Similarly,

(11.29) $$|\sigma_x(\sigma_y\delta_y^4\psi_{i,j}^* - \sigma_y(\partial_y^4\psi)_{i,j}^*)| \le Ch^4\|\partial_y^8\psi\|_{L^\infty}, \quad 2 \le i,j \le N-2.$$

To estimate the truncation error for the mixed term $\partial_x^2\partial_y^2\psi$, we use (9.12) and rewrite it for $1 \le i \le N-1$, $0 \le j \le N$, as

(11.30) $$\delta_x^2\psi_{i,j}^* - (\partial_x^2\psi)_{i,j}^* = \frac{h^2}{12}(\partial_x^4\psi)_{i,j}^* + r_{i,j},$$
$$|r_{i,j}| \le Ch^4\|\partial_x^6\psi\|_{L^\infty}.$$

Operating with δ_y^2 on this equation yields for $1 \le i,j \le N-1$,

(11.31) $$\delta_y^2[\delta_x^2\psi_{i,j}^* - (\partial_x^2\psi)_{i,j}^*] = \frac{h^2}{12}\delta_y^2(\partial_x^4\psi)_{i,j}^* + s_{i,j},$$
$$|s_{i,j}| \le Ch^2\|\partial_x^6\psi\|_{L^\infty}.$$

Using a Taylor expansion for the terms $\delta_y^2(\partial_x^2\psi)^*$ and $\delta_y^2(\partial_x^4\psi)^*$, we get for $1 \le i,j \le N-1$,

(11.32)
$$(a) \quad \delta_y^2(\partial_x^2\psi)_{i,j}^* = (\partial_y^2\partial_x^2\psi)_{i,j}^* + \frac{h^2}{12}(\partial_y^4\partial_x^2\psi)(x_i,\eta_j),$$
$$(b) \quad \delta_y^2(\partial_x^4\psi)_{i,j}^* = \partial_y^2(\partial_x^4\psi)^*(x_i,\tilde{\eta}_j),$$

where $\eta_j, \tilde{\eta}_j \in (y_{j-1}, y_{j+1})$. Inserting (11.32) into (11.31), we obtain for $1 \le i, j \le N-1$,

$$\delta_y^2\delta_x^2\psi^*_{i,j} - (\partial_y^2\partial_x^2\psi)^*_{i,j} = \frac{h^2}{12}\left[(\partial_y^4\partial_x^2\psi)(x_i,\eta_j) + (\partial_y^2\partial_x^4\psi)(x_i,\tilde{\eta}_j)\right] + s_{i,j},$$

$$|s_{i,j}| \le Ch^2\|\partial_x^6\psi\|_{L^\infty}.$$

We may estimate therefore the mixed term in (11.27) for $2 \le i, j \le N-2$ by

$$\begin{aligned}&|\sigma_x\sigma_y(\delta_y^2\delta_x^2\psi^*_{i,j} - (\partial_y^2\partial_x^2\psi)^*_{i,j})|\\ &\le Ch^2\big(\|\partial_y^4\partial_x^2\psi\|_{L^\infty} + \|\partial_y^2\partial_x^4\psi\|_{L^\infty} + \|\partial_x^6\psi\|_{L^\infty}\big).\end{aligned} \tag{11.33}$$

Similarly,

$$\begin{aligned}&|\sigma_x\sigma_y(\delta_x^2\delta_y^2\psi^*_{i,j} - (\partial_x^2\partial_y^2\psi)^*_{i,j})|\\ &\le Ch^2\big(\|\partial_x^4\partial_y^2\psi\|_{L^\infty} + \|\partial_x^2\partial_y^4\psi\|_{L^\infty} + \|\partial_y^6\psi\|_{L^\infty}\big).\end{aligned} \tag{11.34}$$

Combining (11.28), (11.29), (11.33) and (11.34), we obtain (11.24).

At near-boundary points $i = 1$ or $i = N-1$ we have for any fixed $j = 0, \ldots, N$ [see (10.65)]

$$|\sigma_x\delta_x^4\psi^*_{i,j} - \sigma_x(\partial_x^4\psi)^*_{i,j}| \le Ch\|\partial_x^5\psi\|_{L^\infty}. \tag{11.35}$$

Operating with σ_y on (11.35), we have, for $i = 1, N-1$, $1 \le j \le N-1$,

$$|\sigma_y\sigma_x\delta_x^4\psi^*_{i,j} - \sigma_y\sigma_x(\partial_x^4\psi)^*_{i,j}| \le Ch\|\partial_x^5\psi\|_{L^\infty}. \tag{11.36}$$

Similarly, for $j = 1, N-1, 1 \le i \le N-1$,

$$|\sigma_x\sigma_y\delta_y^4\psi^*_{i,j} - \sigma_x\sigma_y(\partial_y^4\psi)^*_{i,j}| \le Ch\|\partial_y^5\psi\|_{L^\infty}. \tag{11.37}$$

Combining (11.36) and (11.37) with (11.33) and (11.34), we obtain (11.25). ■

In analogy with the one-dimensional case [see (10.84], we define the truncation error by

$$\mathfrak{t}_{i,j} = \Delta_h^2\psi^*_{i,j} - (\Delta^2\psi)^*_{i,j}, \qquad 1 \le i, j \le N-1. \tag{11.38}$$

As in Proposition 10.8, we can estimate this error in the energy norm as follows.

Proposition 11.6. *In the energy norm the truncation error satisfies*

$$\begin{aligned}|\mathfrak{t}|_h \le Ch^{3/2}\Big(&\|\partial_x^5\psi\|_{L^\infty} + \|\partial_x^6\psi\|_{L^\infty} + \|\partial_x^8\psi\|_{L^\infty}\\ &+ \|\partial_x^4\partial_y^2\psi\|_{L^\infty} + \|\partial_y^4\partial_x^2\psi\|_{L^\infty}\\ &+\|\partial_y^5\psi\|_{L^\infty} + \|\partial_y^6\psi\|_{L^\infty} + \|\partial_y^8\psi\|_{L^\infty}\Big).\end{aligned} \tag{11.39}$$

Proof. We have, because of Claim 11.5,

$$|\sigma_x\sigma_y \mathfrak{t}|_h^2 \le Ch^2\Big\{4(N-1)h^2\left[\|\partial_x^5\psi\|_{L^\infty} + \|\partial_y^5\psi\|_{L^\infty}\right]^2$$
$$+ (N-2)^2h^4\left[\|\partial_x^2\partial_y^4\psi\|_{L^\infty} + \|\partial_x^4\partial_y^2\psi\|_{L^\infty} + \|\partial_x^6\psi\|_{L^\infty} + \|\partial_y^6\psi\|_{L^\infty}\right]^2$$
$$+ (N-2)^2h^8\left[\|\partial_x^8\psi\|_{L^\infty} + \|\partial_y^8\psi\|_{L^\infty}\right]^2\Big\}.$$

Since $N = 1/h$ and $|(\sigma_x\sigma_y)^{-1}| \le 9$ [see the discussion around (10.87)], we obtain (11.39). ■

11.1.4 *Coercivity and convergence properties of the discrete biharmonic operator*

In Section 10.5 we established the identity (10.97) for the scalar product $\langle\cdot,\cdot\rangle_h$. It led to the coercivity property (Proposition 10.11) and served as a fundamental tool in the proof of the convergence of the discrete solution to the exact one.

Our goal here is to establish a similar result for the nine-point biharmonic operator Δ_h^2. Let $\psi,\varphi \in L_{h,0}^2$ and assume in addition that all four Hermitian derivatives $\psi_x, \psi_y, \varphi_x$ and φ_y are also in $L_{h,0}^2$. We define the operators $\delta_x^+, \delta_y^+, \mu_x^+$ and μ_y^+ as

(11.40)
$$\delta_x^+\psi_{i,j} = \frac{\psi_{i+1,j}-\psi_{i,j}}{h},\ 0 \le i \le N-1, \quad 0 \le j \le N,$$
$$\delta_x^+\psi_{N,j} = 0, \qquad 0 \le j \le N,$$

and

(11.41)
$$\mu_x^+\psi_{i,j} = \frac{1}{2}\frac{\psi_{i+1,j}+\psi_{i,j}}{h},\ 0 \le i \le N-1, \quad 0 \le j \le N,$$
$$\mu_x^+\psi_{N,j} = 0, \qquad 0 \le j \le N,$$

with analogous definitions for δ_y^+ and μ_y^+. For any fixed $j = 1,\ldots,N-1$, we denote by $\psi_{:,j} \in l_{h,0}^2$ the one-dimensional grid function

(11.42)
$$\psi_{:,j} = (\psi_{1,j},\ldots,\psi_{N-1,j}),$$

so that the two-dimensional $L_{h,0}^2$ scalar product [see (8.23)] can be expressed as a sum of the one-dimensional products

(11.43)
$$(\psi,\varphi)_h = h\sum_{j=1}^{N-1}(\psi_{:,j},\varphi_{:,j})_h.$$

Using (10.97) we obtain

$$(\delta_x^4 \boldsymbol{\psi}, \boldsymbol{\varphi})_h = h \sum_{j=1}^{N-1} (\delta_x^4 \boldsymbol{\psi}_{:,j}, \boldsymbol{\varphi}_{:,j})_h$$

(11.44)
$$= h \sum_{j=1}^{N-1} \left[\left(\delta_x^+ (\boldsymbol{\psi}_x)_{:,j}, \delta_x^+ (\boldsymbol{\varphi}_x)_{:,j} \right)_h \right.$$
$$\left. + \frac{12}{h^2} \left(\delta_x^+ \boldsymbol{\psi}_{:,j} - \mu_x^+ \boldsymbol{\psi}_{x,:,j}, \delta_x^+ \boldsymbol{\varphi}_{:,j} - \mu_x^+ \boldsymbol{\varphi}_{x,:,j} \right)_h \right].$$

We note that $(\delta_x^+ \boldsymbol{\psi}_x)_{0,j} = (1/h)(\boldsymbol{\psi}_x)_{1,j}$ so that $\delta_x^+ \boldsymbol{\psi}_x$ is not in $L_{h,0}^2$ and the scalar product $\left(\delta_x^+ (\boldsymbol{\psi}_x)_{:,j}, \delta_x^+ (\boldsymbol{\varphi}_x)_{:,j} \right)_h$ includes also the term $h(\delta_x^+ \boldsymbol{\psi}_x)_{0,j}(\delta_x^+ \boldsymbol{\varphi}_x)_{0,j}$. The same comment applies also to the other inner product. Hence, switching back to the two-dimensional scalar product on the right-hand side of the equation above,

(11.45) $(\delta_x^4 \boldsymbol{\psi}, \boldsymbol{\varphi})_h = (\delta_x^+ \boldsymbol{\psi}_x, \delta_x^+ \boldsymbol{\varphi}_x)_h + \dfrac{12}{h^2} \left(\delta_x^+ \boldsymbol{\psi} - \mu_x^+ \boldsymbol{\psi}_x, \delta_x^+ \boldsymbol{\varphi} - \mu_x^+ \boldsymbol{\varphi}_x \right)_h.$

Similarly, in the y direction,

(11.46) $(\delta_y^4 \boldsymbol{\psi}, \boldsymbol{\varphi})_h = (\delta_y^+ \boldsymbol{\psi}_y, \delta_y^+ \boldsymbol{\varphi}_y)_h + \dfrac{12}{h^2} \left(\delta_y^+ \boldsymbol{\psi} - \mu_y^+ \boldsymbol{\psi}_y, \delta_y^+ \boldsymbol{\varphi} - \mu_y^+ \boldsymbol{\varphi}_y \right)_h.$

In addition, for the mixed term we have [compare with (9.32)],

(11.47) $\qquad (\delta_x^2 \delta_y^2 \boldsymbol{\psi}, \boldsymbol{\varphi})_h = (\delta_x^+ \delta_y^+ \boldsymbol{\psi}, \delta_x^+ \delta_y^+ \boldsymbol{\varphi})_h.$

Combining (11.45), (11.46), and (11.47), we obtain the identity

(11.48)
$$(\Delta_h^2 \boldsymbol{\psi}, \boldsymbol{\varphi})_h = (\delta_x^+ \boldsymbol{\psi}_x, \delta_x^+ \boldsymbol{\varphi}_x)_h + (\delta_y^+ \boldsymbol{\psi}_y, \delta_y^+ \boldsymbol{\varphi}_y)_h$$
$$+ \frac{12}{h^2} \left(\delta_x^+ \boldsymbol{\psi} - \mu_x^+ \boldsymbol{\psi}_x, \delta_x^+ \boldsymbol{\varphi} - \mu_x^+ \boldsymbol{\varphi}_x \right)_h$$
$$+ \frac{12}{h^2} \left(\delta_y^+ \boldsymbol{\psi} - \mu_y^+ \boldsymbol{\psi}_y, \delta_y^+ \boldsymbol{\varphi} - \mu_y^+ \boldsymbol{\varphi}_y \right)_h$$
$$+ 2(\delta_x^+ \delta_y^+ \boldsymbol{\psi}, \delta_x^+ \delta_y^+ \boldsymbol{\varphi})_h = (\boldsymbol{\psi}, \Delta_h^2 \boldsymbol{\varphi})_h.$$

The symmetric character of this equation allows us to define, as in the one-dimensional case, the scalar product

(11.49) $\qquad\qquad \langle \boldsymbol{\psi}, \boldsymbol{\varphi} \rangle_h = (\Delta_h^2 \boldsymbol{\psi}, \boldsymbol{\varphi})_h.$

The identity (11.48) implies the coercivity of Δ_h^2, namely, that this scalar product yields a *positive definite* bilinear form. The presence of the mixed term here makes the proof somewhat more complicated than that of Proposition 10.11.

Proposition 11.7. *Let $\psi, \psi_x, \psi_y \in L^2_{h,0}$ (respectively $\varphi, \varphi_x, \varphi_y \in L^2_{h,0}$), where (ψ_x, ψ_y) (respectively (φ_x, φ_y)) is the Hermitian gradient of ψ (respectively φ). Then, the scalar product $\langle \psi, \varphi \rangle_h$ is positive definite. In particular, the discrete operator Δ_h^2 is symmetric positive definite in $L^2_{h,0}$ and it satisfies*

$$\langle \psi, \psi \rangle_h \geq C\Big(|\delta_x^+ \psi_x|_h^2 + |\delta_y^+ \psi_y|_h^2 + |\delta_x^+ \psi_y|_h^2 + |\delta_y^+ \psi_x|_h^2 \Big), \tag{11.50}$$

as well as

$$\langle \psi, \psi \rangle_h \geq C\Big(|\delta_x^+ \psi|_h^2 + |\delta_y^+ \psi|_h^2 \Big), \tag{11.51}$$

$$\langle \psi, \psi \rangle_h \geq C |\psi|_h^2, \tag{11.52}$$

where C is a positive constant independent of h and ψ.

Proof. It follows from (11.48) that $\langle \psi, \varphi \rangle_h$ as defined in (11.49) is symmetric on $L^2_{h,0}$ and satisfies

$$(\Delta_h^2 \psi, \psi)_h \geq |\delta_x^+ \psi_x|_h^2 + |\delta_y^+ \psi_y|_h^2 + 2 |\delta_x^+ \delta_y^+ \psi|_h^2. \tag{11.53}$$

In order to establish (11.50) we will show that the mixed term $\delta_x^+ \delta_y^+ \psi$ satisfies

$$|\delta_x^+ \delta_y^+ \psi|_h \geq \frac{1}{6} |\delta_x^+ \psi_y|_h. \tag{11.54}$$

Using the identity $\delta_y^+ \psi_{i,j} = \delta_y \psi_{i,j} + \frac{h}{2} \delta_y^2 \psi_{i,j}$ for $i = 0, ..., N-1$ and $j = 1, \ldots, N-1$, and the relation $\sigma_y \psi_y = \delta_y \psi$ for $j = 1, \ldots, N-1$ and $i = 0, \ldots, N-1$, we deduce that

$$\begin{aligned} \delta_y^+ \psi_{i,j} &= \delta_y \psi_{i,j} + \frac{h}{2} \delta_y^2 \psi_{i,j} \\ &= \left(I + \frac{h^2}{6} \delta_y^2 \right) \psi_{y,i,j} + \frac{h}{2} \delta_y^2 \psi_{i,j}, \quad i = 0, \ldots, N, \quad j = 1, \ldots, N-1. \end{aligned}$$

Therefore, for $i = 0, \ldots, N-1$ and $j = 1, \ldots, N-1$,

$$\delta_x^+ \delta_y^+ \psi_{i,j} = \delta_x^+ \psi_{y,i,j} + \frac{h^2}{6} \delta_y^2 \delta_x^+ \psi_{y,i,j} + \frac{h}{2} \delta_y^2 \delta_x^+ \psi_{i,j}. \tag{11.55}$$

In addition, we have the following two estimates

$$\sum_{i=0}^{N-1} \sum_{j=1}^{N-1} |\delta_y^2 \delta_x^+ \psi_{y,i,j}|^2 h^2 \leq \frac{16}{h^4} \sum_{i=0}^{N-1} \sum_{j=1}^{N-1} |\delta_x^+ \psi_{y,i,j}|^2 h^2 = \frac{16}{h^4} |\delta_x^+ \psi_y|_h^2,$$

$$\sum_{i=0}^{N-1} \sum_{j=1}^{N-1} |\delta_y^2 \delta_x^+ \psi_{i,j}|^2 h^2 \leq \frac{4}{h^2} \sum_{i=0}^{N-1} \sum_{j=0}^{N-1} |\delta_y^+ \delta_x^+ \psi_{i,j}|^2 h^2 = \frac{4}{h^2} |\delta_y^+ \delta_x^+ \psi|_h^2.$$

Indeed, the first inequality follows from the definition of δ_y^2 and the inequality $(x + 2y + z)^2 \leq 4(x^2 + 2y^2 + z^2)$. The second is obtained by noting that $\delta_y^2\delta_x^+\boldsymbol{\psi}_{i,j} = \left(\delta_y^+\delta_x^+\boldsymbol{\psi}_{i,j} - \delta_y^+\delta_x^+\boldsymbol{\psi}_{i,j-1}\right)/h$ for $j = 1, \ldots, N-1$ and $i = 0, \ldots, N-1$. Therefore, reverting to norms in L_h^2, we infer from (11.55) that

$$|\delta_x^+\delta_y^+\boldsymbol{\psi}|_h \geq |\delta_x^+\boldsymbol{\psi}_y|_h - \frac{2}{3}|\delta_x^+\boldsymbol{\psi}_y|_h - |\delta_x^+\delta_y^+\boldsymbol{\psi}|_h,$$

which yields $2|\delta_x^+\delta_y^+\boldsymbol{\psi}|_h^2 \geq \frac{1}{3}|\delta_x^+\boldsymbol{\psi}_y|_h^2$. Therefore,

$$|\delta_x^+\delta_y^+\boldsymbol{\psi}|_h \geq \frac{1}{6}|\delta_x^+\boldsymbol{\psi}_y|_h, \tag{11.56}$$

as claimed in (11.54). Similarly, by symmetry we have $|\delta_x^+\delta_y^+\boldsymbol{\psi}|_h \geq \frac{1}{6}|\delta_y^+\boldsymbol{\psi}_x|_h$. Thus, the estimate (11.53) yields

$$\langle\boldsymbol{\psi}, \boldsymbol{\psi}\rangle_h \geq |\delta_x^+\boldsymbol{\psi}_x|_h^2 + |\delta_y^+\boldsymbol{\psi}_y|_h^2 + \frac{1}{36}(|\delta_x^+\boldsymbol{\psi}_y|_h^2 + |\delta_y^+\boldsymbol{\psi}_x|_h^2), \tag{11.57}$$

as stated in (11.50). The proof of estimate (11.51) follows directly by applying (10.101) in the x and y directions. Finally, inequality (11.52) follows from (11.51) and the Poincaré inequality (9.35). ∎

We can now establish an error estimate for the scheme (Definition 11.2) analogous to Theorem 10.14.

Proposition 11.8. *Let ψ^* be the grid function corresponding to the exact solution $\psi(x, y)$ of (11.1) and let $\boldsymbol{\psi} \in L_h^2$ be the discrete solution of (11.21), then the following estimate holds,*

$$\begin{aligned}
&\langle\psi^* - \boldsymbol{\psi}, \psi^* - \boldsymbol{\psi}\rangle_h^{1/2} \\
&\quad \leq Ch^{3/2}\Big(\|\partial_x^5\psi\|_{L^\infty} + \|\partial_x^6\psi\|_{L^\infty} + \|\partial_x^8\psi\|_{L^\infty} + \|\partial_x^4\partial_y^2\psi\|_{L^\infty} \\
&\qquad + \|\partial_y^5\psi\|_{L^\infty} + \|\partial_y^6\psi\|_{L^\infty} + \|\partial_y^8\psi\|_{L^\infty} + \|\partial_y^4\partial_x^2\psi\|_{L^\infty}\Big).
\end{aligned} \tag{11.58}$$

Proof. The proof follows the same lines as that of Theorem 10.14. ∎

11.2 The biharmonic problem in an irregular domain

In the preceding section, we treated the biharmonic equation in a rectangular domain. This section considers biharmonic operators in general simply connected domains, such as circles, ellipses etc. As a by-product, we get an extension of the nine-point biharmonic operator to obtain fourth-order accuracy in the regular case [see (11.81)].

Let Ω be a simply connected domain with smooth boundary. Let ψ be a smooth function defined on $\overline{\Omega}$. We consider again the problem (11.1). Our goal is to define a discrete approximation of the problem, which will be obtained, as in the rectangular case, by means of an interpolating polynomial [compare with Equation (11.19)]. However, such a polynomial will now depend on the geometric character of the grid. As we shall see, the "underlying philosophy" is that of a *compact nine-point stencil.* The corresponding polygon of eight vertices surrounding a central point will be a regular one (as in the rectangular case) for interior points. On the other hand, near the boundary, the polygon will be "distorted," with some of its vertices lying on the boundary, at varying distances from the center. Nonetheless, the most important feature of our geometric construction is that *in all cases* the vertices are located either on the vertical/horizontal lines through the center (parallel to the coordinate axes) or on the principal diagonals through the center (bisecting the right angles between the former lines). Thus, even though the interpolating polynomials are of sixth order, hence different from the fourth-order polynomials discussed in the previous section, they will be mostly "canonical," independent of the particular geometric setting (see Remark 11.20).

11.2.1 *Embedding a Cartesian grid in an irregular domain*

Consider[1] a domain embedded in a large uniform grid of mesh size h. A grid point is a point (ih, jh) for $i, j \in \mathbb{Z}$. Most of the nodes, which are interior to Ω, will be designated as "calculated nodes," i.e. nodes where the approximate functional values are actually calculated. These values are ψ, ψ_x and ψ_y, which serve as approximate values to the analytical values of ψ and $\nabla\psi$ at the nodes. A small number of interior nodes close to $\partial\Omega$ are not calculated nodes and are only used in the construction of the scheme. There are no approximate values associated with these nodes and we label them as "edge nodes." In Figure 11.3 these nodes are marked with an "x".

The division of the interior nodes between edge nodes and calculated nodes is a parameter of the scheme. Some exterior nodes that are close to $\partial\Omega$ are used in the geometric phase of the scheme, as explained below. They are not used in the calculations, i.e. they do not carry approximate

[1]The reader who is interested in the convergence proof appearing in Chapter 12 can skip this section.

values or serve as ghost points. In Figure 11.3 they are shown as simple dots "·".

The essential step is to determine the nodes that are designated as "edge nodes." They are selected to limit the distortion of the irregular polygons. More specifically, only nodes that are sufficiently close to the boundary are marked as "edge nodes," so that all the resulting irregular polygons (as constructed in the following paragraphs) are subject to geometry constraints [see (11.92)]. This also ensures that the distance of the edge nodes from the boundary is $O(h)$. Our scheme is a compact scheme; all approximate

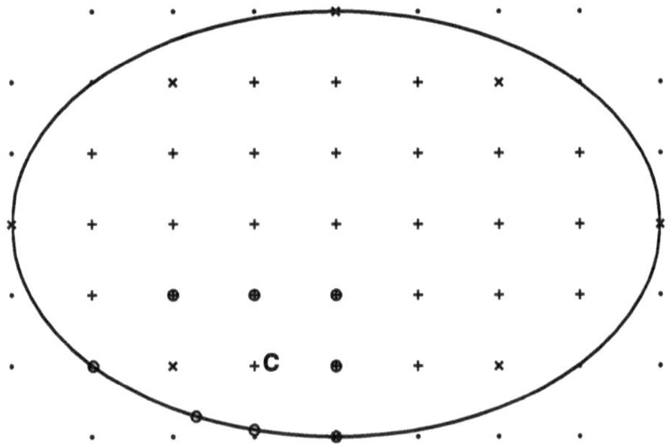

Fig. 11.3 An ellipse embedded in the grid. "+" denotes calculated nodes, "x" denotes edge nodes, and "·" denotes exterior nodes. The circled nodes are the eight neighbors of the calculated node C.

values of high-order derivatives use values of ψ, ψ_x and ψ_y at immediate neighbors. More specifically, given a node $\mathbf{p_0} = (ih, jh)$ we consider the eight grid points:

$$\tilde{\mathbf{p_1}} = ((i-1)h, (j+1)h),\ \tilde{\mathbf{p_2}} = (ih, (j+1)h),\ \tilde{\mathbf{p_3}} = ((i+1)h, (j+1)h),$$
$$\tilde{\mathbf{p_4}} = ((i-1)h, jh),\ \tilde{\mathbf{p_5}} = ((i+1)h, jh),$$
$$\tilde{\mathbf{p_6}} = ((i-1)h, (j-1)h),\ \tilde{\mathbf{p_7}} = (ih, (j-1)h),\ \tilde{\mathbf{p_8}} = ((i+1)h, (j-1)h).$$

Our goal is to construct suitable neighboring points $\mathbf{p_1}, ..., \mathbf{p_8}$, that are either calculated nodes or boundary points. Only the values of ψ, ψ_x and ψ_y at these points are needed to calculate the various approximate derivatives at $\mathbf{p_0}$.

Suppose we wish to approximate $\Delta^2\psi(x_0, y_0)$ at a calculated node $\mathbf{p_0} = (x_0, y_0)$, such that one of its neighbors $\tilde{\mathbf{p_i}} = (x_i, y_i)$ is either an edge node or an exterior node. Take the ray that begins at $\mathbf{p_0}$ and goes through $\tilde{\mathbf{p}}_\mathbf{i}$. This ray intersects the boundary at a point, which we define as $\mathbf{p_i}$ (in the case of a convex domain this point is well defined). The calculation of the approximate value to $\Delta^2\psi(x_0, y_0)$ relies on the data at $\mathbf{p_i}$ rather than $\tilde{\mathbf{p}}_\mathbf{i}$. This idea is demonstrated in Figures 11.3 and 11.4. Consider a calculated node designated by C. We construct eight neighboring points $\mathbf{p_1}, ..., \mathbf{p_8}$, which carry the data needed for the calculation at C. The four neighbors $\tilde{\mathbf{p}}_\mathbf{1}, \tilde{\mathbf{p}}_\mathbf{2}, \tilde{\mathbf{p}}_\mathbf{3}$ and $\tilde{\mathbf{p}}_\mathbf{5}$ are also calculated nodes so we use them in the calculation, i.e. $\tilde{\mathbf{p}}_\mathbf{i} = \mathbf{p_i}$. The other four neighbors $\tilde{\mathbf{p}}_\mathbf{4}, \tilde{\mathbf{p}}_\mathbf{6}, \tilde{\mathbf{p}}_\mathbf{7}$ and $\tilde{\mathbf{p}}_\mathbf{8}$ are either edge or exterior nodes so they are replaced by points on the boundary as described above. We therefore obtain the eight points $\mathbf{p_i}$ (marked with circles), which are the actual points used in the calculation.

Once the eight points $\mathbf{p_i}$ are determined and approximate values of ψ, ψ_x and ψ_y are assigned to them, we can proceed to evaluate an approximate value for $\Delta^2\psi$ at the point $\mathbf{p_0}$. This is described in the following subsection.

11.2.2 *The biharmonic $\Delta_h^2\psi$ operator*

Here we define a scheme for the approximation of the biharmonic operator. Figure 11.4 shows the nine-point irregular stencil used for the approximation of $\Delta^2\psi$ at $\mathbf{p_0} = (0, 0)$. Each of the nine grid points $\mathbf{p_i}$ carries three values ψ, ψ_x and ψ_y. These are calculated values if $\mathbf{p_i}$ is a calculated node and are given boundary data if $\mathbf{p_i}$ is a boundary point. The first step is

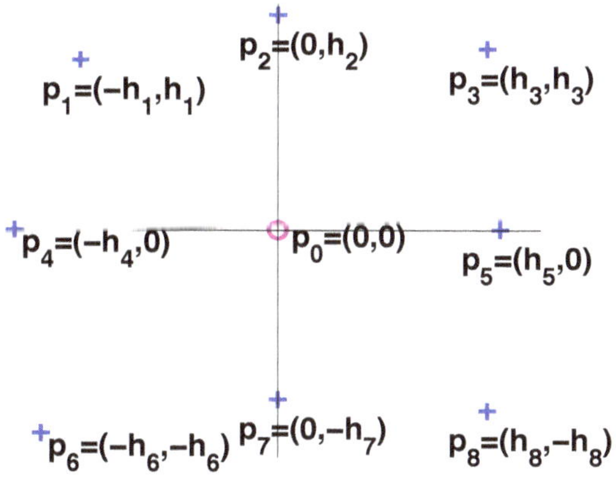

Fig. 11.4 The eight neighbors.

to obtain from ψ, ψ_x and ψ_y the data to be interpolated on the stencil $\mathbf{p_0}, ..., \mathbf{p_8}$. Our goal is to interpolate these data by a polynomial of degree six. To do so we define at $\mathbf{p_i}$, $1 \le i \le 8$ the directional derivative $\psi_d(\mathbf{p_i})$ as the derivative in the direction $\mathbf{p_i} \to \mathbf{p_0}$. Thus, for example, $\psi_d(\mathbf{p_2}) = -\psi_y(\mathbf{p_2})$ while $\psi_d(\mathbf{p_3}) = -(1/\sqrt{2})(\psi_x(\mathbf{p_3}) + \psi_y(\mathbf{p_3}))$. We thus obtain the 19 data

$$\begin{cases} \psi(\mathbf{p_i}), & 0 \le i \le 8, \\ \psi_d(\mathbf{p_i}), & 1 \le i \le 8, \\ \psi_x(\mathbf{p_0}),\ \psi_y(\mathbf{p_0}). \end{cases} \tag{11.59}$$

These are the only data used in the definition of the interpolating polynomial. To deal with an irregular stencil we set

$$\vec{\mathbf{h}} = (h_1, ..., h_8) \tag{11.60}$$

and denote by $Q_{\psi,\vec{\mathbf{h}}}$ the interpolating polynomial. The eight vertices of the irregular polygon around $\mathbf{p_0}$ are now $\mathbf{p_1} = (-h_1, h_1), \ldots, \mathbf{p_8} = (h_8, -h_8)$ (see Figure 11.4).

Definition 11.9. *The finite difference operator $\Delta^2_{\vec{\mathbf{h}}}\psi$, for the approximation of $\Delta^2\psi$ at $\mathbf{p_0} = (0,0)$, is $\Delta^2 Q_{\psi,\vec{\mathbf{h}}}(0,0)$.*

11.2.2.1 *Approximating the data using a sixth-order polynomial*

Let $\mathbb{P}^i$ be the linear space of polynomials in two variables of degree $\le i$. Let P^i be their single variable counterparts. There are 28 monomials in $\mathbb{P}^6$, so that $\dim \mathbb{P}^6 = 28$.

We designate by Dat_{19} the 19-dimensional space of the data in (11.59). The goal is to find a polynomial in $\mathbb{P}^6$ that interpolates the 19 data in the linear space $\mathrm{Dat}_{19} \cong \mathbb{R}^{19}$.

Let $A : \mathbb{P}^6 \to \mathrm{Dat}_{19}$ be the linear transformation that is the evaluation operator. That is, given $Q \in \mathbb{P}^6$,

$$A(Q) = \left\{ Q(\mathbf{p_0}), ..., Q(\mathbf{p_8}), \frac{\partial Q}{\partial d}(\mathbf{p_1}), ..., \frac{\partial Q}{\partial d}(\mathbf{p_8}), \frac{\partial Q}{\partial x}(\mathbf{p_0}), \frac{\partial Q}{\partial y}(\mathbf{p_0}) \right\}.$$

Proposition 11.10. *A is surjective, that is,* $\dim(\mathrm{Im}(A)) = 19$. *In addition,* $\dim(\ker(A)) = 9$.

Proof. Let $\mathbf{d} \in \mathrm{Dat}_{19}$ be a set of data as defined in (11.59). We will construct a polynomial $Q \in \mathbb{P}^6$ such that $A(Q) = \mathbf{d}$. Let $Q_1 \in P^5$ be the unique single-variable fifth-order polynomial such that $Q_1(x)$ interpolates the six data on the x axis,

$$\psi(\mathbf{p_4}), \psi_x(\mathbf{p_4}), \psi(\mathbf{p_0}), \psi_x(\mathbf{p_0}), \psi(\mathbf{p_5}), \psi_x(\mathbf{p_5}). \tag{11.61}$$

Let $Q_2 \in P^4$ be the unique polynomial such that $Q_1(0) + yQ_2(y)$ interpolates the following five data on the y axis,

$$\psi(\mathbf{p_2}), \psi_y(\mathbf{p_2}), \psi_y(\mathbf{p_0}), \psi(\mathbf{p_7}), \psi_y(\mathbf{p_7}). \tag{11.62}$$

In a similar fashion, let $Q_3 \in P^3$ be the unique polynomials such that $xyQ_3(x-y) + yQ_2(y) + Q_1(x)$ interpolates the values along the $y = -x$ diagonal,

$$\psi(\mathbf{p_1}), \psi_{x-y}(\mathbf{p_1}), \psi(\mathbf{p_8}), \psi_{x-y}(\mathbf{p_8}). \tag{11.63}$$

Finally, let $Q_4 \in P^3$ be the unique polynomials such that $xy(x+y)Q_4(x+y) + xyQ_3(x-y) + yQ_2(y) + Q_1(x)$ interpolates the values along the $y = x$ diagonal,

$$\psi(\mathbf{p_3}), \psi_{x+y}(\mathbf{p_3}), \psi(\mathbf{p_6}), \psi_{x+y}(\mathbf{p_6}). \tag{11.64}$$

Let $Q(x, y) = Q_1(x) + yQ_2(y) + xyQ_3(x-y) + xy(x+y)Q_4(x+y)$. At each step in the construction above, the interpolations already made are not modified. Therefore, $Q(x, y)$ interpolates all the required 19 data: $A(Q) = \mathbf{d}$. Finally in order to show that $\dim(\ker(A)) = 9$, we note that $\dim(\ker(A)) = \dim(\mathbb{P}^6) - \dim(\operatorname{Im}(A)) = 28 - 19 = 9$. ■

Remark 11.11. Note that by restricting $Q_4 \in P^2$ in the proof above, we obtain $Q(x, y) \in \mathbb{P}^5$, which can interpolate all but one of the 19 data (leaving out at most one value on the $x = y$ diagonal). This remark will be useful in the proof of Lemma 11.15.

Proposition 11.10 shows that it is always possible to interpolate the data in Dat_{19} using a polynomial in $\mathbb{P}^6$. However, this interpolation is not unique and therefore we construct a canonical (unique) interpolating polynomial. To do so, we must first disregard the polynomials in $\ker(A)$ as they are not influenced by the data (even though they might influence the biharmonic operator). Therefore, if a polynomial space Θ is chosen such that

$$\mathbb{P}^6 = \ker(A) \oplus \Theta. \tag{11.65}$$

then the linear transformation $A|_\Theta$ is one to one onto Dat_{19}. Thus, using $(A|_\Theta)^{-1}$, any set of data uniquely defines an interpolating polynomial in $\Theta \subseteq \mathbb{P}^6$.

Unfortunately, the equation $\mathbb{P}^6 = \ker(A) \oplus \Theta$ does not uniquely define Θ. Moreover, a different choice of Θ might produce different discrete biharmonic operators $\Delta^2_{\vec{\mathbf{h}}}$ [see Definition 11.9 and Example 11.19]. In other words, it is possible to have a polynomial $Q \in \mathbb{P}^6$ such that $A(Q) = 0$ (meaning that all 19 data for Q vanish) and yet $\Delta^2 Q(0,0) \neq 0$. This is shown in the following example.

Example 11.12. In the regular grid (i.e. $h_i = h$, $0 \leq i \leq 8$) the following two polynomials are in $\ker(A)$ and are not null-biharmonic:

$$\begin{aligned} G_1(x,y) &= x^2(x-h)^2(x+h)^2, \\ G_2(x,y) &= y^2(y-h)^2(y+h)^2. \end{aligned} \tag{11.66}$$

In order to construct Θ we first characterize $\ker(A)$ using the notion of indifferent polynomials.

Definition 11.13.

(i) A ***null-biharmonic*** *polynomial $Q \in \mathbb{P}^6$ is a polynomial such that $\Delta^2 Q(0,0) = 0$.*

(ii) An ***indifferent*** *polynomial is a null-biharmonic polynomial in $\ker(A) \subseteq \mathbb{P}^6$.*

(iii) An indifferent (respectively null-biharmonic) subspace L is a linear space such that $\forall\, Q \in L$, Q is indifferent (respectively null-biharmonic).

The biharmonic operator is simply a linear functional operating on $\mathbb{P}^6$. The kernel of this linear functional is the maximal null-biharmonic subspace. The importance of indifferent polynomials is demonstrated by the following proposition, which shows that Θ needs to be defined up to an indifferent subspace.

Proposition 11.14. *The finite difference operator $\Delta^2_{\vec{\mathbf{h}}}$ is identical for different Θ's [satisfying (11.65)] which are equivalent up to an indifferent subspace. To be precise, assume $\mathbb{P}^6 = \ker(A) \oplus \Theta_1$, $\mathbb{P}^6 = \ker(A) \oplus \Theta_2$ where $\Theta_2 \subseteq \Theta_1 \oplus L$, where L is an indifferent subspace, then the finite difference operators $\Delta^2_{\vec{\mathbf{h}}}$ using either Θ_1 or Θ_2 are identical.*

Proof. Let $\mathbf{d} \in \mathrm{Dat}_{19}$ be a set of data. Define $Q_1 = A|^{-1}_{\Theta_1}(\mathbf{d})$, $Q_2 = A|^{-1}_{\Theta_2}(\mathbf{d})$, and their difference by $Q = Q_1 - Q_2$. Notice that $A(Q_1) = A(Q_2) = \mathbf{d}$, hence $Q \in \ker(A)$.

Since $Q \in \Theta_1 \oplus L$ and $L \subseteq \ker(A)$, where L is an indifferent subspace, then $Q \in L$ and therefore Q is an indifferent polynomial.

This shows that $\Delta^2 Q(0,0) = 0$ and therefore $\Delta^2 Q_1(0,0) = \Delta^2 Q_2(0,0)$ proving that the scheme is equivalent for Θ_1 and Θ_2. ■

If $\ker(A)$ were an indifferent subspace then the proposition shows that any Θ satisfying (11.65) would yield an equivalent scheme for the biharmonic operator. However, this is not the case in general, as was seen in Example 11.12.

11.2.2.2 *Choice of* Θ

We need to construct Θ as a complementary subspace to $\ker(A)$ in $\mathbb{P}^6$ [see (11.65)]. We first distinguish a subspace of $\mathbb{P}^6$ which is independent of the specific structure of the polygon (namely, independent of $\vec{\mathbf{h}}$) and yet is indifferent, namely, null-biharmonic and contained in $\ker(A)$. This is the subspace

$$B = \{Q \in \mathbb{P}^6 \mid Q(x,y) = xy(x-y)(x+y)R(x,y),\ R \in \mathbb{P}^2\}. \tag{11.67}$$

The easy verification that $B \subseteq \ker(A)$ is indeed indifferent is left to the reader (compare the proof of Proposition 11.10). Restricting R to linear polynomials we obtain

$$\widetilde{B} = B \cap \mathbb{P}^5. \tag{11.68}$$

Let $\widetilde{A} = A|_{\mathbb{P}^5}$, the restriction of A to the polynomials of order five. Note that the proof of Proposition 11.10 (see also Remark 11.11) implied that by using only fifth-order polynomials we can interpolate 18 (out of 19) given data in Dat_{19}. In other words, the image of $\widetilde{A}$ (in Dat_{19}) is at least 18-dimensional. On the other hand, $\widetilde{B} \subseteq \ker(\widetilde{A})$ and $\dim(\widetilde{B}) = 3$. Since $\dim(\mathbb{P}^5) = 21$ we must have $\dim(\mathrm{Im}(\widetilde{A})) = 18$. These observations are summarized in the following lemma.

Lemma 11.15. $\widetilde{B} = \ker(\widetilde{A})$ *and in particular all fifth-order polynomials in* $\ker(\widetilde{A})$ *are null-biharmonic.*

Turning back to $\mathbb{P}^6$, we note that $\dim(\ker(A)) = 9$ while for the indifferent subspace B we have $\dim(B) = 6$. Let

$$\ker(A) = \mathcal{L}^{ind} \oplus \mathcal{L}^{res} \tag{11.69}$$

be a decomposition of $\ker(A)$ where $\mathcal{L}^{ind}$ is the maximal indifferent subspace (in particular $B \subseteq \mathcal{L}^{ind}$).

Notice that in the regular case $B \neq \mathcal{L}^{\text{ind}}$ as $xy(x^2+y^2-2h^2)^2 \in \ker(A)$ is an indifferent polynomial. Let

$$\tilde{\Theta} = \text{Span}\{1, x, x^2, x^3, x^4, x^5, y, y^2, y^3, y^4, y^5, xy, xy(x+y), xy(x+y)^2, xy(x+y)^3, xy(x-y), xy(x-y)^2, xy(x-y)^3\}. \tag{11.70}$$

$\tilde{\Theta}$ is spanned by 18 polynomials. Note that $\ker(\tilde{A}) \cap \tilde{\Theta} = 0$. Indeed, if $Q(x,y) \in \ker(\tilde{A}) \cap \tilde{\Theta} \subset \mathbb{P}^5$ then it should be divisible by $xy(x+y)(x-y)$. It is readily seen, from the definition of $\tilde{\Theta}$ (11.70) that no non-trivial Q in $\tilde{\Theta}$ is divisible by $xy(x+y)(x-y)$. Therefore, we conclude:

$$\mathbb{P}^5 = \ker(\tilde{A}) \oplus \tilde{\Theta} = \tilde{B} \oplus \tilde{\Theta}. \tag{11.71}$$

We construct Θ by adding one more polynomials of degree six to $\tilde{\Theta}$. This yields the desired 19-dimensional space. The choice of Θ determines the finite difference operator (as presented in Definition 11.9).

We note that Proposition 11.14 and Lemma 11.15 show that any choice of $\tilde{\Theta}$ satisfying (11.71) would result in an equivalent scheme.

The polynomial that should be added to $\tilde{\Theta}$ must not be in $\tilde{\Theta} \oplus \ker(A)$. In the regular case, $h_i = h$, $0 \le i \le 8$, we have

$$\mathcal{L}^{\text{ind}} = \text{Span}\{B, xy(x^2+y^2-2h^2)^2\}, \tag{11.72}$$

$$\mathcal{L}^{\text{res}} = \text{Span}\{G_1, G_2\}, \tag{11.73}$$

where $\mathcal{L}^{\text{ind}}$ and $\mathcal{L}^{\text{res}}$ were defined in (11.69) and G_1 and G_2 were defined in (11.66). It can then be easily verified that,

$$\{x^6, x^5y, x^3y^3, xy^5, y^6\} \subseteq \tilde{\Theta} \oplus \mathcal{L}^{\text{ind}} \oplus \mathcal{L}^{\text{res}}. \tag{11.74}$$

This leaves just two monomials of degree six: x^4y^2 and x^2y^4. Note that $x^4y^2 - x^2y^4 = x^2y^2(x-y)(x+y) \in B$ is an indifferent polynomial. Thus, Proposition 11.14 tells us that adding x^2y^4 or x^4y^2 to $\tilde{\Theta}$ yields an equivalent scheme. Therefore, the scheme is essentially determined for the regular case $h_i = h$, $0 \le i \le 8$. For the sake of symmetry we define

$$\Theta = \text{Span}\{\tilde{\Theta}, x^2y^2(x^2+y^2)\}. \tag{11.75}$$

Proposition 11.16. *The subspace Θ as defined in (11.75) satisfies our requirement in (11.65), namely, the direct sum $\Theta \oplus \ker(A) = \mathbb{P}^6$.*

Proof. For the regular case the claim follows from the preceding discussion. For the general case we must show that $x^2y^2(x^2+y^2) \notin \text{Span}\{\tilde{\Theta}, \ker(A)\}$.

The comment after (11.74) explains that it is equivalent to show that $x^4y^2 \notin \mathrm{Span}\{\tilde{\Theta}, \ker(A)\}$.

Assume to the contrary that

$$x^4y^2 = Q_1 + Q_2, \quad Q_1 \in \tilde{\Theta}, \; Q_2 \in \ker(A). \tag{11.76}$$

We have $A(x^4y^2) = A(Q_1)$. Consider the restrictions of these polynomials on the $x = y$ diagonal. The six data on the $x = y$ diagonal satisfy

$$\begin{aligned} \psi(\mathbf{p_i}) &= Q_1(\mathbf{p_i}), \quad i = 0, 3, 6, \\ \psi_d(\mathbf{p_i}) &= \frac{\partial Q_1}{\partial d}(\mathbf{p_i}), \quad i = 0, 3, 6. \end{aligned} \tag{11.77}$$

Therefore, the polynomial $x^6 - Q_1(x,x)$ has three double zeros and must be of the form $Cx^2(x-h_3)^2(x+h_6)^2$ where C is an arbitrary constant. We have $C = 1$ by comparing the x^6 coefficient in

$$x^6 - Q_1(x,x) = Cx^2(x-h_3)^2(x+h_6)^2. \tag{11.78}$$

A similar argument for the $x = -y$ diagonal gives

$$x^6 - Q_1(x,-x) = x^2(x-h_8)^2(x+h_1)^2. \tag{11.79}$$

Considering the restriction of (11.76) to the x axis, we have

$$\begin{aligned} Q_1(\mathbf{p_4}) &= Q_1(\mathbf{p_0}) = Q_1(\mathbf{p_5}) = 0, \\ \frac{\partial}{\partial x}Q_1(\mathbf{p_4}) &= \frac{\partial}{\partial x}Q_1(\mathbf{p_0}) = \frac{\partial}{\partial x}Q_1(\mathbf{p_5}) = 0. \end{aligned} \tag{11.80}$$

Therefore $Q_1(x,0)$ has three double zeros. Since $Q_1(x,0)$ is a polynomial of degree five we have $Q_1(x,0) \equiv 0$. A similar argument shows that $Q_1(0,y) \equiv 0$. In particular, there are no pure powers of x or y in $Q_1(x,y)$. Thus, the only second-order term in $Q_1(x,y)$ is αxy. However, Equation (11.78) [respectively (11.79)] shows that α is positive (respectively negative). This contradiction implies that such a polynomial $Q_1(x,y)$ does not exist. ■

In Subsection 11.2.4 we will give all the coefficients of the interpolating polynomial $Q \in \Theta$ in the regular case $h_i = h$, $1 \leq i \leq 8$. In particular, the coefficients of the finite difference scheme for $\Delta^2\psi$ are equivalent to the fourth-order scheme presented in [13],

$$
\begin{aligned}
\Delta^2_{\vec{\mathbf{h}}}\psi(0,0) &= \Delta^2 Q_{\psi,\vec{\mathbf{h}}}(0,0)\\
&= \frac{6}{h^4}\Big(12\psi(0,0)+\psi(-h,h)-4\psi(0,h)\\
&\quad+\psi(h,h)-4\psi(-h,0)-4\psi(h,0)\\
&\quad+\psi(-h,-h)-4\psi(0,-h)+\psi(h,-h)\Big)\\
&\quad+\frac{1}{h^3}\Big(\psi_x(-h,h)-\psi_x(h,h)-8\psi_x(-h,0)\\
&\quad+8\psi_x(h,0)+\psi_x(-h,-h)-\psi_x(h,-h)\\
&\quad-\psi_y(-h,h)+8\psi_y(0,h)-\psi_y(h,h)\\
&\quad+\psi_y(-h,-h)-8\psi_y(0,-h)+\psi_y(h,-h)\Big).
\end{aligned}
\tag{11.81}
$$

Remark 11.17. It should be emphasized that the discrete operator $\Delta^2_{\vec{\mathbf{h}}}\psi(0,0)$ given in this equation is different from the nine-point biharmonic operator $\Delta^2_h\psi$ of Definition 11.1. In fact, the latter was shown (Claim 11.5) to be second-order accurate and we shall see below (Subsection 11.2.4) that the former is fully fourth-order accurate, due to the higher accuracy in the discretization of the mixed term $\partial_x^2\partial_y^2$. For notational simplicity, we denote in this regular case $\tilde{\Delta}^2_h\psi = \Delta^2_{\vec{\mathbf{h}}}\psi(0,0)$. In fact, we shall show [see Equations (11.125) and (11.130)] that

$$
\tilde{\Delta}^2_h\psi = \Delta^2_h\psi - \frac{h^2}{6}(\delta_x^2\delta_y^4+\delta_y^2\delta_x^4)\psi.
\tag{11.82}
$$

In the irregular case we are still left with a degree of freedom in the definition of Θ. However, we proceed to show that $\Delta^2_{\vec{\mathbf{h}}}\psi$ is second-order accurate if Θ is taken as in (11.75) even in the irregular case.

11.2.3 *Accuracy of the finite difference biharmonic operator*

In this subsection we analyze the accuracy of the finite difference operator $\Delta^2_{\vec{\mathbf{h}}}$ (Definition 11.9). We consider the geometric setting as in Figure 11.4. We denote by Λ the polygon whose vertices are $\{\mathbf{p_i},\ 1\le i\le 8\}$. Let ψ be a regular function in the neighborhood of Λ, differentiable as much as needed. Taylor's formula around $(0,0)$ reads

$$\psi(x,y) = \sum_{0\le|\alpha|\le 6} \frac{D^\alpha\psi(0,0)}{\alpha!} x^{\alpha_1} y^{\alpha_2} + O(h^7)E(x,y)$$

(11.83)

$$= \tilde{\psi} + O(h^7)E(x,y),\ \tilde{\psi}\in\mathbb{P}^6, \quad \alpha=(\alpha_1,\alpha_2),$$

where

$$|E(x,y)| \le C \sup_{\Lambda} |\psi^{(7)}|.$$

Combining (11.65) and (11.69), we have

$$\mathbb{P}^6 = \Theta \oplus \mathcal{L}^{\text{ind}} \oplus \mathcal{L}^{\text{res}}. \tag{11.84}$$

We assume that all the derivatives of ψ of order ≤ 7 are bounded with a universal bound. We now express $\tilde{\psi}$ in terms of the direct sum (11.84),

$$\tilde{\psi} = \psi^\Theta + \psi^{\text{ind}} + \psi^{\text{res}},\ \ \psi^\Theta\in\Theta,\ \ \psi^{\text{ind}}\in\mathcal{L}^{\text{ind}},\ \psi^{\text{res}}\in\mathcal{L}^{\text{res}}. \tag{11.85}$$

Let $Q_{\boldsymbol{\psi},\vec{\mathbf{h}}}\in\Theta$ be the interpolating polynomial of $\boldsymbol{\psi}$ using the 19 data in Dat_{19} on the $\vec{\mathbf{h}}$ stencil [see (11.59) and (11.60)]. Recall (Definition 11.9) that $\Delta^2 Q_{\boldsymbol{\psi},\vec{\mathbf{h}}}(0,0)$ is the finite difference approximation for the biharmonic operator. Note that in general $\psi^\Theta - Q_{\boldsymbol{\psi},\vec{\mathbf{h}}}$ is not identically zero since it is the polynomial interpolating the 19 data of the term $O(h^7)E(x,y)$.

The truncation error for the biharmonic operator is given by

$$K_{\vec{\mathbf{h}}}(\psi) = \Delta^2\psi(0,0) - \Delta^2 Q_{\boldsymbol{\psi},\vec{\mathbf{h}}}(0,0). \tag{11.86}$$

We note that

$$S\in\Theta \Rightarrow K_{\vec{\mathbf{h}}}(S)=0, \tag{11.87}$$

$$R\in\mathcal{L}^{\text{ind}} \Rightarrow K_{\vec{\mathbf{h}}}(R)=0. \tag{11.88}$$

Indeed, if $S\in\Theta$ then $S = Q_{\mathbf{S},\vec{\mathbf{h}}}$ by definition. If $R\in\mathcal{L}^{\text{ind}}$ then $R\in\ker(A)$ is an indifferent polynomial and therefore $Q_{\mathbf{R},\vec{\mathbf{h}}}=0$ and $\Delta^2 R(0,0) = 0$.

Thus we are left with

$$K_{\vec{\mathbf{h}}}(\psi) = K_{\vec{\mathbf{h}}}(\psi^{\text{res}}) + O(h^7)K_{\vec{\mathbf{h}}}(E). \tag{11.89}$$

This equation shows us something we already suspected: the error of the scheme results from polynomials which are in $\ker(A)$ and are not indifferent, and from the negligible error resulting from the Taylor remainder term of order ≥ 7.

For the irregular mesh analysis we will continue to use a *grid parameter* $h > 0$ and define the ratios

$$c_i = \frac{h_i}{h}, \ \ 0 < c_i, \tag{11.90}$$

$$\vec{\mathbf{c}} = \{c_1, ..., c_8\}. \tag{11.91}$$

Notice that c_i can be greater than, less than, or equal to 1. This can be seen in Figure 11.3, where a boundary point used in the calculation of the scheme can be closer or farther apart than a neighboring grid point in the same direction.

Assume that there exists a constant $M \geq 1$ such that

$$\frac{1}{M} \leq c_i \leq M, \ 1 \leq i \leq 8. \tag{11.92}$$

Our fundamental result is the following theorem.

Theorem 11.18. *Under the assumption (11.92) the finite difference discrete operator $\Delta^2_{\vec{\mathbf{h}}}$ approximating the biharmonic operator using (11.75) is second-order accurate.*

Proof. The proof consists of four steps.

Step I. Clearly, $\Delta^2_{\vec{\mathbf{h}}}\psi$ is a linear combination of the data in (11.59). We claim that the coefficients of the ψ terms in this combination are $O(\frac{1}{h^4})$ while the coefficients of the ψ_x and ψ_y terms are $O(\frac{1}{h^3})$.

This claim is obvious in the regular case [see (11.81)]. For an irregular stencil the proof is obtained by a scaling argument. The details are given in Subsection 11.2.3.1.

Step II. We claim that

$$K_{\vec{\mathbf{h}}}(\psi) = \Delta^2\psi^{\text{res}}(0,0) + O(h^3). \tag{11.93}$$

Applying (11.86) to $E(x,y)$ we get

$$K_{\vec{\mathbf{h}}}(E) = \Delta^2 E(0,0) - \Delta^2 Q_{E,\vec{\mathbf{h}}}(0,0), \tag{11.94}$$

and from Step I,

$$\Delta^2 Q_{E,\vec{\mathbf{h}}}(0,0) = \sum_i O\left(\frac{1}{h^4}\right) E(\mathbf{p_i}) + \sum_i O\left(\frac{1}{h^3}\right) E_x(\mathbf{p_i})$$
$$+ \sum_i O\left(\frac{1}{h^3}\right) E_y(\mathbf{p_i}) = O\left(\frac{1}{h^4}\right).$$

Since $\Delta^2 E(0,0)$ is a constant we conclude

$$K_{\vec{\mathbf{h}}}(E) = O\left(\frac{1}{h^4}\right).$$

Equation (11.93) now follows from (11.89), noting that $K_{\vec{\mathbf{h}}}(\psi^{\mathrm{res}}) = \Delta^2\psi^{\mathrm{res}}(0,0)$ by definition.

Step III. We construct a smooth (with respect to $\vec{\mathbf{h}}$) basis $\left\{G_1(\vec{\mathbf{h}};x,y),...,G_9(\vec{\mathbf{h}};x,y)\right\}$ of $\ker(A_{\vec{\mathbf{h}}})$. Note that we add the $\vec{\mathbf{h}}$ subscript to A to clarify the stencil in which the calculation is done.

To this end we let

$$\begin{aligned}\Theta^c &= \tilde{B} \oplus \mathrm{Span}\{x^6, x^5y, x^4y^2 - x^2y^4, x^3y^3, xy^5, y^6\} \\ &= \mathrm{Span}\{x^3y - xy^3, x^4y - x^2y^3, x^3y^2 - xy^4, x^6, x^5y, \\ &\qquad x^4y^2 - x^2y^4, x^3y^3, xy^5, y^6\}.\end{aligned} \tag{11.95}$$

In view of (11.75),

$$\mathbb{P}^6 = \Theta \oplus \Theta^c. \tag{11.96}$$

Indeed, $\mathbb{P}^5 = \tilde{\Theta} \oplus \tilde{B}$ by (11.71) and Θ includes only one polynomial of degree six while the six others are in Θ^c.

Define

$$\begin{aligned}F &: \mathbb{R}^8_+ \times \Theta^c \times \Theta \to \mathrm{Dat}_{19} \\ H &: \mathbb{R}^8_+ \times \Theta^c \to \Theta\end{aligned}$$

as

$$F(\vec{\mathbf{h}}, T_1, T_2) = A_{\vec{\mathbf{h}}}(T_1 + T_2), \qquad \vec{\mathbf{h}} \in \mathbb{R}^8_+,\ T_1 \in \Theta^c,\ T_2 \in \Theta, \tag{11.97}$$

$$H(\vec{\mathbf{h}}, T_1) = -Q_{T_1,\vec{\mathbf{h}}}, \qquad \vec{\mathbf{h}} \in \mathbb{R}^8_+,\ T_1 \in \Theta^c, \tag{11.98}$$

where $Q_{T_1,\vec{\mathbf{h}}} \in \Theta$ is the interpolating polynomial of T_1 with respect to the $\vec{\mathbf{h}}$ stencil. Recalling Proposition 11.16, we know that this interpolating polynomial is uniquely determined in Θ. Moreover, $\forall \vec{\mathbf{h}} \in \mathbb{R}^8_+$, $T_1 \in \Theta^c$, $T \in \Theta$,

$$H(\vec{\mathbf{h}}, T_1) = T \ \Leftrightarrow\ F(\vec{\mathbf{h}}, T_1, T) = 0 \ \Leftrightarrow\ T_1 + T \in \ker(A_{\vec{\mathbf{h}}}). \tag{11.99}$$

Given $\vec{\mathbf{h}} \in \mathbb{R}^8_+$ and $T_1 \in \Theta^c$, it is obvious from (11.98) that $T = H(\vec{\mathbf{h}}, T_1)$ is the unique solution to $F(\vec{\mathbf{h}}, T_1, T) = 0$. Since $F(\vec{\mathbf{h}}, T_1, T)$ is smooth with respect to all its variables, the implicit function theorem implies that $H(\vec{\mathbf{h}}, T)$ is continuously differentiable with respect to all its variables.

From (11.99) it follows that

$$\ker(A_{\vec{\mathbf{h}}}) = \{T_1 + H(\vec{\mathbf{h}}, T_1) \mid T_1 \in \Theta^c\}. \tag{11.100}$$

Thus, denoting by $e_i(x,y)$, $1 \le i \le 9$, the nine homogeneous polynomials spanning Θ^c and defining

$$\begin{aligned}G_i(\vec{\mathbf{h}}; x, y) &= e_i(x,y) + H(\vec{\mathbf{h}}, e_i(x,y)) \\ &= e_i(x,y) - Q_{e_i,\vec{\mathbf{h}}}(x,y), \quad 1 \le i \le 9,\end{aligned} \tag{11.101}$$

we obtain the desired smooth basis of $\ker(A_{\vec{\mathbf{h}}})$. Note that the linear independence of the G_i's follows from the fact that the e_i's form a basis for Θ^c while $Q_{e_i,\vec{\mathbf{h}}} \in \Theta$ is the complementary subspace.

Step IV. We can now conclude the proof of the theorem. Combining (11.65) and (11.83),

$$\psi(x,y) = \psi^\Theta + \sum_{i=1}^{9} b_i G_i(\vec{\mathbf{h}}) + O(h^7)E(x,y),$$
$$= \psi^\Theta + \sum_{i=1}^{9} b_i H(\vec{\mathbf{h}}, e_i) + \sum_{i=1}^{9} b_i e_i + O(h^7)E(x,y),$$

where $\psi^\Theta \in \Theta$ and $G_i(\vec{\mathbf{h}}) \in \ker(A_{\vec{\mathbf{h}}})$.

Notice that $\psi^\Theta + \sum_{i=1}^9 b_i H(\vec{\mathbf{h}}, e_i) \in \Theta$. In terms of the decompositions (11.83) and (11.96), the Θ^c component of $\tilde{\psi}(x,y)$ is $\sum_{i=1}^9 b_i e_i$. Since $\{e_i,\ 1 \le i \le 9\}$ are fixed polynomials (independent of $\vec{\mathbf{h}}$), the coefficients $\{b_i,\ 1 \le i \le 9\}$ are constants, which only depend on the derivatives of ψ of degree ≤ 6.

Using (11.93), we have

$$K_{\vec{\mathbf{h}}}(\psi) = \sum_{i=1}^{9} b_i\ \Delta^2 G_i(\vec{\mathbf{h}}; x, y)\Big|_{x=0,y=0} + O(h^3). \tag{11.102}$$

Recalling the assumption (11.92), we now show that

$$\left| \Delta^2 G_i(\vec{\mathbf{h}}; x, y)\Big|_{(0,0)} \right| \le h^2 J, \quad \vec{\mathbf{h}} \in \mathbb{R}^8_+, \quad 1 \le i \le 9, \tag{11.103}$$

where

$$J = \sup_{1 \le i \le 9} \left| \Delta^2 G_i(\vec{\mathbf{c}}; x, y)\big|_{(0,0)} \right|, \quad \vec{\mathbf{c}} \in \left[\frac{1}{M}, M\right]^8 \subset \mathbb{R}^8. \tag{11.104}$$

Since $\Delta^2 G_i(\vec{\mathbf{h}}; x, y)\Big|_{(0,0)}$ is a continuous function of $\vec{\mathbf{h}}$ it attains a maximum in the compact cube $\left[\frac{1}{M}, M\right]^8 \subseteq \mathbb{R}^8$.

Using the definition (11.95) of Θ^c,

$$\Theta^c \cap \mathbb{P}^5 = \tilde{B}.$$

Let $1 \le i \le 9$. If $e_i \in \Theta^c \cap \mathbb{P}^5$ then $e_i \in \tilde{B} \subseteq \ker(A_{\vec{\mathbf{h}}})$. Hence, $H(\vec{\mathbf{h}}, e_i) = 0$ and $G_i(\vec{\mathbf{h}}; x, y) = e_i$. It follows that $\Delta^2 G_i(\vec{\mathbf{h}}; x, y)\Big|_{(0,0)} = 0$ (since $\tilde{B}$ is an indifferent subspace). Otherwise, e_i is one of the six homogeneous polynomial of degree six in (11.95). Consider the decomposition (11.96), applied to the polynomial $R(x,y) = h^6 G_i(\vec{\mathbf{c}}; \frac{x}{h}, \frac{y}{h}) = h^6 \left(e_i(\frac{x}{h}, \frac{y}{h}) + H(\vec{\mathbf{c}}, e_i(\frac{x}{h}, \frac{y}{h}))\right)$. Its Θ^c component is $h^6 \left(e_i(\frac{x}{h}, \frac{y}{h})\right) = e_i(x,y)$. Also, notice that $R(x,y)$ vanishes at all points on the $\vec{\mathbf{h}}$ stencil. Hence, $R \in \ker(A_{\vec{\mathbf{h}}})$ and $R(x,y) = e_i(x,y) - Q_{e_i,\vec{\mathbf{h}}}(x,y) = G_i(\vec{\mathbf{h}}; x, y)$ by (11.101). It follows that

$$G_i(\vec{\mathbf{h}}; x, y) = h^6 G_i(\vec{\mathbf{c}}; \frac{x}{h}, \frac{y}{h}). \tag{11.105}$$

Hence using (11.104),

$$\left| \Delta^2 G_i(\vec{\mathbf{h}}; x, y)\Big|_{(0,0)} \right| = \left| \frac{h^6}{h^4} \Delta^2 G_i(\vec{\mathbf{c}}; x, y)\Big|_{(0,0)} \right| \leq h^2 J. \tag{11.106}$$

Finally, invoking Equation (11.102) we obtain

$$|K_{\vec{\mathbf{h}}}(\psi)| \leq \sum_{i=1}^{9} |b_i| h^2 J + O(h^3) = O(h^2). \tag{11.107}$$

■

Example 11.19. We calculate $\Delta^2 \psi^{\text{res}}(0,0)$ for a regular grid. Using G_1 and G_2 as in (11.66), we have $\mathcal{L}^{\text{res}} = \text{Span}\{G_1, G_2\}$ [see (11.73)]. Using our choice (11.75) of Θ, there are no polynomials containing the x^6 or y^6 monomials in Θ. Also, there are no polynomials containing these monomials in $\mathcal{L}^{\text{ind}}$ [see (11.72)]. Comparing the coefficients of x^6 and y^6 in $\mathcal{L}^{\text{res}}$ and in (11.83), we have

$$\psi^{\text{res}} = \frac{1}{6!} \left(\frac{\partial^6 \psi}{\partial x^6}(0,0) G_1 + \frac{\partial^6 \psi}{\partial y^6}(0,0) G_2 \right), \tag{11.108}$$

and therefore

$$\begin{aligned} |\Delta^2 \psi^{\text{res}}(0,0)| &\leq \frac{1}{6!} \left(48 \sup_{\Lambda} |\psi^{(6)}| h^2 + 48 \sup_{\Lambda} |\psi^{(6)}| h^2 \right) \\ &= O(h^2), \end{aligned}$$

where the domain Λ is defined at the beginning of this section. In fact, due to the explicit formula for the discrete biharmonic operator on a regular grid (11.81), the approximation is actually fourth-order accurate.

A different choice of Θ could give us different results. Consider the space $\overline{\Theta} = \Theta \backslash \{y^4\} \cup \{y^6\}$. Using $G_2 \in \ker(A_{\vec{\mathbf{h}}}))$, we know that

$$y^4 - \frac{y^6}{2h^2} + \frac{h^2 y^2}{2} \in \ker(A_{\vec{\mathbf{h}}}).$$

Therefore, $\overline{\Theta}$ satisfies (11.65). We rewrite (11.85), using $\overline{\Theta}$,

$$\tilde{\psi} = \psi^{\overline{\Theta}} + \psi^{\overline{\text{ind}}} + \psi^{\overline{\text{res}}},\ \psi^{\overline{\Theta}} \in \overline{\Theta},\ \psi^{\overline{\text{ind}}} \in \mathcal{L}^{\text{ind}},\ \psi^{\overline{\text{res}}} \in \mathcal{L}^{\text{res}}.$$

In this case, we must compare the coefficients of x^6 and y^4 in $\mathcal{L}^{\text{res}}$ and in (11.83)

$$\begin{aligned}\psi^{\overline{\text{res}}} &= \frac{1}{6!}\frac{\partial^6\psi}{\partial x^6}(0,0)G_1 - \frac{1}{4!}\frac{\partial^4\psi}{\partial y^4}(0,0)\frac{G_2}{2h^2},\\ |\Delta^2\psi^{\overline{\text{res}}}(0,0)| &\le \frac{48}{6!}\sup_{\Lambda}|\psi^{(6)}|h^2 + \frac{24}{4!}\sup_{\Lambda}|\psi^{(4)}|\\ &= O(h^2) + O(1) = O(1).\end{aligned}$$

Summary. Our goal is to approximate the biharmonic operator at $(0,0)$ given the data from the nine-point irregular stencil (Figure 11.4). We use 19 data from the stencil, as defined in (11.59). We first approximate the data by a polynomial in $\Theta \subseteq \mathbb{P}^6$, which is defined in (11.70) and (11.75). Proposition 11.16 shows that there exists a unique polynomial, $Q \in \Theta$, which interpolates the 19 data. $\Delta^2 Q(0,0)$ is the finite difference approximation of the biharmonic operator at $(0,0)$. The discussion preceding Proposition 11.16 explains the logic in the construction of Θ. Theorem 11.18 shows that the resulting scheme has second-order accuracy.

Remark 11.20. Note that we can interpret the nine-point biharmonic operator (11.1) [15,168] using our polynomial approach. Let $\hat{A} : \mathbb{P}^4 \to \text{Dat}_{13}$ where Dat_{13} is the 13-dimensional linear space spanned by the 13 data used in (11.15). It is easy to show that

$$\ker(\hat{A}) = \text{Span}\{xy(x-y)(x+y), xy(x-h)(x+h)\}$$

showing that $\ker(\hat{A})$ is an indifferent subspace (for a regular stencil). Using Proposition 11.14 we know that any space Θ_{13}, such that $\mathbb{P}^4 = \ker(\hat{A}) \oplus \Theta_{13}$ yields an equivalent scheme. We can define Θ_{13} by omitting the two polynomials x^3y and xy^3, obtaining an equivalent scheme to the one given in [168].

Remark 11.21. We note that in the regular case, the discrete biharmonic operator relied *solely* on the values of ψ at the grid points [see (11.20)]. The derivative values were taken as the *Hermitian derivatives*, which are obtained from the ψ values.

In contrast, our construction of the interpolating polynomial in the irregular case (Definition 11.9) and our analysis of its accuracy follow a more traditional pattern of interpolation theory; the values of both ψ and its gradient are given at the grid points. In the practical implementation [11] the gradient values are calculated from the values of ψ using a suitable extension of the Hermitian derivation approach. Refer also to the Notes at the end of this chapter.

11.2.3.1 *Proof of Step I of Theorem 11.18*

We will prove Step I of Theorem 11.18. Let

$$Q_{\psi,\vec{\mathbf{h}}} = Q^0_{\psi,\vec{\mathbf{h}}} + Q^d_{\psi,\vec{\mathbf{h}}},$$

where $Q^0_{\psi,\vec{\mathbf{h}}}$ interpolates the ψ data, while substituting 0 for the ψ_x and ψ_y data. In a similar fashion, $Q^d_{\psi,\vec{\mathbf{h}}}$ satisfies the following (as in (11.59)) after substituting 0 for all the ψ data,

$$\begin{aligned}
&Q^d_{\psi,\vec{\mathbf{h}}}(\mathbf{p_i}) = 0, && 0 \le i \le 8,\\
&\frac{\partial Q^d_{\psi,\vec{\mathbf{h}}}}{\partial d}(\mathbf{p_i}) = \psi_d(\mathbf{p_i}), && 1 \le i \le 8,\\
&\frac{\partial Q^d_{\psi,\vec{\mathbf{h}}}}{\partial x}(\mathbf{p_0}) = \psi_x(\mathbf{p_0}), &&\\
&\frac{\partial Q^d_{\psi,\vec{\mathbf{h}}}}{\partial y}(\mathbf{p_0}) = \psi_y(\mathbf{p_0}). &&
\end{aligned}$$

First, we show that the coefficients of $Q^0_{\psi,\vec{\mathbf{h}}}$ are $O(\frac{1}{h^4})$.

Note that the elements of Θ are all homogeneous polynomials [see (11.70) and (11.75)] of degree k where $0 \le k \le 6$. We designate by $\{S^k_j\}_{j=1}^{n(k)}$ the elements of Θ having degree k. Let

$$Q^0_{\psi,\vec{\mathbf{h}}}(x,y) = \sum_{i=0}^{8}\sum_{k=0}^{6}\sum_{j=1}^{n(k)} a^0_{j,k,i}\left(\vec{\mathbf{h}}\right)\psi(\mathbf{p_i})S^k_j(x,y). \tag{11.109}$$

Equation (11.109) is a polynomial equality, which holds for *every choice* of the eight points $\mathbf{p_i}$ (as in Figure 11.4). In other words, we can choose the vector $\vec{\mathbf{h}}$ arbitrarily in $\mathbb{R}^8_+$ and the equation will give the unique polynomial (in the subspace spanned by Θ) interpolating the data $\psi(\mathbf{p_i})$ (with zero derivatives).

Proposition 11.16 ensures that there is always a unique representation as in (11.109). Clearly, the $a^0_{j,k,i}\left(\vec{\mathbf{h}}\right)$ are continuous functions of $\vec{\mathbf{h}} \in \mathbb{R}^8_+$. Since the cube $\left[\frac{1}{M}, M\right]^8 \subseteq \mathbb{R}^8_+$ is compact we have a uniform bound

$$\Gamma = \max_{\vec{\mathbf{c}}\in\left[\frac{1}{M},M\right]^8} \left|a^0_{j,k,i}\left(\vec{\mathbf{c}}\right)\right|. \tag{11.110}$$

We proceed by a homogeneity argument to prove that the non-vanishing coefficients $a^0_{j,k,i}\left(\vec{\mathbf{h}}\right)$ are $O(\frac{1}{h^4})$. Let

$$\psi_\lambda(x,y) = \psi(\lambda x, \lambda y), \quad \lambda > 0. \tag{11.111}$$

Evaluate the polynomial $R(x,y) = Q^0_{\psi_\lambda, \frac{\vec{\mathbf{h}}}{\lambda}}(\frac{x}{\lambda}, \frac{y}{\lambda})$ at $\mathbf{p_i}$. Since $Q^0_{\psi_\lambda, \frac{\vec{\mathbf{h}}}{\lambda}}$ is the interpolating polynomial of ψ_λ on the $\frac{\vec{\mathbf{h}}}{\lambda}$ stencil we have

$$R(\mathbf{p_i}) = Q^0_{\psi_\lambda, \frac{\vec{\mathbf{h}}}{\lambda}}\left(\frac{\mathbf{p_i}}{\lambda}\right) = \psi_\lambda\left(\frac{\mathbf{p_i}}{\lambda}\right) = \psi(\mathbf{p_i}). \tag{11.112}$$

Notice also that all the derivative data is 0,

$$\frac{\partial R}{\partial d}(\mathbf{p_i}) = 0, \ 0 \le i \le 8, \ \text{including both derivatives at } \mathbf{p}_0. \tag{11.113}$$

Combining (11.112) and (11.113) and using the uniqueness of the interpolating polynomial, we get

$$R(x,y) = Q^0_{\psi, \vec{\mathbf{h}}}(x,y),$$

hence

$$Q^0_{\psi_\lambda, \frac{\vec{\mathbf{h}}}{\lambda}}\left(\frac{x}{\lambda}, \frac{y}{\lambda}\right) = Q^0_{\psi, \vec{\mathbf{h}}}(x,y). \tag{11.114}$$

From (11.109), we can write the last equation as

$$\begin{aligned} &\sum_{i=0}^{8}\sum_{k=0}^{6}\sum_{j=1}^{n(k)} a^0_{j,k,i}\left(\frac{\vec{\mathbf{h}}}{\lambda}\right)\psi_\lambda\left(\frac{\mathbf{p_i}}{\lambda}\right) S^k_j\left(\frac{x}{\lambda}, \frac{y}{\lambda}\right) \\ &\quad = \sum_{i=0}^{8}\sum_{k=0}^{6}\sum_{j=1}^{n(k)} a^0_{j,k,i}\left(\vec{\mathbf{h}}\right)\psi(\mathbf{p_i}) S^k_j(x,y). \end{aligned} \tag{11.115}$$

Using the assumed homogeneity of $S^k_j(x,y)$ and Equation (11.111), we further obtain

$$\begin{aligned} &\sum_{i=0}^{8}\sum_{k=0}^{6}\sum_{j=1}^{n(k)} \frac{1}{\lambda^k} a^0_{j,k,i}\left(\frac{\vec{\mathbf{h}}}{\lambda}\right)\psi(\mathbf{p_i}) S^k_j(x,y) \\ &\quad = \sum_{i=0}^{8}\sum_{k=0}^{6}\sum_{j=1}^{n(k)} a^0_{j,k,i}\left(\vec{\mathbf{h}}\right)\psi(\mathbf{p_i}) S^k_j(x,y). \end{aligned} \tag{11.116}$$

Thus, we have, by the uniqueness of the expansion

$$a^0_{j,k,i}\left(\vec{\mathbf{h}}\right) = \frac{1}{\lambda^k} a^0_{j,k,i}\left(\frac{\vec{\mathbf{h}}}{\lambda}\right). \tag{11.117}$$

Choosing $\lambda = h$ and noting (11.92) and (11.110), we obtain

$$\left|a_{j,k,i}^0\left(\vec{\mathbf{h}}\right)\right| = \frac{1}{h^k}\left|a_{j,k,i}^0\left(\frac{\vec{\mathbf{h}}}{h}\right)\right| \le \frac{1}{h^k}\Gamma = O\left(\frac{1}{h^k}\right).$$

Recall that we are interested in estimating the order of magnitude of $a_{j,k,i}^0\left(\vec{\mathbf{h}}\right)$ for those coefficients that appear in the evaluation of $\Delta^2 Q_{\boldsymbol{\psi},\vec{\mathbf{h}}}^0(0,0)$.

$$\Delta^2 Q_{\boldsymbol{\psi},\vec{\mathbf{h}}}^0(0,0) = \Delta^2\left(\sum_{i=0}^{8}\sum_{k=0}^{6}\sum_{j=1}^{n(k)} a_{j,k,i}^0\left(\vec{\mathbf{h}}\right)\psi(\mathbf{p_i})S_j^k(x,y)\right)_{x=0,y=0}$$

$$(11.118)\qquad = \sum_{i=0}^{8}\sum_{k=0}^{6}\sum_{j=1}^{n(k)} a_{j,k,i}^0\left(\vec{\mathbf{h}}\right)\psi(\mathbf{p_i})\ \Delta^2 S_j^k(x,y)\Big|_{x=0,y=0}$$

$$= \sum_{i=0}^{8}\sum_{k=0}^{6}\sum_{j=1}^{n(k)} b_{j,k}\ a_{j,k,i}^0\left(\vec{\mathbf{h}}\right)\psi(\mathbf{p_i}),$$

where $b_{j,k} = \Delta^2 S_j^k(x,y)\big|_{x=0,y=0}$. Taking into account the homogeneity of $S_j^k(x,y)$, it is clear that $b_{j,k} = 0$ if $k \neq 4$. We therefore obtain

$$(11.119)\qquad \Delta^2 Q_{\boldsymbol{\psi},\vec{\mathbf{h}}}^0(0,0) = \sum_{i=0}^{8} O\left(\frac{1}{h^4}\right)\psi(\mathbf{p_i}).$$

We proceed to show that the coefficients of $Q_{\boldsymbol{\psi},\vec{\mathbf{h}}}^d$ are $O(\frac{1}{h^3})$. Let

$$(11.120)\qquad \begin{aligned} Q_{\boldsymbol{\psi},\vec{\mathbf{h}}}^d(x,y) &= \sum_{i=1}^{8}\sum_{k=0}^{6}\sum_{j=1}^{n(k)} a_{j,k,i}^d\left(\vec{\mathbf{h}}\right)\psi_d(\mathbf{p_i})S_j^k(x,y) \\ &+ \sum_{k=0}^{6}\sum_{j=1}^{n(k)}[a_{j,k,0}^x\psi_x(\mathbf{p_0})S_j^k(x,y) + a_{j,k,0}^y\psi_y(\mathbf{p_0})S_j^k(x,y)], \end{aligned}$$

and assume that the approximate values of the derivatives satisfy

$$(11.121)\qquad (\psi_\lambda)_d = (\psi(\lambda x, \lambda y))_d = \lambda\psi_d(\lambda x, \lambda y).$$

This equation is a natural requirement for the connection between ψ and ψ_x, ψ_y.

In order to estimate the coefficients $a_{j,k,i}^d$ in Equation (11.120), we evaluate the polynomial $R^d(x,y) = Q_{\boldsymbol{\psi}_\lambda,\frac{\vec{\mathbf{h}}}{\lambda}}^d(\frac{x}{\lambda},\frac{y}{\lambda})$ at $\mathbf{p_i}$. In a similar fashion to (11.112),

$$\begin{aligned} \frac{\partial R^d}{\partial d}(\mathbf{p_i}) &= \frac{\partial Q_{\boldsymbol{\psi}_\lambda,\frac{\vec{\mathbf{h}}}{\lambda}}^d(x/\lambda, y/\lambda)}{\partial d}\left(\frac{\mathbf{p_i}}{\lambda}\right) \\ &= \frac{1}{\lambda}\frac{\partial Q_{\boldsymbol{\psi}_\lambda,\frac{\vec{\mathbf{h}}}{\lambda}}^d(x,y)}{\partial d}\left(\frac{\mathbf{p_i}}{\lambda}\right) = \frac{1}{\lambda}(\psi_\lambda)_d\left(\frac{\mathbf{p_i}}{\lambda}\right) = \psi_d(\mathbf{p_i}), \end{aligned}$$

where the final equality follows from (11.121). Notice also that

$$R^d(\mathbf{p_i}) = 0,$$

and therefore, by uniqueness,

$$R^d(x,y) = Q^d_{\boldsymbol{\psi},\vec{\mathbf{h}}}(x,y).$$

We expand the equality above using (11.120),

$$\sum_{i=1}^{8}\sum_{k=0}^{6}\sum_{j=1}^{n(k)} a^d_{j,k,i}\left(\frac{\vec{\mathbf{h}}}{\lambda}\right)(\boldsymbol{\psi}_\lambda)_d\left(\frac{\mathbf{p_i}}{\lambda}\right)S^k_j\left(\frac{x}{\lambda},\frac{y}{\lambda}\right)$$
$$+\sum_{k=0}^{6}\sum_{j=1}^{n(k)}\Big[a^x_{j,k,0}\left(\frac{\vec{\mathbf{h}}}{\lambda}\right)(\boldsymbol{\psi}_\lambda)_x\left(\frac{\mathbf{p_0}}{\lambda}\right)S^k_j\left(\frac{x}{\lambda},\frac{y}{\lambda}\right)$$
$$+\,a^y_{j,k,0}\left(\frac{\vec{\mathbf{h}}}{\lambda}\right)(\boldsymbol{\psi}_\lambda)_y\left(\frac{\mathbf{p_0}}{\lambda}\right)S^k_j\left(\frac{x}{\lambda},\frac{y}{\lambda}\right)\Big]$$
$$=\sum_{i=1}^{8}\sum_{k=0}^{6}\sum_{j=1}^{n(k)} a^d_{j,k,i}\left(\vec{\mathbf{h}}\right)\boldsymbol{\psi}_d(\mathbf{p_i})S^k_j(x,y)$$
$$+\sum_{k=0}^{6}\sum_{j=1}^{n(k)}\Big[a^x_{j,k,0}(\vec{\mathbf{h}})\boldsymbol{\psi}_x(\mathbf{p_0})S^k_j(x,y)+a^y_{j,k,0}(\vec{\mathbf{h}})\boldsymbol{\psi}_y(\mathbf{p_0})S^k_j(x,y)\Big],$$

By the homogeneity of $S^k_j(x,y)$, we deduce that

$$\sum_{i=1}^{8}\sum_{k=0}^{6}\sum_{j=1}^{n(k)}\left(\frac{1}{\lambda}\right)^{n(k)-1} a^d_{j,k,i}\left(\frac{\vec{\mathbf{h}}}{\lambda}\right)\boldsymbol{\psi}_d(\mathbf{p_i})S^k_j(x,y)$$
$$+\sum_{k=0}^{6}\sum_{j=1}^{n(k)}\left(\frac{1}{\lambda}\right)^{n(k)-1}\Big[a^x_{j,k,0}\left(\frac{\vec{\mathbf{h}}}{\lambda}\right)\boldsymbol{\psi}_x(\mathbf{p_0})S^k_j(x,y)$$
$$+\,a^y_{j,k,0}\left(\frac{\vec{\mathbf{h}}}{\lambda}\right)\boldsymbol{\psi}_y(\mathbf{p_0})S^k_j(x,y)\Big]$$
$$=\sum_{i=1}^{8}\sum_{k=0}^{6}\sum_{j=1}^{n(k)} a^d_{j,k,i}\left(\vec{\mathbf{h}}\right)\boldsymbol{\psi}_d(\mathbf{p_i})S^k_j(x,y)$$
$$+\sum_{k=0}^{6}\sum_{j=1}^{n(k)}\Big[a^x_{j,k,0}(\vec{\mathbf{h}})\boldsymbol{\psi}_x(\mathbf{p_0})S^k_j(x,y)+a^y_{j,k,0}(\vec{\mathbf{h}})\boldsymbol{\psi}_y(\mathbf{p_0})S^k_j(x,y)\Big].$$

The homogeneity of the coefficients now follows from (11.117) and a calculation as in (11.118) leads to

$$\Delta^2 Q^d_{\boldsymbol{\psi},\vec{\mathbf{h}}}(0,0) = \sum_i O\left(\frac{1}{h^3}\right)\boldsymbol{\psi}_d(\mathbf{p_i}),$$

where $\psi_d(\mathbf{p_0})$ represents the two values $\psi_x(\mathbf{p_0})$ and $\psi_y(\mathbf{p_0})$. The proof is now complete.

11.2.4 *Fourth-order improvement of the nine-point biharmonic operator*

As mentioned in the paragraph preceding Equation (11.81) we will give here all of the coefficients of the interpolating polynomial $Q(x,y) \in \Theta$ [see (11.75)]. Using the simplified notation $Q(x,y)$ instead of $Q_{\psi,\vec{\mathrm{h}}}$ and taking $\mathbf{p_0} = (0,0)$ we have (see Figure 11.5)

$$\begin{aligned} Q(x,y) = {} & a_{0,0} + a_{1,0}x + a_{0,1}y \\ & + a_{2,0}x^2 + a_{1,1}xy + a_{0,2}y^2 \\ & + a_{3,0}x^3 + a_{2,1}x^2y + a_{1,2}xy^2 + a_{0,3}y^3 \\ & + a_{4,0}x^4 + a_{3,1}xy(x+y)^2 + a_{1,3}xy(x-y)^2 + a_{0,4}y^4 \\ & + a_{5,0}x^5 + a_{3,2}xy(x+y)^3 + a_{2,3}xy(x-y)^3 + a_{0,5}y^5 \\ & + a_{3,3}x^2y^2(x^2+y^2). \end{aligned} \tag{11.122}$$

Based on the data (11.59), we consider the system of linear equations

$$\begin{cases} Q(0,0) = \psi(0,0), \quad \partial_x Q(0,0) = \psi_x(0,0), \quad \partial_y Q(0,0) = \psi_y(0,0), \\ Q(h,0) = \psi(h,0), \quad Q(-h,0) = \psi(-h,0), \\ Q(0,h) = \psi(0,h), \quad Q(0,-h) = \psi(0,-h), \\ Q(h,h) = \psi(h,h), \quad Q(-h,h) = \psi(-h,h), \\ Q(-h,h) = \psi(-h,h), \quad Q(-h,-h) = \psi(-h,-h), \\ \partial_d Q(h,0) = \psi_d(h,0), \quad \partial_d Q(-h,0) = \psi_d(-h,0), \\ \partial_d Q(0,h) = \psi_d(0,h), \quad \partial_d Q(0,-h) = \psi_d(0,-h), \\ \partial_d Q(h,h) = \psi_d(h,h), \quad \partial_d Q(-h,h) = \psi_d(-h,h), \\ \partial_d Q(h,-h) = \psi_d(h,-h), \quad \partial_d Q(-h,-h) = \psi_d(-h,-h), \end{cases}$$

Solving this system using Maple gives a unique solution,

(11.123)

$$\begin{aligned} &(a) \quad a_{0,0} = \psi(0,0) \\ &(b) \quad a_{1,0} = \psi_x(0,0), \quad a_{0,1} = \psi_y(0,0), \\ &(c_1) \quad a_{2,0} = \delta_x^2\psi(0,0) - \frac{1}{2}(\delta_x\psi_x)(0,0), \\ &(c_2) \quad a_{0,2} = \delta_y^2\psi(0,0) - \frac{1}{2}(\delta_y\psi_y)(0,0), \\ &(c_3) \quad a_{1,1} = 2\delta_x\delta_y\psi(0,0) - \frac{1}{2}(\delta_x\mu_y\psi_y)(0,0) - \frac{1}{2}(\delta_y\mu_x\psi_x)(0,0), \end{aligned}$$

$$
\begin{aligned}
(d_1)\quad a_{3,0} &= \frac{1}{4h^2}(10(\delta_x\boldsymbol{\psi}(0,0)-\boldsymbol{\psi}_x(0,0))-h^2\delta_x^2\boldsymbol{\psi}_x(0,0)),\\
(d_2)\quad a_{0,3} &= \frac{1}{4h^2}(10(\delta_y\boldsymbol{\psi}(0,0)-\boldsymbol{\psi}_y(0,0))-h^2\delta_y^2\boldsymbol{\psi}_y(0,0)),\\
(d_3)\quad a_{2,1} &= \frac{1}{4}\left(5\delta_y\delta_x^2\boldsymbol{\psi}(0,0)-2(\delta_x\delta_y\boldsymbol{\psi}_x)(0,0)-(\delta_x^2\mu_y\boldsymbol{\psi}_y)(0,0)\right),\\
(d_4)\quad a_{1,2} &= \frac{1}{4}\left(5(\delta_x\delta_y^2\boldsymbol{\psi})(0,0)-2\delta_x\delta_y\boldsymbol{\psi}_y(0,0)-(\delta_y^2\mu_x\boldsymbol{\psi}_x)(0,0)\right),
\end{aligned}
$$

$$
\begin{aligned}
(e_1)\quad a_{4,0} &= \frac{1}{2h^2}\left((\delta_x\boldsymbol{\psi}_x)(0,0)-\delta_x^2\boldsymbol{\psi}(0,0)\right),\\
(e_2)\quad a_{0,4} &= \frac{1}{2h^2}\left((\delta_y\boldsymbol{\psi}_y)(0,0)-\delta_y^2\boldsymbol{\psi}(0,0)\right),\\
(e_3)\quad a_{3,1} &= \frac{1}{16h^4}\Big[-6h^2\delta_x^2\boldsymbol{\psi}(0,0)+(2\boldsymbol{\psi}(h,h)+4\boldsymbol{\psi}(-h,h))\\
&\quad+(-6\boldsymbol{\psi}(0,h)+2\boldsymbol{\psi}(-h,-h)+4\boldsymbol{\psi}(h,-h)-6\boldsymbol{\psi}(0,-h))\\
&\quad+h\left(2h\delta_x\boldsymbol{\psi}_x(0,0)+(\boldsymbol{\psi}_x(-h,h)-\boldsymbol{\psi}_x(h,-h))\right)\\
&\quad+h\left(2h\delta_y\boldsymbol{\psi}_y(0,0)-(\boldsymbol{\psi}_y(-h,h)-\boldsymbol{\psi}_y(h,-h))\right)\Big],\\
(e_4)\quad a_{1,3} &= -\frac{1}{16h^4}\Big[-6h^2\delta_x^2\boldsymbol{\psi}(0,0)+(2\boldsymbol{\psi}(-h,h)+4\boldsymbol{\psi}(h,h))\\
&\quad+(-6\boldsymbol{\psi}(0,h)+2\boldsymbol{\psi}(h,-h)+4\boldsymbol{\psi}(-h,-h)-6\boldsymbol{\psi}(0,-h))\\
&\quad+h\left(2h\delta_x\boldsymbol{\psi}_x(0,0)+(\boldsymbol{\psi}_x(-h,-h)-\boldsymbol{\psi}_x(h,h))\right)\\
&\quad+h\left(2h\delta_y\boldsymbol{\psi}_y(0,0)-(\boldsymbol{\psi}_y(h,h)-\boldsymbol{\psi}_y(-h,-h))\right)\Big],
\end{aligned}
$$

$$
\begin{aligned}
(f_1)\quad a_{5,0} &= \frac{3}{2h^4}\left(\boldsymbol{\psi}_x(0,0)+\frac{h^2}{6}\delta_x^2\boldsymbol{\psi}_x(0,0)-\delta_x\boldsymbol{\psi}(0,0)\right),\\
(f_2)\quad a_{0,5} &= \frac{3}{2h^4}\left(\boldsymbol{\psi}_y(0,0)+\frac{h^2}{6}\delta_y^2\boldsymbol{\psi}_y(0,0)-\delta_y\boldsymbol{\psi}(0,0)\right),
\end{aligned}
$$

$$
\begin{aligned}
(f_3)\quad a_{3,2} &= \frac{1}{32h^5}\Big(-3(\boldsymbol{\psi}(h,-h)-\boldsymbol{\psi}(0,h)-\boldsymbol{\psi}(-h,h)+\boldsymbol{\psi}(-h,-h))\\
&\quad+6h\delta_x\boldsymbol{\psi}(0,0)\\
&\quad+h\left(\boldsymbol{\psi}_x(h,h)+\boldsymbol{\psi}_x(-h,-h)-2\boldsymbol{\psi}_x(0,0)-h^2\delta_x^2\boldsymbol{\psi}_x(0,0)\right)\\
&\quad+h\left(\boldsymbol{\psi}_y(h,h)+\boldsymbol{\psi}_y(-h,-h)-2\boldsymbol{\psi}_y(0,0)-h^2\delta_y^2\boldsymbol{\psi}_y(0,0)\right)\Big),\\
(f_4)\quad a_{2,3} &= \frac{1}{32h^5}\Big(3(\boldsymbol{\psi}(0,h)-\boldsymbol{\psi}(-h,h)+\boldsymbol{\psi}(h,-h)-\boldsymbol{\psi}(0,-h))\\
&\quad-6\,h\delta_x\boldsymbol{\psi}(0,0)\\
&\quad+h\left(-\boldsymbol{\psi}_x(-h,h)-\boldsymbol{\psi}_x(h,-h)+2\boldsymbol{\psi}_x(0,0)+h^2\delta_x^2\boldsymbol{\psi}_x(0,0)\right)\\
&\quad+h\left(\boldsymbol{\psi}_y(-h,h)+\boldsymbol{\psi}_y(h,-h)-2\boldsymbol{\psi}_y(0,0)-h^2\delta_y^2\boldsymbol{\psi}_y(0,0)\right)\Big),
\end{aligned}
$$

(g) $a_{3,3} = \frac{1}{8h^2}\left(\delta_x^2(\delta_y\psi_y(0,0) - \delta_y^2\psi(0,0)) + \delta_y^2(\delta_x\psi_x(0,0) - \delta_y^2\psi(0,0))\right)$.

The discrete biharmonic operator is defined by $\Delta^2 Q(0,0)$ (see Definition 11.9)

$$\Delta^2 Q(0,0) = 24a_{4,0} + 24a_{0,4} + 16(a_{3,1} - a_{1,3}), \tag{11.124}$$

which is

$$\begin{aligned}\Delta^2 Q(0,0) &= \frac{12}{h^2}\left(\delta_x\psi_x - \delta_x^2\psi\right)(0,0) + \frac{12}{h^2}\left(\delta_y\psi_y - \delta_y^2\psi\right)(0,0) \\ &\quad + 2\big(3\delta_x^2\delta_y^2\psi(0,0) - \delta_y^2\delta_x\psi_x(0,0) - \delta_x^2\delta_y\psi_y(0,0)\big).\end{aligned} \tag{11.125}$$

It may be readily verified that this expression is identical to (11.81). We conclude that $\Delta^2 Q(0,0)$ is $\tilde{\Delta}_h^2$, where we define

$$\tilde{\Delta}_h^2\psi := \delta_x^4\psi + \delta_y^4\psi + 2\ (3\delta_x^2\delta_y^2\psi - \delta_x^2\delta_y\psi_y - \delta_y^2\delta_x\psi_x). \tag{11.126}$$

For simplicity we omit the indices i, j in the following. Since [by (11.12)]

$$\delta_x^2\delta_y^2\psi - \delta_x^2\delta_y\psi_y = -\frac{h^2}{12}\delta_x^2\delta_y^4\psi \tag{11.127}$$

and

$$\delta_x^2\delta_y^2\psi - \delta_y^2\delta_x\psi_x = -\frac{h^2}{12}\delta_y^2\delta_x^4\psi, \tag{11.128}$$

we infer that

$$\tilde{\Delta}_h^2\psi = \delta_x^4\psi + \delta_y^4\psi + 2\ \delta_x^2\delta_y^2\psi - 2\ \frac{h^2}{12}(\delta_x^2\delta_y^4\psi + \delta_y^2\delta_x^4\psi). \tag{11.129}$$

Using the definition of the nine-point biharmonic operator, introduced in (11.11),

$$\Delta_h^2\psi = \delta_x^4\psi + \delta_y^4\psi + 2\delta_x^2\delta_y^2\psi,$$

we get, [see (11.82)],

$$\tilde{\Delta}_h^2\psi = \Delta_h^2\psi - \frac{h^2}{6}(\delta_x^2\delta_y^4\psi + \delta_y^2\delta_x^4\psi). \tag{11.130}$$

We now study the accuracy of $\tilde{\Delta}_h^2$. We start with some formal considerations. Recall that the truncation error of $\delta_x^2\delta_y^2$ is [see (11.31) and (11.32)]

$$\delta_y^2\delta_x^2\psi^* - (\partial_y^2\partial_x^2\psi)^* = \frac{h^2}{12}\left((\partial_x^2\partial_y^4\psi)^* + (\partial_y^2\partial_x^4\psi)^*\right) + O(h^4). \tag{11.131}$$

Therefore, if we approximate $\partial_x^2\partial_y^4 + \partial_y^2\partial_x^4$ to second-order accuracy, then we obtain a fourth-order approximation of the mixed term. Since δ_x^4 and δ_y^4 are fourth-order discretizations of ∂_x^4 and ∂_y^4, respectively, and since δ_x^2

and δ_y^2 are second-order approximations of ∂_x^2 and ∂_y^2, respectively, therefore $\delta_x^2\delta_y^4 + \delta_y^2\delta_x^4$ is a second-order approximation of $\partial_x^2\partial_y^4 + \partial_y^2\partial_x^4$. Hence, $\delta_x^2\delta_y^2 - \frac{h^2}{12}(\delta_x^2\delta_y^4 + \delta_y^2\delta_x^4)$ is a fourth-order approximation of the mixed term. Thus, $\tilde{\Delta}_h^2$ approximates Δ^2 within fourth-order accuracy.

The following claim contains a rigorous statement of these considerations.

Claim 11.22 (Accuracy of the improved discrete biharmonic operator). *Let ψ be a smooth function in Ω and suppose that it vanishes along with its gradient on $\partial\Omega$. Then the improved discrete biharmonic operator $\tilde{\Delta}_h^2$ satisfies the following accuracy properties:*

- *At interior points $2 \le i, j \le N-2$, we have*

$$(11.132)\quad \begin{aligned} |\sigma_x\sigma_y\tilde{\Delta}_h^2\psi_{i,j}^* - \sigma_x\sigma_y(\Delta^2\psi)_{i,j}^*| &\le Ch^4\big(\|\partial_x^8\psi\|_{L^\infty} + \|\partial_y^8\psi\|_{L^\infty} \\ &+\|\partial_x^2\partial_y^6\psi\|_{L^\infty} + \|\partial_y^2\partial_x^6\psi\|_{L^\infty} + \|\partial_x^4\partial_y^4\psi\|_{L^\infty}\big). \end{aligned}$$

- *At near-boundary points $i = 1, N-1,\ \ 1 \le j \le N-1$ or $j = 1, N-1$, $1 \le i \le N-1$, we have*

$$(11.133)\quad \begin{aligned} |\sigma_x\sigma_y\tilde{\Delta}_h^2\psi_{i,j}^* - \sigma_x\sigma_y(\Delta^2\psi)_{i,j}^*| &\le Ch(\|\partial_x^5\psi\|_{L^\infty} + \|\partial_y^5\psi\|_{L^\infty}) \\ &+Ch^4(\|\partial_x^8\psi\|_{L^\infty} + \|\partial_y^8\psi\|_{L^\infty} \\ &+\|\partial_x^2\partial_y^6\psi\|_{L^\infty} + \|\partial_y^2\partial_x^6\psi\|_{L^\infty} + \|\partial_x^4\partial_y^4\psi\|_{L^\infty}). \end{aligned}$$

In the above estimates $C > 0$ is a constant independent of h and ψ.

Proof. The left-hand side of (11.132) can be expanded as in (11.27), yielding for $2 \le i, j \le N-2$

(11.134)

$$\begin{aligned} \sigma_x\sigma_y(\tilde{\Delta}_h^2\psi_{i,j}^* - (\Delta^2\psi)_{i,j}^*) &= \sigma_y(\sigma_x\delta_x^4\psi_{i,j}^* - \sigma_x(\partial_x^4\psi)_{i,j}^*) \\ &+ \sigma_x(\sigma_y\delta_y^4\psi_{i,j}^* - \sigma_y(\partial_y^4\psi)^*)_{i,j} \\ &+ 2\sigma_x\sigma_y\left(\delta_x^2\delta_y^2\psi_{i,j}^* - \frac{h^2}{12}(\delta_x^2\delta_y^4\psi_{i,j}^* + \delta_y^2\delta_x^4\psi_{i,j}^*) - (\partial_x^2\partial_y^2\psi)_{i,j}^*\right). \end{aligned}$$

It was shown in (11.28) and (11.29) that

$$(11.135)\quad |\sigma_y(\sigma_x\delta_x^4\psi_{i,j}^* - \sigma_x(\partial_x^4\psi)_{i,j}^*)| \le Ch^4\|\partial_x^8\psi\|_{L^\infty},\ \ 2 \le i, j \le N-2,$$

$$(11.136)\quad |\sigma_x(\sigma_y\delta_y^4\psi_{i,j}^* - \sigma_y(\partial_y^4\psi)_{i,j}^*)| \le Ch^4\|\partial_y^8\psi\|_{L^\infty},\ \ 2 \le i, j \le N-2.$$

To estimate the truncation error in the mixed term $\partial_x^2\partial_y^2\psi$, we rewrite (9.12) as

$$\begin{aligned} &\delta_x^2\psi_{i,j}^* - (\partial_x^2\psi)_{i,j}^* = \frac{h^2}{12}(\partial_x^4\psi)_{i,j}^* + \frac{2h^4}{6!}(\partial_x^6\psi)_{i,j}^* + r_{i,j}, \\ &|r_{i,j}| \le Ch^6\|\partial_x^8\psi\|_{L^\infty}. \end{aligned} \tag{11.137}$$

Operating with δ_y^2 on this equation yields

$$\begin{aligned} &\delta_y^2[\delta_x^2\psi_{i,j}^* - (\partial_x^2\psi)_{i,j}^*] = \frac{h^2}{12}\delta_y^2(\partial_x^4\psi)_{i,j}^* + \frac{2h^4}{6!}\delta_y^2(\partial_x^6\psi)_{i,j}^* + s_{i,j}, \\ &|s_{i,j}| \le Ch^4\|\partial_x^8\psi\|_{L^\infty}. \end{aligned} \tag{11.138}$$

Using a Taylor expansion for the terms $\delta_y^2(\partial_x^2\psi)^*$, $\delta_y^2(\partial_x^4\psi)^*$ and $\delta_y^2(\partial_x^6\psi)^*$, we find that

(11.139)

$$(a)\quad \delta_y^2(\partial_x^2\psi)_{i,j}^* = (\partial_y^2\partial_x^2\psi)_{i,j}^* + \frac{h^2}{12}(\partial_y^4\partial_x^2\psi)_{i,j}^* + \frac{2h^4}{6!}(\partial_y^6\partial_x^2\psi)(x_i,\eta_j),$$

$$(b)\quad \frac{2h^2}{4!}\delta_y^2(\partial_x^4\psi)_{i,j}^* = \frac{2h^2}{4!}\partial_y^2(\partial_x^4\psi)_{i,j}^* + \frac{4h^4}{(4!)^2}(\partial_y^4\partial_x^4\psi)(x_i,\tilde{\eta}_j),$$

$$(c)\quad \frac{2h^4}{6!}\delta_y^2(\partial_x^6\psi)_{i,j}^* = \frac{2h^4}{6!}\partial_y^2\partial_x^6\psi(x_i,\bar{\eta}_j),$$

where $\eta_j, \tilde{\eta}_j, \bar{\eta}_j \in (y_{j-1}, y_{j+1})$. Inserting (11.139) into (11.138), we obtain for $1 \le i,j \le N-1$,

$$\begin{aligned} &\delta_y^2\delta_x^2\psi_{i,j}^* - (\partial_y^2\partial_x^2\psi)_{i,j}^* - \frac{h^2}{12}[(\partial_y^4\partial_x^2\psi)_{i,j}^* + (\partial_y^2\partial_x^4\psi)_{i,j}^*] = t_{i,j}, \\ &|t_{i,j}| \le Ch^4\big(\|\partial_x^8\psi\|_{L^\infty} + \|\partial_x^6\partial_y^2\psi\|_{L^\infty} \\ &\qquad + \|\partial_x^2\partial_y^6\psi\|_{L^\infty} + \|\partial_x^4\partial_y^4\psi\|_{L^\infty}\big). \end{aligned} \tag{11.140}$$

We may estimate therefore the mixed term in (11.134) by

$$\begin{aligned} &\left|\sigma_x\sigma_y\left(\delta_y^2\delta_x^2\psi_{i,j}^* - (\partial_y^2\partial_x^2\psi)_{i,j}^* - \frac{h^2}{12}[\partial_y^4\partial_x^2\psi_{i,j}^* + \partial_y^2\partial_x^4\psi_{i,j}^*]\right)\right| \\ &\le Ch^4[\|\partial_x^8\psi\|_{L^\infty} + \|\partial_x^6\partial_y^2\psi\|_{L^\infty} + \|\partial_x^2\partial_y^6\psi\|_{L^\infty} + \|\partial_x^4\partial_y^4\psi\|_{L^\infty}]. \end{aligned} \tag{11.141}$$

Similarly,

$$\left|\sigma_x\sigma_y\left(\delta_x^2\delta_y^2\psi_{i,j}^* - (\partial_x^2\partial_y^2\psi)_{i,j}^* - \frac{h^2}{12}(\partial_y^4\partial_x^2\psi_{i,j}^* + \partial_y^2\partial_x^4\psi_{i,j}^*)\right)\right| \tag{11.142}$$
$$\leq Ch^4[\|\partial_y^8\psi\|_{L^\infty} + \|\partial_x^6\partial_y^2\psi\|_{L^\infty} + \|\partial_x^2\partial_y^6\psi\|_{L^\infty} + \|\partial_x^4\partial_y^4\psi\|_{L^\infty}].$$

Combining (11.135) and (11.136), and (11.141) and (11.142), we obtain (11.132).

At near-boundary points $i = 1$ or $i = N-1$ we have for any fixed $j = 0, \cdots, N$ [see (10.64)]

$$|\sigma_x\delta_x^4\psi_{i,j}^* - \sigma_x(\partial_x^4\psi)_{i,j}^*| \leq Ch\|\partial_x^5\psi\|_{L^\infty}. \tag{11.143}$$

Operating with σ_y on (11.143), we have, for $i = 1, N-1,\ 1 \leq j \leq N-1$,

$$|\sigma_y\sigma_x\delta_x^4\psi_{i,j}^* - \sigma_y\sigma_x(\partial_x^4\psi)_{i,j}^*| \leq Ch\|\partial_x^5\psi\|_{L^\infty}. \tag{11.144}$$

Similarly, for $j = 1, N-1,\ 1 \leq i \leq N-1$,

$$|\sigma_x\sigma_y\delta_y^4\psi_{i,j}^* - \sigma_x\sigma_y(\partial_y^4\psi)_{i,j}^*| \leq Ch\|\partial_y^5\psi\|_{L^\infty}. \tag{11.145}$$

Combining (11.144) and (11.145) with (11.141) and (11.142), we obtain (11.133). ∎

As for the second-order discrete biharmonic operator Δ_h^2, the evaluation of $\tilde{\Delta}_h^2$ does not require any artificial boundary conditions but only natural boundary data.

The corresponding scheme for solving the problem (11.1) is the same as (11.21) with Δ_h^2 replaced by $\tilde{\Delta}_h^2$,

$$\begin{cases} \tilde{\Delta}_h^2\psi_{i,j} = f^*(x_i, y_j), \quad 1 \leq i,j \leq N-1, \\ \psi_{i,j} = g_1^*(x_i, y_j), \quad \{i = 0, N, \quad 0 \leq j \leq N\} \\ \qquad\qquad\qquad\qquad or\ \{j = 0, N, \quad 0 \leq i \leq N\}, \\ \psi_{x,i,j} = -g_2^*(x_i, y_j), \quad i = 0, \quad 0 \leq j \leq N, \\ \psi_{x,i,j} = g_2^*(x_i, y_j), \quad i = N, \quad 0 \leq j \leq N, \\ \psi_{y,i,j} = -g_2^*(x_i, y_j), \quad j = 0, \quad 0 \leq i \leq N, \\ \psi_{y,i,j} = g_2^*(x_i, y_j), \quad j = N, \quad 0 \leq i \leq N. \end{cases} \tag{11.146}$$

In Proposition 11.7 we gave the coercivity of the nine-point biharmonic operator Δ_h^2. In the following proposition we prove the coercivity of the improved discrete biharmonic operator $\tilde{\Delta}_h^2$.

Proposition 11.23. *Let $\psi, \psi_x, \psi_y \in L_{h,0}^2$, where (ψ_x, ψ_y) is the Hermitian gradient of ψ. Then, the fourth-order discrete biharmonic operator $\tilde{\Delta}_h^2$*

satisfies the coercivity property

$$(\tilde{\Delta}_h^2\psi,\psi)_h \geq (\Delta_h^2\psi,\psi)_h$$

(11.147)

$$\geq C\Big(|\delta_x^+\psi_x|_h^2 + |\delta_y^+\psi_y|_h^2 + |\delta_x^+\psi_y|_h^2 + |\delta_y^+\psi_x|_h^2\Big).$$

In addition we have

$$(\tilde{\Delta}_h^2\psi,\psi)_h \geq C'|\psi|_h^2, \tag{11.148}$$

where C *and* C' *are constants independent of* h.

Proof. Let $\varphi,\varphi_x,\varphi_y \in L^2_{h,0}$ where (φ_x,φ_y) is the Hermitian gradient of φ. Using (11.130) we have

$$(\tilde{\Delta}_h^2\psi,\varphi)_h = (\Delta_h^2\psi,\varphi)_h - \frac{h^2}{6}(\delta_x^2\delta_y^4\psi,\varphi)_h - \frac{h^2}{6}(\delta_y^2\delta_4^4\psi,\varphi)_h. \tag{11.149}$$

Discrete integration by parts of the third term of (11.149) yields [using (11.45)]

$$(\delta_y^2\delta_x^4\psi,\varphi)_h = -(\delta_y^+\delta_x^4\psi,\delta_y^+\varphi)_h$$

$$= -(\delta_x^+\delta_y^+\psi_x,\delta_x^+\delta_y^+\varphi_x)_h - \frac{12}{h^2}(\delta_x^+\delta_y^+\psi - \mu_x\delta_y^+\psi_x,\delta_x^+\delta_y^+\varphi - \mu_x\delta_y^+\varphi_x)_h,$$

where δ_x^+ and δ_y^+ are defined in Subsection 11.1.4. Similarly, for the second term of (11.149), we have

$$(\delta_x^2\delta_y^4\psi,\varphi)_h = -(\delta_x^+\delta_y^4\psi,\delta_x^+\varphi)_h$$

$$= -(\delta_x^+\delta_y^+\psi_y,\delta_x^+\delta_y^+\varphi_y)_h - \frac{12}{h^2}(\delta_x^+\delta_y^+\psi - \mu_y\delta_x^+\psi_y,\delta_x^+\delta_y^+\varphi - \mu_y\delta_x^+\varphi_y)_h.$$

Combining these three equalities, we have

$$\begin{aligned}(\tilde{\Delta}_h^2\psi,\varphi)_h &= (\Delta_h^2\psi,\varphi)_h \\ &\quad + \frac{h^2}{6}\big((\delta_x^+\delta_y^+\psi_x,\delta_x^+\delta_y^+\varphi_x)_h + (\delta_x^+\delta_y^+\psi_y,\delta_x^+\delta_y^+\varphi_y)_h\big) \\ &\quad + 2(\delta_y^+(\delta_x^+\psi - \mu_x\psi_x),\delta_y^+(\delta_x^+\varphi - \mu_x\varphi_x))_h \\ &\quad + 2(\delta_x^+(\delta_y^+\psi - \mu_y\psi_y),\delta_x^+(\delta_y^+\varphi - \mu_y\varphi_y))_h.\end{aligned}$$

This allows us to estimate $(\tilde{\Delta}_h^2\psi,\psi)$ from below

$$(\tilde{\Delta}_h^2\psi,\psi)_h \geq (\Delta_h^2\psi,\psi)_h. \tag{11.150}$$

Using (11.150) and (11.50), we have

$$(\tilde{\Delta}_h^2\psi,\psi)_h \geq C\Big(|\delta_x^+\psi_x|_h^2 + |\delta_y^+\psi_y|_h^2 + |\delta_x^+\psi_y|_h^2 + |\delta_y^+\psi_x|_h^2\Big). \tag{11.151}$$

By (11.150) and (11.52) we also obtain

$$(\tilde{\Delta}_h^2\psi,\psi)_h \geq C'|\psi|_h, \tag{11.152}$$

where C and C' are constants independent of h. ■

Finally, Proposition 11.8 extends to the fourth-order biharmonic operator $\tilde{\Delta}_h^2$ in a straightforward manner.

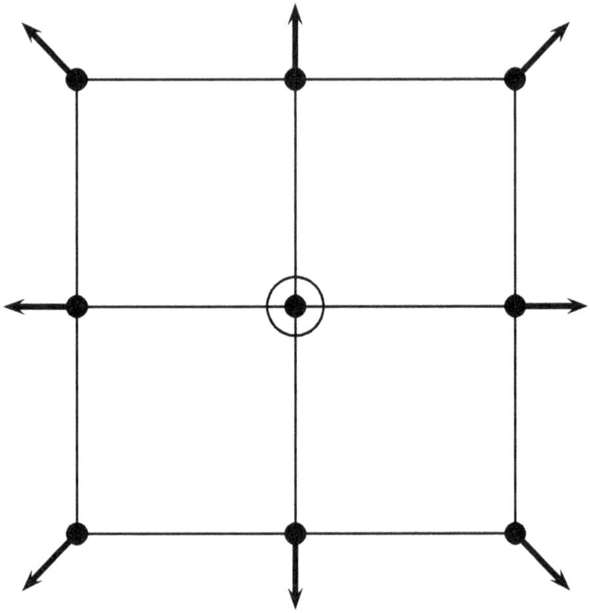

Fig. 11.5 Stencil of the 19-data compact biharmonic operator $\tilde{\Delta}_h^2$. There are nine degrees of freedom for the values of ψ, eight degrees of freedom for the directional derivatives at neighboring points, and two degrees of freedom for the full gradient at the center.

11.3 Notes for Chapter 11

- Equation (11.3), together with its truncation error, appears in the early numerical treatise [46]. Refer to [21] and the references therein for a detailed study of (11.3).
- The literature on polynomial interpolations based on functional values and their gradients, the so-called "Hermite–Birkhoff interpolations," is quite extensive (see [62] and references therein). In the one-dimensional case it is also referred as a "Lagrangian interpolation problem with repeated arguments." A general formula for this kind of problem is available [118]. However, the *unique character* of the interpolation approach presented in Section 11.2 is due to the following two ingredients: (i) A *subspace* $\Theta \subseteq \mathbb{P}^6$ is used, rather than the whole polynomial space. This subspace is spanned by *canonical* polynomials, independent of the underlying geometry of the irregular polygon.

(ii) The geometry of the polygon is connected to the interpolated data, that is, only the directional derivative towards the center is required at each vertex.

- The material of Section 11.2 is taken from [11]. That paper also contains a suitable Hermitian construction of approximate derivatives for irregular grids. The numerical implementation makes use of these derivatives so that the whole algorithm for the discrete biharmonic operator relies solely on the values of the function ψ at the grid points.
- Refer to [59, 60] and references therein for recent treatments of the biharmonic problem using finite element and finite volume methods.

Chapter 12

Compact Approximation of the Navier–Stokes Equations in Streamfunction Formulation

In this chapter we arrive at our ultimate goal, namely, the approximation of the full time-dependent Navier–Stokes system. The previous chapters have prepared the tools needed for this endeavor. We have at hand compact discrete Laplacian and biharmonic difference operators with high-order accuracy and coercivity properties. These properties allow us to prove convergence to the exact solution along the same line of proof as that used in establishing well-posedness of solutions in the continuous case (via *energy estimates*).

Our treatment relies solely on the pure streamfunction formulation of the system as discussed in the Introduction to this part. We briefly recall this formulation below, highlighting the fact that we shall be dealing with one scalar unknown function - the streamfunction. The boundary conditions are naturally expressed in terms of this function and its gradient. The price to be paid is the presence of the fourth-order biharmonic operator in the evolution equation. However, the discrete biharmonic operator at our disposal will prove to be a very good answer to this added difficulty.

12.1 The Navier–Stokes equations in streamfunction formulation

In the Introduction to this part we saw that the Navier–Stokes system is equivalent to the pure streamfunction Equation (7.2). In our treatment we use the "no-slip" boundary condition, namely, vanishing velocity on the boundary $\partial\Omega$. We thus arrive at the initial boundary-value problem for the streamfunction ψ,

$$\begin{cases} \partial_t(\Delta\psi) + (\boldsymbol{\nabla}^\perp\psi) \cdot \boldsymbol{\nabla}(\Delta\psi) - \nu\Delta^2\psi = 0, \quad (x,y) \in \Omega, \ t > 0, \\[2mm] \psi = \dfrac{\partial\psi}{\partial n} = 0, \quad (x,y) \in \partial\Omega, \ t > 0, \\[2mm] \psi(x,y,0) = \psi_0(x,y), \quad (x,y) \in \Omega. \end{cases}$$

(12.1)

Recall that in terms of the streamfunction $\psi(x,y,t)$, the velocity is given by $\mathbf{u} = \boldsymbol{\nabla}^\perp\psi$ and the vorticity is given by $\omega = \Delta\psi$.

As observed already, the (planar) Navier–Stokes system is thus reduced to an evolution equation for a scalar function ψ. This equation will also serve as the starting point of our discrete approximation.

Our method of proof for the convergence of the discrete approximation to the exact solution is an *energy method*. In fact, it is closely related to the energy method employed in the stability analysis of Equation (12.1). We therefore give the following theorem and its proof as a prelude to the convergence proof expounded in the following sections.

Theorem 12.1 (Stability and Uniqueness). *Let $\psi, \tilde{\psi} \in H_0^2(\Omega)$ be solutions of (12.1) having the same initial data. Then $\psi \equiv \tilde{\psi}$.*[1]

Proof. Let $\mathbf{u} = \boldsymbol{\nabla}^\perp\psi$ and $\mathbf{v} = \boldsymbol{\nabla}^\perp\tilde{\psi}$ be the corresponding velocity fields, and $\xi = \Delta\psi$ and $\eta = \Delta\tilde{\psi}$ the corresponding vorticities. Both ψ and $\tilde{\psi}$ satisfy Equation (12.1). Taking the difference and multiplying by $\psi - \tilde{\psi}$ we get

(12.2)
$$\int_\Omega (\psi - \tilde{\psi})\partial_t(\xi - \eta)dxdy - \int_\Omega \xi((\mathbf{u} - \mathbf{v}) \cdot \boldsymbol{\nabla})(\psi - \tilde{\psi})dxdy$$
$$- \int_\Omega (\xi - \eta)(\mathbf{v} \cdot \boldsymbol{\nabla})(\psi - \tilde{\psi})dxdy = \nu \int_\Omega (\xi - \eta)^2 dxdy.$$

But clearly $(\mathbf{u} - \mathbf{v}) \cdot \boldsymbol{\nabla}(\psi - \tilde{\psi}) \equiv 0$. The first term on the left-hand side of (12.2) can be rewritten as

(12.3) $$\int_\Omega (\psi - \tilde{\psi})\partial_t(\xi - \eta)dxdy = -\frac{1}{2}\frac{d}{dt}\int_\Omega |\boldsymbol{\nabla}(\psi - \tilde{\psi})|^2 dxdy.$$

For the third integral on the left-hand side of (12.2), we use Hölder's inequality, and an interpolation inequality for the $L^4(\Omega)$ norm [171, Sec.

---

[1] Readers unfamiliar with the basic theory of Sobolev spaces may skip this proof. All the necessary facts can be found in [58, Chapter 5].

3.3.3],

$$
(12.4)\quad
\begin{aligned}
&\Big|\int_\Omega (\xi-\eta)(\mathbf{v}\cdot\boldsymbol{\nabla})(\psi-\tilde{\psi})\,dxdy\Big| \\
&\quad\le \|\xi-\eta\|_{L^2(\Omega)}\ \|\mathbf{v}\|_{L^4(\Omega)}\ \|\boldsymbol{\nabla}(\psi-\tilde{\psi})\|_{L^4(\Omega)} \\
&\quad\le C\|\xi-\eta\|_{L^2(\Omega)}\ \Big[\|\mathbf{v}\|^{1/2}_{L^2(\Omega)}\ \|\boldsymbol{\nabla}\mathbf{v}\|^{1/2}_{L^2(\Omega)}\Big] \\
&\qquad\cdot\Big[\|\boldsymbol{\nabla}(\psi-\tilde{\psi})\|^{1/2}_{L^2(\Omega)}\ \|\Delta(\psi-\tilde{\psi})\|^{1/2}_{L^2(\Omega)}\Big],
\end{aligned}
$$

where C is a *generic* constant depending only on Ω. We note that, by definition and standard elliptic estimates

$$
\begin{aligned}
\|\mathbf{v}\|_{L^2(\Omega)} &\le C\|\tilde{\psi}\|_{H^1_0(\Omega)} \le C\|\Delta\tilde{\psi}\|_{L^2(\Omega)}, \\
\|\boldsymbol{\nabla}\mathbf{v}\|_{L^2(\Omega)} &\le C\|\Delta\tilde{\psi}\|_{L^2(\Omega)},
\end{aligned}
$$

where we have used the assumption $\tilde{\psi}\in H^0_2(\Omega)$ in substituting $\|\Delta\tilde{\psi}\|_{L^2(\Omega)}$ for the $H^0_2(\Omega)$ norm of $\tilde{\psi}$.

It follows that (12.4) can be rewritten as

$$
(12.5)\quad
\begin{aligned}
&\Big|\int_\Omega (\xi-\eta)(\mathbf{v}\cdot\boldsymbol{\nabla})(\psi-\tilde{\psi})\,dxdy\Big| \\
&\quad\le C\|\xi-\eta\|^{3/2}_{L^2(\Omega)}\ \|\boldsymbol{\nabla}(\psi-\tilde{\psi})\|^{1/2}_{L^2(\Omega)}\|\Delta\tilde{\psi}\|_{L^2(\Omega)}.
\end{aligned}
$$

The right-hand side of (12.5) can be further estimated as

$$
(12.6)\quad
\begin{aligned}
&\|\xi-\eta\|^{3/2}_{L^2(\Omega)}\ \|\boldsymbol{\nabla}(\psi-\tilde{\psi})\|^{1/2}_{L^2(\Omega)}\|\Delta\tilde{\psi}\|_{L^2(\Omega)} \\
&\quad\le \epsilon\|\xi-\eta\|^2_{L^2(\Omega)} + \frac{27}{256\epsilon^3}\|\Delta\tilde{\psi}\|^4_{L^2(\Omega)}\ \|\boldsymbol{\nabla}(\psi-\tilde{\psi})\|^2_{L^2(\Omega)}.
\end{aligned}
$$

Here we have used Young's inequality $AB\le(1/p)A^p+(1/q)B^q$, where $A,B>0$ and $(1/p)+(1/q)=1$, with $p=4/3$, $q=4$ and

$$
\begin{aligned}
A &= \Lambda\epsilon^{3/4}\|\xi-\eta\|^{3/2}_{L^2(\Omega)},\quad \Lambda=(4/3)^{3/4}, \\
B &= \frac{1}{\Lambda\epsilon^{3/4}}\|\boldsymbol{\nabla}(\psi-\tilde{\psi})\|^{1/2}_{L^2(\Omega)}\|\Delta\tilde{\psi}\|_{L^2(\Omega)}.
\end{aligned}
$$

We take $\epsilon=\frac{\nu}{2C}$ in this estimate and insert it into (12.2). In conjunction with (12.3) and (12.5) we get

$$
(12.7)\quad \frac{d}{dt}\|\boldsymbol{\nabla}(\psi-\tilde{\psi})\|^2_{L^2(\Omega)}\le C\|\boldsymbol{\nabla}(\psi-\tilde{\psi})\|^2_{L^2(\Omega)},
$$

where $C>0$ depends on $\tilde{\psi}$ (in addition to ν and Ω) but not on ψ. Since $\psi(x,y,0)=\tilde{\psi}(x,y,0)$, Gronwall's inequality yields $\psi\equiv\tilde{\psi}$. ■

The convergence proof of the semidiscrete version of (12.1) (see Section 12.3 below) follows the same lines as the proof of Theorem 12.1, in a suitable discrete sense.

12.2 Discretizing the streamfunction equation

In Subsection 11.1.2 we considered the discretization of the biharmonic equation in a rectangle. As seen in Equation (12.1), the biharmonic operator is the highest-order differential operator appearing in the streamfunction formulation. In this section we assume that our domain is $\Omega = [0,1] \times [0,1]$ and discretize Equation (12.1) in this domain.

We use the notation developed in Section 11.1. In particular, we approximate the unknown streamfunction $\psi(x,y,t)$ by functions $\boldsymbol{\psi}_{i,j}(t)$ at nodes (x_i, y_j). Note that the dependence on time remains *continuous*. Such an approximation is called a *semi-discrete approximation*. We therefore obtain a system of ordinary differential equations for the unknown (time-dependent) grid function $\boldsymbol{\psi}(t) = \{\boldsymbol{\psi}_{i,j}(t)\}$, given by

$$
(12.8)\qquad \begin{cases} \dfrac{d}{dt}\Delta_h \boldsymbol{\psi}_{i,j}(t) + C_h(\boldsymbol{\psi}(t))_{i,j} - \nu \Delta_h^2 \boldsymbol{\psi}_{i,j}(t) = 0, \\ \qquad\qquad\qquad 1 \le i,j \le N-1, \\ \boldsymbol{\psi}_{i,j}(t) = 0, \quad \{i = 0, N, \quad 0 \le j \le N\} \\ \qquad\qquad\qquad or \quad \{j = 0, N, \quad 0 \le i \le N\}, \\ \boldsymbol{\psi}_{x,i,j}(t) = 0, \quad i = 0, N, \quad 0 \le j \le N, \\ \boldsymbol{\psi}_{y,i,j}(t) = 0, \quad j = 0, N, \quad 0 \le i \le N, \end{cases}
$$

where $\Delta_h = \delta_x^2 + \delta_y^2$ and Δ_h^2 is the nine-point biharmonic operator given by (11.11). It remains to define the term $C_h(\boldsymbol{\psi}(t))_{i,j}$, which is a discrete version of the convective term $\nabla^{\perp}\psi \cdot \nabla\Delta\psi$. Recall that we already have approximations to first-order derivatives, namely, the Hermitian derivatives defined in (11.9). Thus, $\nabla^{\perp}\psi$ is naturally replaced by $(-\boldsymbol{\psi}_y, \boldsymbol{\psi}_x)$ and $\nabla\Delta\psi = \Delta\nabla\psi$ is replaced by $\Delta_h(\boldsymbol{\psi}_x, \boldsymbol{\psi}_y)$. We can therefore take

$$
(12.9)\qquad C_h(\boldsymbol{\psi})_{i,j} = -\boldsymbol{\psi}_{y,i,j}(\Delta_h \boldsymbol{\psi}_x)_{i,j} + \boldsymbol{\psi}_{x,i,j}(\Delta_h \boldsymbol{\psi}_y)_{i,j},
$$

as a discrete analog of the convective term. The coupled system of equations (12.8) is supplemented by initial data $\boldsymbol{\psi}_{i,j}(0) = \boldsymbol{\psi}^0_{i,j} = \psi_0(x_i, y_j)$ for $0 \le i,j \le N$.

Remark 12.2. In a full analogy with the continuous case, the (discrete) incompressibility condition is automatically satisfied, as noted in Proposition 11.4.

12.3 Convergence of the scheme

We assume that $\psi(x,y,t)$ is a smooth solution to (12.1) and denote by $\psi^*_{i,j}(t)$ its value at the node (x_i,y_j). At time t we have two sets of grid functions expressed as

$$\begin{aligned}&\{\psi^*_{i,j}(t), \quad 0\le i,j\le N\}\in L^2_0,\\ &\{\psi^*_{x,i,j}(t), \quad 0\le i,j\le N\}\in L^2_0,\\ &\{\psi^*_{y,i,j}(t), \quad 0\le i,j\le N\}\in L^2_0,\end{aligned} \tag{12.10}$$

$$\begin{aligned}&\{\boldsymbol{\psi}_{i,j}(t), \quad 0\le i,j\le N\}\in L^2_0,\\ &\{\boldsymbol{\psi}_{x,i,j}(t), \quad 0\le i,j\le N\}\in L^2_0,\\ &\{\boldsymbol{\psi}_{y,i,j}(t), \quad 0\le i,j\le N\}\in L^2_0.\end{aligned} \tag{12.11}$$

Here $(\psi^*_{x,i,j},\psi^*_{y,i,j})$ is the Hermitian gradient of ψ^*, satisfying

$$\sigma_x\psi^*_x=\delta_x\psi^*, \qquad \sigma_y\psi^*_y=\delta_y\psi^*. \tag{12.12}$$

Similarly for $(\boldsymbol{\psi}_{x,i,j},\boldsymbol{\psi}_{y,i,j})$, the Hermitian derivatives of $\boldsymbol{\psi}$. For the finite difference operators δ_x, δ_y, σ_x and σ_y, as well as δ_x^+ and δ_y^+, refer to Subsections 11.1.2 and 11.1.4.

We are interested in estimating the error, which is the grid function $\mathfrak{e}$ given by

$$\mathfrak{e}_{i,j}(t)=\boldsymbol{\psi}_{i,j}(t)-\psi^*_{i,j}(t). \tag{12.13}$$

Recall the definition of δ_x^+ from (11.40). Our fundamental convergence result is the following theorem.

Theorem 12.3. *Let $T>0$. Then there exist constants $C,h_0>0$, depending possibly on T,ν and the exact solution ψ, such that, for all $0\le t\le T$,*

$$\begin{cases}|\delta_x^+\mathfrak{e}|_h^2+|\delta_y^+\mathfrak{e}|_h^2\le Ch^3, \quad 0<h\le h_0,\\ \\ \qquad |\mathfrak{e}|_h^2\le Ch^3, \quad 0<h\le h_0.\end{cases} \tag{12.14}$$

Proof. The grid function $\boldsymbol{\psi}$ satisfies the semi-discrete system (12.8), which we rewrite as

$$\partial_t\Delta_h\boldsymbol{\psi}=-\boldsymbol{\nabla}_h^{\perp}\boldsymbol{\psi}\cdot(\Delta_h\boldsymbol{\nabla}_h\boldsymbol{\psi})+\nu\Delta_h^2\boldsymbol{\psi}, \tag{12.15}$$

subject to the initial condition

$$\boldsymbol{\psi}_{i,j}(0)=\psi_0(x_i,y_j). \tag{12.16}$$

The grid function ψ^* satisfies the approximate system

$$\partial_t\Delta_h\psi^*=-\boldsymbol{\nabla}_h^{\perp}\psi^*\cdot\left[\Delta_h\boldsymbol{\nabla}_h(\psi^*)\right]+\nu\Delta_h^2\psi^*+F, \tag{12.17}$$

where F is the truncation error of the scheme depending on the regularity of the exact solution. Let $\mathbf{u}^*$ and $\mathbf{u}$ be the discrete velocities associated with ψ^* and $\boldsymbol{\psi}$, namely,

$$\mathbf{u}^* = (-\psi^*_y, \psi^*_x), \quad \mathbf{u} = (-\boldsymbol{\psi}_y, \boldsymbol{\psi}_x). \tag{12.18}$$

The error $\mathfrak{e}(t)$ evolves according to

$$\partial_t \Delta_h \mathfrak{e} - \nu \Delta_h^2 \mathfrak{e} = -\left[\mathbf{u} \cdot \Delta_h(\boldsymbol{\psi}_x, \boldsymbol{\psi}_y) - \mathbf{u}^* \cdot \Delta_h(\psi^*_x, \psi^*_y)\right] - F. \tag{12.19}$$

The right-hand side of (12.19) is decomposed as

$$\begin{aligned}
&\left[\mathbf{u} \cdot \Delta_h(\boldsymbol{\psi}_x, \boldsymbol{\psi}_y) - \mathbf{u}^* \cdot \Delta_h(\psi^*_x, \psi^*_y)\right] + F \\
&\quad = (\mathbf{u} - \mathbf{u}^*) \cdot \Delta_h\left[(\boldsymbol{\psi} - \psi^*)_x, (\boldsymbol{\psi} - \psi^*)_y\right] + (\mathbf{u} - \mathbf{u}^*) \cdot \Delta_h\left[(\psi^*_x, \psi^*_y)\right] \\
&\qquad + \mathbf{u}^* \cdot \Delta_h\left[(\boldsymbol{\psi} - \psi^*)_x, (\boldsymbol{\psi} - \psi^*)_y\right] + F.
\end{aligned}$$

Taking the discrete scalar product [see (8.25)] with $\mathfrak{e}(t)$, we obtain

$$\begin{aligned}
(\partial_t \Delta_h \mathfrak{e} - \nu \Delta_h^2 \mathfrak{e}, \mathfrak{e})_h = &-\Big((\mathbf{u} - \mathbf{u}^*) \cdot \Delta_h\left[(\boldsymbol{\psi} - \psi^*)_x, (\boldsymbol{\psi} - \psi^*)_y\right], \mathfrak{e}\Big)_h \\
&-\Big((\mathbf{u} - \mathbf{u}^*) \cdot \Delta_h(\psi^*_x, \psi^*_y), \mathfrak{e}\Big)_h \\
&-\Big(\mathbf{u}^* \cdot \Delta_h\left[(\boldsymbol{\psi} - \psi^*)_x, (\boldsymbol{\psi} - \psi^*)_y\right], \mathfrak{e}\Big)_h \\
&-\big(F, \mathfrak{e}\big)_h.
\end{aligned} \tag{12.20}$$

We denote the four terms of the right-hand side by J_1, J_2, J_3 and J_4,

$$\begin{aligned}
J_1 &= \big((\mathbf{u} - \mathbf{u}^*) \cdot \Delta_h[(\boldsymbol{\psi} - \psi^*)_x, (\boldsymbol{\psi} - \psi^*)_y], \mathfrak{e}\big)_h, \\
J_2 &= \big((\mathbf{u} - \mathbf{u}^*) \cdot \Delta_h(\psi^*_x, \psi^*_y), \mathfrak{e}\big)_h, \\
J_3 &= \big(\mathbf{u}^* \cdot \Delta_h[(\boldsymbol{\psi} - \psi^*)_x, (\boldsymbol{\psi} - \psi^*)_y], \mathfrak{e}\big)_h, \\
J_4 &= (F, \mathfrak{e})_h.
\end{aligned}$$

We estimate separately the four terms J_1, J_2, J_3 and J_4.

Estimating the term J_1.

We have

$$\mathbf{u} - \mathbf{u}^* = \Big(-(\boldsymbol{\psi} - \psi^*)_y, (\boldsymbol{\psi} - \psi^*)_x\Big) = (-\mathfrak{e}_y, \mathfrak{e}_x), \tag{12.21}$$

so that

$$
\begin{aligned}
J_1 &= \Big((\mathbf{u}-\mathbf{u}^*)\cdot \Delta_h(\mathfrak{e}_x,\mathfrak{e}_y),\mathfrak{e}\Big)_h \\
&= \big(-\mathfrak{e}_y(\delta_x^2\mathfrak{e}_x+\delta_y^2\mathfrak{e}_x)+\mathfrak{e}_x(\delta_x^2\mathfrak{e}_y+\delta_y^2\mathfrak{e}_y),\mathfrak{e}\big)_h \\
&= \big(-\mathfrak{e}_y(\delta_x^2\mathfrak{e}_x+\delta_y^2\mathfrak{e}_x),\mathfrak{e}\big)_h + \big(\mathfrak{e}_x(\delta_x^2\mathfrak{e}_y+\delta_y^2\mathfrak{e}_y),\mathfrak{e}\big)_h \\
&= -\big(\delta_x^2\mathfrak{e}_x,\mathfrak{e}\mathfrak{e}_y\big)_h - \big(\delta_y^2\mathfrak{e}_x,\mathfrak{e}\mathfrak{e}_y\big)_h + \big(\delta_x^2\mathfrak{e}_y,\mathfrak{e}\mathfrak{e}_x\big)_h + \big(\delta_y^2\mathfrak{e}_y,\mathfrak{e}\mathfrak{e}_x\big)_h \\
&= \big(\delta_x^+\mathfrak{e}_x,\delta_x^+(\mathfrak{e}\mathfrak{e}_y)\big)_h + \big(\delta_y^+\mathfrak{e}_x,\delta_y^+(\mathfrak{e}\mathfrak{e}_y)\big)_h \\
&\quad -\big(\delta_x^+\mathfrak{e}_y,\delta_x^+(\mathfrak{e}\mathfrak{e}_x)\big)_h - \big(\delta_y^+\mathfrak{e}_y,\delta_y^+(\mathfrak{e}\mathfrak{e}_x)\big)_h.
\end{aligned}
\tag{12.22}
$$

The integration by parts in the last equality was carried out as in (9.32).

In order to formulate a discrete Leibniz rule for $\mathfrak{w},\mathfrak{z}\in L^2_{h,0}$ we use the "shift operators"

$$
\begin{aligned}
(S_x\mathfrak{w})_{i,j} &= \mathfrak{w}_{i+1,j}, \quad 1\le i,j\le N-1,\\
(S_y\mathfrak{z})_{i,j} &= \mathfrak{z}_{i,j+1}, \quad 1\le i,j\le N-1.
\end{aligned}
\tag{12.23}
$$

The boundary conditions give $\mathfrak{w}_{N,j}=\mathfrak{z}_{i,N}=0$ for $1\le i,j\le N-1$. It may be easily checked that

$$
\delta_x^+(\mathfrak{w}\mathfrak{z}) = (S_x\mathfrak{w})\delta_x^+\mathfrak{z} + \mathfrak{z}\delta_x^+\mathfrak{w}. \tag{12.24}
$$

Using (12.24), we expand J_1 as a sum of eight terms

$$
\begin{aligned}
J_1 = {} & \big(\delta_x^+\mathfrak{e}_x,(S_x\mathfrak{e}_y)\delta_x^+\mathfrak{e}\big)_h + \big(\delta_x^+\mathfrak{e}_x,\mathfrak{e}\delta_x^+\mathfrak{e}_y\big)_h \\
&+\big(\delta_y^+\mathfrak{e}_x,(S_y\mathfrak{e}_y)\delta_y^+\mathfrak{e}\big)_h + \big(\delta_y^+\mathfrak{e}_x,\mathfrak{e}\delta_y^+\mathfrak{e}_y\big)_h \\
&-\big(\delta_x^+\mathfrak{e}_y,(S_x\mathfrak{e}_x)\delta_x^+\mathfrak{e}\big)_h - \big(\delta_x^+\mathfrak{e}_y,\mathfrak{e}\delta_x^+\mathfrak{e}_x\big)_h \\
&-\big(\delta_y^+\mathfrak{e}_y,(S_y\mathfrak{e}_x)\delta_y^+\mathfrak{e}\big)_h - \big(\delta_y^+\mathfrak{e}_y,\mathfrak{e}\delta_y^+\mathfrak{e}_x\big)_h.
\end{aligned}
$$

The second and fourth terms are equal, respectively, to the sixth and eighth terms with opposite signs. Thus,

$$
\begin{aligned}
J_1 = {} & \big(\delta_x^+\mathfrak{e}_x,(S_x\mathfrak{e}_y)\delta_x^+\mathfrak{e}\big)_h + \big(\delta_y^+\mathfrak{e}_x,(S_y\mathfrak{e}_y)\delta_y^+\mathfrak{e}\big)_h \\
&-\big(\delta_x^+\mathfrak{e}_y,(S_x\mathfrak{e}_x)\delta_x^+\mathfrak{e}\big)_h - \big(\delta_y^+\mathfrak{e}_y,(S_y\mathfrak{e}_x)\delta_y^+\mathfrak{e}\big)_h.
\end{aligned}
$$

We now observe that if $\mathfrak{w}\in L^2_{h,0}$ then $|\mathfrak{w}|_\infty\le \frac{1}{h}|\mathfrak{w}|_h$. Since $|S_x\mathfrak{w}|_\infty = |\mathfrak{w}|_\infty$

and $|S_y\mathfrak{w}|_\infty = |\mathfrak{w}|_\infty$ we can estimate J_1 as

$$\begin{aligned}|J_1| &= \left|\Big((\mathbf{u}-\mathbf{u}^*)\cdot\Delta_h(\mathfrak{e}_x,\mathfrak{e}_y),\mathfrak{e}\Big)_h\right| \\ &\le \varepsilon\Big[|\delta_x^+\mathfrak{e}_x|_h^2 + |\delta_y^+\mathfrak{e}_x|_h^2 + |\delta_x^+\mathfrak{e}_y|_h^2 + |\delta_y^+\mathfrak{e}_y|_h^2\Big] \\ &\quad + \frac{1}{4\varepsilon}\Big[|(\mathfrak{e}_x,\mathfrak{e}_y)|_\infty^2\Big(|\delta_x^+\mathfrak{e}|_h^2 + |\delta_y^+\mathfrak{e}|_h^2\Big)\Big] \\ &\le \varepsilon\Big[|\delta_x^+\mathfrak{e}_x|_h^2 + |\delta_y^+\mathfrak{e}_x|_h^2 + |\delta_x^+\mathfrak{e}_y|_h^2 + |\delta_y^+\mathfrak{e}_y|_h^2\Big] + \frac{C}{\varepsilon h^2}\Big[|\delta_x^+\mathfrak{e}|_h^2 + |\delta_y^+\mathfrak{e}|_h^2\Big]^2.\end{aligned}$$

Note that in the last step we used (10.11) and (10.24) to estimate $|\mathfrak{e}_x|_\infty \le C|\delta_x\mathfrak{e}|_\infty \le C|\delta_x^+\mathfrak{e}|_\infty$ and $|\mathfrak{e}_y|_\infty \le C|\delta_y\mathfrak{e}|_\infty \le C|\delta_y^+\mathfrak{e}|_\infty$, where C is a constant independent of h. The factor $\varepsilon > 0$ will be specified later.

Estimating the term J_2.

The term J_2 is estimated by

(12.25) $$|J_2| = \left|\big((\mathbf{u}-\mathbf{u}^*)\cdot\Delta_h(\psi_x^*,\psi_y^*),\mathfrak{e}\big)_h\right| \le C\big[|\mathbf{u}-\mathbf{u}^*|_h^2 + |\mathfrak{e}|_h^2\big],$$

where C is a generic constant (depending on ψ). Here, we used the fact that $\Delta_h(\psi_x^*,\psi_y^*)$ is the discrete operator Δ_h operating on the Hermitian gradient of the exact solution. Thus, due to Lemma 10.1 (the high accuracy of the Hermitian derivatives), it is bounded if the exact solution is sufficiently regular. In addition, using that $\mathbf{u}-\mathbf{u}^* = \big[-(\boldsymbol{\psi}_y - \psi_y^*), \boldsymbol{\psi}_x - \psi_x^*\big]$, we have

(12.26) $$|\mathbf{u}-\mathbf{u}^*|_h^2 = |\mathfrak{e}_x|_h^2 + |\mathfrak{e}_y|_h^2.$$

Furthermore, in view of (10.11) and (10.87) we have

(12.27) $$|\mathfrak{e}_x|_h \le C|\delta_x^+\mathfrak{e}|_h,\quad |\mathfrak{e}_y|_h \le C|\delta_y^+\mathfrak{e}|_h,$$

and due to the Poincaré inequality (9.35) we obtain

(12.28) $$|J_2| \le C\Big[|\delta_x^+\mathfrak{e}|_h^2 + |\delta_y^+\mathfrak{e}|_h^2\Big].$$

Estimating the term J_3.

We have

$$\begin{aligned}J_3 = \Big(\mathbf{u}^*\cdot\Delta_h(\mathfrak{e}_x,\mathfrak{e}_y),\mathfrak{e}\Big)_h &= \underbrace{(-\psi_y^*\delta_x^2\mathfrak{e}_x,\mathfrak{e})_h}_{J_{3,1}} + \underbrace{(-\psi_y^*\delta_y^2\mathfrak{e}_x,\mathfrak{e})_h}_{J_{3,2}} \\ &\quad + \underbrace{(\psi_x^*\delta_x^2\mathfrak{e}_y,\mathfrak{e})_h}_{J_{3,3}} + \underbrace{(\psi_x^*\delta_y^2\mathfrak{e}_y,\mathfrak{e})_h}_{J_{3,4}}.\end{aligned}$$

In particular,

$$J_{3,1} = (-\psi_y^* \delta_x^2 \mathfrak{e}_x, \mathfrak{e})_h = -(\delta_x^2 \mathfrak{e}_x, \psi_y^* \mathfrak{e})_h = \Big(\delta_x^+ \mathfrak{e}_x, \delta_x^+(\psi_y^* \mathfrak{e})\Big)_h. \tag{12.29}$$

Using (12.24), this term can be estimated as

$$\begin{aligned}|J_{3,1}| = \left|\big(\delta_x^+ \mathfrak{e}_x, \delta_x^+(\psi_y^* \mathfrak{e})\big)_h\right| &\le |\delta_x^+ \mathfrak{e}_x|_h |\delta_x^+(\psi_y^* \mathfrak{e})|_h \\ &\le |\delta_x^+ \mathfrak{e}_x|_h \Big[|(S_x \psi_y^*)\delta_x^+ \mathfrak{e}|_h + |\mathfrak{e}\delta_x^+ \psi_y^*|_h\Big] \\ &\le |\delta_x^+ \mathfrak{e}_x|_h \Big[|\psi_y^*|_\infty |\delta_x^+ \mathfrak{e}|_h + |\delta_x^+ \psi_y^*|_\infty |\mathfrak{e}|_h\Big].\end{aligned}$$

Therefore, using the Poincaré inequality (9.35), we can further estimate

$$\begin{aligned}|J_{3,1}| &\le \max\big(|\psi_y^*|_\infty, |\delta_x^+ \psi_y^*|_\infty\big)\left[\varepsilon|\delta_x^+ \mathfrak{e}_x|_h^2 + \frac{1}{4\varepsilon}(|\delta_x^+ \mathfrak{e}|_h + |\mathfrak{e}|_h)^2\right] \\ &\le \max\big(|\psi_y^*|_\infty, |\delta_x^+ \psi_y^*|_\infty\big)\left[\varepsilon|\delta_x^+ \mathfrak{e}_x|_h^2 + \frac{C}{\varepsilon}(|\delta_x^+ \mathfrak{e}|_h^2 + |\delta_y^+ \mathfrak{e}|_h^2)\right].\end{aligned}$$

Similarly, in the y direction, we obtain for the term $J_{3,2}$

$$\begin{aligned}|J_{3,2}| &= |(\psi_y^* \delta_y^2 \mathfrak{e}_x, \mathfrak{e})_h| \\ &\le \max\big(|\psi_y^*|_\infty, |\delta_y^+ \psi_y^*|_\infty\big)\left[\varepsilon|\delta_y^+ \mathfrak{e}_x|_h^2 + \frac{C}{\varepsilon}(|\delta_x^+ \mathfrak{e}|_h^2 + |\delta_y^+ \mathfrak{e}|_h^2)\right].\end{aligned} \tag{12.30}$$

Therefore, with $m_1 = \max\left[|\psi_y^*|_\infty, |\delta_x^+ \psi_y^*|_\infty, |\delta_y^+ \psi_y^*|_\infty\right]$, the estimate of the term $J_{3,1} + J_{3,2}$ is

$$\begin{aligned}|J_{3,1} + J_{3,2}| &\le |J_{3,1}| + |J_{3,2}| \\ &\le m_1\left[\varepsilon\left\{|\delta_x^+ \mathfrak{e}_x|_h^2 + |\delta_y^+ \mathfrak{e}_x|_h^2\right\} + \frac{C}{\varepsilon}\left\{|\delta_x^+ \mathfrak{e}|_h^2 + |\delta_y^+ \mathfrak{e}|_h^2\right\}\right].\end{aligned} \tag{12.31}$$

Treating the term $J_{3,3} + J_{3,4}$ in a similar way, we obtain

$$\begin{aligned}|J_{3,3} + J_{3,4}| &\le |J_{3,3}| + |J_{3,4}| \\ &\le m_2\left[\varepsilon\{|\delta_x^+ \mathfrak{e}_y|_h^2 + |\delta_y^+ \mathfrak{e}_y|_h^2\} + \frac{C}{\varepsilon}\{|\delta_x^+ \mathfrak{e}|_h^2 + |\delta_y^+ \mathfrak{e}|_h^2\}\right],\end{aligned}$$

where $m_2 = \max\left[|\psi_x^*|_\infty, |\delta_x^+ \psi_x^*|_\infty, |\delta_y^+ \psi_x^*|_\infty\right]$.

Finally, defining $M = \max(m_1, m_2)$, the estimate for the term J_3 is

$$\begin{aligned}|J_3| \le M\bigg[&\varepsilon\left\{|\delta_x^+ \mathfrak{e}_x|_h^2 + |\delta_y^+ \mathfrak{e}_x|_h^2 + |\delta_x^+ \mathfrak{e}_y|_h^2 + |\delta_y^+ \mathfrak{e}_y|_h^2\right\} \\ &+ \frac{2C}{\varepsilon}\left\{|\delta_x^+ \mathfrak{e}|_h^2 + |\delta_y^+ \mathfrak{e}|_h^2\right\}\bigg].\end{aligned} \tag{12.32}$$

Here M depends on the exact solution using the accuracy of the Hermitian derivatives (see Lemma 10.1).

Estimating the term J_4.
The term J_4 is associated with the truncation error. From Lemma 10.1 and Equation (11.39), it is of order 3/2 in the $|\cdot|_h$ norm. For any time $T > 0$ the term J_4 is estimated by

$$|J_4| \leq C(T)|\mathfrak{e}|_h h^{3/2} \leq C(T)\left[|\delta_x^+\mathfrak{e}|_h^2 + |\delta_y^+\mathfrak{e}|_h^2 + h^3\right], \tag{12.33}$$

where $C(T)$ is a constant depending only on $T > 0$ and on the regularity of the exact solution $\psi(t)$ on the interval $[0, T]$. Here the discrete Poincaré inequality (9.35) has been used for the estimate of $|\mathfrak{e}|_h$.

Turning back to (12.20), we have, on $[0, T]$,

$$\begin{aligned}(\partial_t\Delta_h\mathfrak{e}, \mathfrak{e})_h - \nu(\Delta_h^2\mathfrak{e}, \mathfrak{e})_h &= -\frac{1}{2}\frac{d}{dt}\left\{|\delta_x^+\mathfrak{e}|_h^2 + |\delta_y^+\mathfrak{e}|_h^2\right\} - \nu(\Delta_h^2\mathfrak{e}, \mathfrak{e})_h \\ &= -J_1 - J_2 - J_3 - J_4,\end{aligned}$$

or

$$\begin{aligned}\frac{1}{2}\frac{d}{dt}\left\{|\delta_x^+\mathfrak{e}|_h^2 + |\delta_y^+\mathfrak{e}|_h^2\right\} &= J_1 + J_2 + J_3 + J_4 - \nu(\Delta_h^2\mathfrak{e}, \mathfrak{e})_h \\ &\leq |J_1| + |J_2| + |J_3| + |J_4| - \nu(\Delta_h^2\mathfrak{e}, \mathfrak{e})_h \\ &\leq |J_1| + |J_2| + |J_3| + |J_4| \\ &\quad - C\nu\left[|\delta_x^+\mathfrak{e}_x|_h^2 + |\delta_y^+\mathfrak{e}_y|_h^2 + |\delta_x^+\mathfrak{e}_y|_h^2 + |\delta_y^+\mathfrak{e}_x|_h^2\right].\end{aligned}$$

In the final inequality, we used the coercivity property (11.50). Collecting the terms containing $|\delta_x^+\mathfrak{e}_x|_h^2 + |\delta_y^+\mathfrak{e}_y|_h^2 + |\delta_x^+\mathfrak{e}_y|_h^2 + |\delta_y^+\mathfrak{e}_x|_h^2$, which appear in the estimates for J_1, J_2, J_3 and J_4, and selecting $\varepsilon > 0$ sufficiently small, we conclude that these terms are absorbed in the right-hand side of the final inequality. We are therefore left with the estimate

$$\begin{aligned}&\frac{d}{dt}\left\{|\delta_x^+\mathfrak{e}|_h^2 + |\delta_y^+\mathfrak{e}|_h^2\right\} \\ &\leq C\left[|\delta_x^+\mathfrak{e}|_h^2 + |\delta_y^+\mathfrak{e}|_h^2\right]\left[1 + \frac{1}{h^2}(|\delta_x^+\mathfrak{e}|_h^2 + |\delta_y^+\mathfrak{e}|_h^2)\right] + C'h^3.\end{aligned} \tag{12.34}$$

Here, C and C' depend on the exact solution ψ and on the viscosity coefficient ν but not on h.

In order to prove convergence of the approximate solution $\widetilde{\psi}$ to the exact solution ψ using (12.34), we proceed as follows. We use the fact that at $t = 0$ the error $\mathfrak{e} = 0$, and prove an estimate for $|\delta_x^+\mathfrak{e}|_h + |\delta_y^+\mathfrak{e}|_h$ up to any given time $T > 0$. Fix some $K > 0$. Since $\mathfrak{e} = 0$ at $t = 0$, also $\delta_x^+ \mathfrak{e} = \delta_y^+ \mathfrak{e} = 0$. Thus, taking $h > 0$, there exists a time $\tau > 0$ (which may depend on h) such that

$$\sup_{0 \le t \le \tau} \left\{ |\delta_x^+ \mathfrak{e}|_h + |\delta_y^+ \mathfrak{e}|_h \right\} \le Kh. \tag{12.35}$$

Inserting (12.35) into (12.34) we have for $t \le \tau$

$$\begin{aligned} &\frac{d}{dt}\left[|\delta_x^+ \mathfrak{e}|_h^2 + |\delta_y^+ \mathfrak{e}|_h^2 \right] \\ &\le C(1+K^2)\left[|\delta_x^+ \mathfrak{e}|_h^2 + |\delta_y^+ \mathfrak{e}|_h^2 \right] + C'h^3, \quad 0 < h \le h_0. \end{aligned} \tag{12.36}$$

Using Gronwall's inequality on (12.36) yields

$$|\delta_x^+ \mathfrak{e}|_h^2 + |\delta_y^+ \mathfrak{e}|_h^2 \le C_1 e^{C(1+K^2)t} h^3, \quad t \le \tau \tag{12.37}$$

with a suitable constant $C_1 > 0$. Observe that in (12.37), τ depends on h. Now, define $\tau_0 = \tau_0(h)$ by

$$\tau_0 = \sup\left\{ t > 0 \text{ such that } |\delta_x^+ \mathfrak{e}|_h + |\delta_y^+ \mathfrak{e}|_h \le Kh \right\}. \tag{12.38}$$

We have $\tau_0 \ge \tau$ and, as in (12.37), we obtain

$$|\delta_x^+ \mathfrak{e}|_h^2 + |\delta_y^+ \mathfrak{e}|_h^2 \le C_1 e^{C(1+K^2)t} h^3, \quad t \le \tau_0. \tag{12.39}$$

We can now select h_0 so small that

$$C_1 e^{C(1+K^2)T} h_0 < K^2. \tag{12.40}$$

Now the definition of τ_0, (12.39) and (12.40) imply that, for any $0 < h \le h_0$ we have $\tau_0(h) \ge T$ and, in particular, for such h, the estimate (12.37) holds for all $t \le T$.

The estimate $|\mathfrak{e}|_h^2 \le Ch^3$ of (12.14) is a result of $|\delta_x^+ \mathfrak{e}|_h^2 + |\delta_y^+ \mathfrak{e}|_h^2 \le Ch^3$ and the discrete Poincaré inequality (9.35). This concludes the proof of the theorem.

■

12.4 Notes for Chapter 12

- The proof of Theorem 12.3 was taken from [12].
- In [55] the fourth-order convergence of an essentially compact scheme for the Navier–Stokes system in vorticity-streamfunction formulation was proved. It uses the Briley formula for the application of vorticity boundary conditions.

- The application of various compact schemes to time-dependent Navier–Stokes equations in pure streamfunction formulation is fairly recent [16, 32, 63, 99, 119]. We mention also the use of the streamfunction formulation for stationary Stokes or Navier–Stokes equations [38, 91, 115, 117, 122, 149, 161].

Appendix B

Eigenfunction Approach for $u_{xxt} - u_{xxxx} = f(x,t)$

The search for a simple, linear, one-dimensional model for Equation (12.1) led Kupferman [119] to the following equation

$$u_{xxt} = u_{xxxx}, \tag{B.1}$$

where $u(x,t)$ is a function defined for $(x,t) \in [0,1] \times [0,\infty)$.

This equation is supplemented with an initial condition

$$u(x,0) = u_0(x), \quad x \in [0,1], \tag{B.2}$$

and boundary conditions

$$u(0,t) = u_x(0,t) = u(1,t) = u_x(0,t) = 0, \quad t \in [0,\infty). \tag{B.3}$$

It is obvious why this equation models the streamfunction equation (12.1). Indeed, on the left-hand side the Laplacian is replaced by the one-dimensional second-order derivative, with a similar replacement for the biharmonic operator on the right-hand side. On the other hand, the nonlinear convective term of (12.1) is not represented in this linear model.

Linearization of nonlinear evolution equations is a classical method [156]. It allows a simple, basic investigation of the properties of the solutions. It is also a useful tool in the investigation of the convergence of discrete approximations. In particular, it is amenable to the von Neumann methodology, namely, the use of a Fourier decomposition (both for the continuous and discrete versions), thus obtaining the "amplification factors" of pure modes. A closely related method relies on the replacement of the e^{ikx} modes by *eigenfunctions of the underlying* spatial operator, again both in the continuous and the discrete versions. Decomposing the initial function in terms of these eigenfunctions, it is easy (due to linearity) to follow the evolution in time of the solution and to estimate the accuracy of the approximation. Indeed, this was the purpose of the treatment in [119].

In contrast, our treatment of the convergence in Section 12.3 was *fully nonlinear* and relied in a substantial way on the coercivity of the discrete biharmonic operator. No use was made of any decomposition, either by Fourier modes or actual eigenfunctions.

However, we feel that it is appropriate to study this model in the context of the streamfunction equation. It gives a simple platform for experimentation with various approximations, where the biharmonic operator plays an important role. Thus, while we shall not use it in this Appendix for the time-dependent study of the equation, we shall carry out a detailed study of the eigenfunctions in both the continuous and the discrete cases. In the latter, we use the (one-dimensional versions of the) finite difference operators used in Chapter 12. We note that due to the special character of the operator (with the mixed space-time derivative $\partial_x^2\partial_t$), the very notion of an "eigenfunction" needs to be clarified. The completeness of such eigenfunctions depends on a suitable modification of the standard theorems of the Sturm–Liouville theory.

In Section 9.3 we studied the discrete analog of the standard second-order Sturm–Liouville problem on an interval. We introduced the complete set of (discrete) eigenfunctions of $-\delta_x^2$. This allowed us to use a "matrix representation" to study the convergence of the grid function $\mathfrak{u}$ satisfying $-\delta_x^2\mathfrak{u} = f^*$ to the continuous solution of $-u'' = f$. In other words, we could "diagonalize" the operator, leading to an easy estimate of the norm of its inverse.

We shall first study the basic properties of the continuous solutions and then focus on the discrete approximation. In both cases we will obtain a complete set of eigenfunctions. In the discrete case, the various features of the three-point biharmonic operator will come into play.

B.1 Some basic properties of the equation

We are interested in time-independent solutions obtained by assuming that the time behavior of the solution is given by $e^{\mu t}$. However, we will begin with a brief discussion of the well-posedness of the evolution equation. This discussion uses an elementary "energy" estimate. We assume that the equation is homogeneous, $f(x,t) \equiv 0$. In fact, due to linearity, Equation (12.1) can be treated by a standard application of the (continuous or discrete) Duhamel principle.

The essential feature of this simple equation is the presence of both second-order and fourth-order (spatial) derivatives. Thus, in spite of its resemblance to the heat equation, it displays some interesting features relevant to the more general treatment in chapter 13. One remarkable feature is the need for *two* boundary conditions, both for the function and its derivative. Thus, "factoring out" the second-order derivative and reducing Equation (B.1) to the heat equation is actually misleading, as we will see.

If Equation (B.1) is multiplied by $u(x,t)$, the homogeneous boundary conditions allow us to integrate by parts, thus giving

$$\frac{1}{2}\frac{d}{dt}\left\|\frac{\partial}{\partial x}u(\cdot,t)\right\|_{L^2}^2 = -\int_0^1\left|\frac{\partial^2}{\partial x^2}u(x,t)\right|^2 dx.$$

Combining this with the familiar inequality

$$\int_0^1 \phi(x)^2 dx \le \frac{1}{2}\int_0^1 \phi'(x)^2 dx,$$

which is valid for any differentiable function $\phi(x)$ vanishing at one of the endpoints, we obtain

$$\frac{d}{dt}\left\|\frac{\partial}{\partial x}u(\cdot,t)\right\|_{L^2}^2 \le -4\left\|\frac{\partial}{\partial x}u(\cdot,t)\right\|_{L^2}^2.$$

This type of estimate is a fundamental building block in establishing the well-posedness of (B.1) (in a suitable function space). We shall not pursue this topic further here but refer the reader to [58]. Instead, we present here the elementary eigenfunction approach to the equation, following [119].

We define an eigenfunction $\phi(x)$ of (B.1) as a function satisfying, for some $\mu \in \mathbb{R}$ (which is the associated eigenvalue),

(B.4) $$\mu\phi''(x) = \phi^{(4)}(x), \quad x \in [0,1],$$

along with the boundary conditions

(B.5) $$\phi(0) = \phi'(0) = \phi(1) = \phi'(1) = 0.$$

Claim B.1. *A complete set of orthogonal eigenfunctions and their associated eigenvalues are given by the union of the following two families*

(B.6) $$\begin{cases} \phi_k^{(1)}(x) = 1 - \cos(2\pi kx), \quad \mu_k^{(1)} = -(2\pi k)^2, \quad k = 1,2,..., \\ \phi_k^{(2)}(x) = \dfrac{1}{\pi q_k}\sin(2q_k\pi x) - \cos(2q_k\pi x) - 2x + 1, \\ \qquad\qquad \mu_k^{(2)} = -(2\pi q_k)^2, \qquad k = 2,3,..., \end{cases}$$

where (for the second family) q_k *is the (unique) solution of*

(B.7) $$\tan(q_k\pi) = q_k\pi, \quad q_k \in (k-1, k-\frac{1}{2}).$$

Proof. It is straightforward to check[1] that these functions are all eigenfunctions (in particular, satisfying all the boundary conditions). Their orthogonality is a consequence of their association with pairwise different eigenvalues. The fact that they constitute a complete set is not as trivial.

We first consider Equation (B.4) in the space $H_0^2[0,1]$, namely, the functions in $H^2[0,1]$ satisfying the boundary conditions (B.5) [58, Chapter 5]. Define the linear operators

$$L_2 u = u'', \quad L_4 u = u^{(4)}, \quad u \in H_0^2.$$

Note that the domain of L_4 is $H^4 \cap H_0^2$.

Equation (B.4) can be written as

$$\mu L_4^{-1} L_2 u = u, \quad u \in H_0^2[0,1].$$

By the Rellich compactness theorem the operator $L = L_4^{-1} L_2$ is compact on $H_0^2[0,1]$, so the Riesz theory applies and we have a complete orthonormal sequence of eigenfunctions with corresponding (negative) eigenvalues such that $\mu_j \to -\infty$.

The difficulty, of course, is to show that the union of the two families $\left\{\phi_k^{(1)}\right\}_{k=1}^{\infty}$ and $\left\{\phi_k^{(2)}\right\}_{k=2}^{\infty}$ is indeed complete, as asserted in the claim.

We proceed by distinguishing two cases:

- Case I: Assume that μ is an eigenvalue for the simpler Sturm–Liouville eigenvalue problem

  $$\mu\phi(x) = \phi''(x), \quad x \in [0,1], \quad \phi(0) = \phi(1) = 0.$$

  Then differentiating this equation twice we recover Equation (B.4). However, we need to verify that the boundary conditions $\phi'(0) = \phi'(1) = 0$ are satisfied in addition to $\phi(0) = \phi(1) = 0$. This imposes a further restriction. A simple inspection reveals that only the family $\left\{\phi_k^{(1)}(x)\right\}_{k=1}^{\infty}$ satisfies all conditions.

- Case II: Assume now that μ is not an eigenvalue for the simpler Sturm–Liouville problem. Using the operators introduced above, Equation (B.4) can be written as

  $$(\mu - L_2)\phi'' = 0.$$

  Using the fact that $\mu < 0$ we have, for any $\tilde{a}, \tilde{b} \in \mathbb{R}$,

  $$(\mu - L_2)(\phi''(x) - \tilde{a}\cos(\sqrt{-\mu}x) - \tilde{b}\sin(\sqrt{-\mu}x)) = 0.$$

[1] Readers unfamiliar with the elementary theory of Sobolev spaces can skip this proof.

Denoting $\psi(x) = \phi''(x) - \tilde{a}\cos(\sqrt{-\mu}x) - \tilde{b}\sin(\sqrt{-\mu}x)$, we take $\tilde{a}$ and $\tilde{b}$ so that $\psi(0) = \psi(1) = 0$. But then the assumption that μ is not an eigenvalue for the simpler Sturm–Liouville problem means that $\psi(x) \equiv 0$. Integrating twice we conclude that

$$\phi(x) = a\cos(\sqrt{-\mu}x) + b\sin(\sqrt{-\mu}x) + cx + d.$$

This function should satisfy all four boundary conditions (B.5). If $d = 0$ then the conditions at $x = 0$ imply $a = 0$ and $c + b\sqrt{-\mu} = 0$. On the other hand the conditions at $x = 1$ imply $c + b\sqrt{-\mu}\cos(\sqrt{-\mu}) = c + b\sqrt{-\mu}\sin(\sqrt{-\mu}) = 0$. This is a contradiction thus we conclude that $d \neq 0$ and we normalize to get $d = 1$. The four conditions (B.5) then determine a, b, c, and μ thus yield the family $\left\{\phi_k^{(2)}\right\}_{k=2}^{\infty}$ as asserted.

It is clear from the above arguments that the two cases exhaust all possible eigenfunctions, thus concluding the proof of the claim. ■

B.2 The discrete approximation

In Chapter 9 and Chapter 10 we introduced the discrete spatial derivatives δ_x^2 and δ_x^4, respectively. Using these operators we may approximate (B.4-B.5) by

(B.8) $$\mu\delta_x^2\mathfrak{v} = \delta_x^4\mathfrak{v}, \quad \mu \in \mathbb{R},$$

subject to the homogeneous boundary conditions

(B.9) $$\mathfrak{v}(x_0) = \mathfrak{v}_x(x_0) = \mathfrak{v}(x_N) = \mathfrak{v}_x(x_N) = 0.$$

We consider the interval $[0,1]$ and use the setup established in Chapter 8, with a uniform grid $\{x_i = ih,\ i = 0,1,...,N\}$. Recall that for functions of the type [see Equation (10.30)]

$$v(x) = A\cos(ax) + B\sin(bx) + Cx + D, \quad x \in [0,1], \quad A,B,C,D,a,b \in \mathbb{R},$$

we computed the Hermitian derivative and the action of δ_x^2 and δ_x^4 on $\mathfrak{v}$, the corresponding grid function, in Equations (10.31),(10.32), and (10.58). In the following claim we show that the discrete eigenfunctions span the space $l_{h,0}^2$. In fact, we give an explicit basis consisting of a union of two families of eigenfunctions, analogous to the continuous case of Claim B.1. The assumption that N is even is made for notational simplicity.

Claim B.2. *Suppose that N is even. Then a complete set of $(N-1)$ linearly independent eigenfunctions (in $l^2_{h,0}$) is given by the union of the following two families*

$$\text{(B.10)} \qquad \begin{cases} \mathfrak{v}_k^{(1)}(x_i) = 1 - \cos(2\pi k x_i), \quad 0 \le i \le N, \quad k = 1, 2, ..., \frac{N}{2}, \\ \mathfrak{v}_k^{(2)}(x_i) = A_k \sin(2 r_k \pi x_i) - \cos(2 r_k \pi x_i) - 2x_i + 1, \\ \qquad\qquad\qquad\qquad 0 \le i \le N, \qquad k = 2, ..., \frac{N}{2}, \end{cases}$$

where (for the second family), A_k and r_k are uniquely determined by the pair of equations

$$\text{(B.11)} \qquad \frac{1}{A_k} = \tan(r_k \pi), \quad r_k \in \left(k-1, k-\frac{1}{2}\right),$$

$$\text{(B.12)} \qquad A_k = \frac{2h}{3} \frac{2 + \cos(2 r_k \pi h)}{\sin(2 r_k \pi h)}.$$

Proof. Clearly $\mathfrak{v}_k^{(1)}(x_0) = \mathfrak{v}_k^{(1)}(x_N) = 0$. Also, we have for the Hermitian derivative

$$(\mathfrak{v}_k^{(1)})_x(x_i) = \frac{\sin(2\pi k h) h^{-1}}{1 - \frac{2}{3}\sin^2(\pi k h)} \sin(2\pi k x_i), \quad 0 \le i \le N,$$

so that $(\mathfrak{v}_k^{(1)})_x(x_0) = (\mathfrak{v}_k^{(1)})_x(x_N) = 0$.

From Equations (10.31) and (10.58) we readily have for $1 \le i \le N-1$,

$$\text{(B.13)} \qquad \delta_x^2 \mathfrak{v}_k^{(1)}(x_i) = 4h^{-2} \cos(2\pi k x_i) \sin^2(\pi k h),$$

and

$$\text{(B.14)} \qquad \delta_x^4 \mathfrak{v}_k^{(1)}(x_i) = -\frac{12}{h^4}\left[4\sin^2(\pi k h) - \frac{\sin^2(2\pi k h)}{1 - \frac{2}{3}\sin^2(\pi k h)}\right] \cos(2\pi k x_i).$$

Thus, $\mathfrak{v}_k^{(1)}$ are eigenfunctions associated with the eigenvalues

$$\text{(B.15)} \qquad \tilde{\mu}_k^{(1)} = -\frac{12}{h^2}\left[1 - \frac{\cos^2(\pi k h)}{1 - \frac{2}{3}\sin^2(\pi k h)}\right] = -\frac{4}{h^2}\frac{\sin^2(\pi k h)}{1 - \frac{2}{3}\sin^2(\pi k h)}.$$

For the discrete eigenfunctions of the second family in the claim, we note first that $\mathfrak{v}_k^{(2)}(x_0) = 0$ while the condition $\mathfrak{v}_k^{(2)}(x_N) = 0$ is satisfied [compare with Equation (B.7)] provided that, for $k = 2, ..., N/2$, Equation (B.11) is satisfied. For the Hermitian derivative we have by Equation (10.32) that for $1 \le i \le N-1$,

$$\begin{aligned} (\mathfrak{v}_k^{(2)})_x(x_i) &= A_k \frac{\sin(2 r_k \pi h) h^{-1}}{1 - \frac{2}{3}\sin^2(r_k \pi h)} \cos(2 r_k \pi x_i) \\ &\quad + \frac{\sin(2 r_k \pi h) h^{-1}}{1 - \frac{2}{3}\sin^2(r_k \pi h)} \sin(2 r_k \pi x_i) - 2 \\ &= \frac{3}{h}\frac{\sin(2 r_k \pi h)}{2 + \cos(2 r_k \pi h)}\left[A_k \cos(2 r_k \pi x_i) + \sin(2 r_k \pi x_i)\right] - 2. \end{aligned}$$

We need to extend this formula to $i = 0, N$, and require that $(\mathfrak{v}_k^{(2)})_x$ vanish there. The condition $(\mathfrak{v}_k^{(2)})_x(x_0) = 0$ holds if Equation (B.12) is satisfied.

Equations (B.11) and (B.12) constitute a pair of equations for the unknowns A_k and r_k. Solving for A_k we get

$$\frac{3}{h}\frac{\sin(2r_k\pi h)}{2+\cos(2r_k\pi h)} = 2\tan(r_k\pi), \quad r_k \in \left(k-1, k-\frac{1}{2}\right).$$

Since the right-hand side ranges (as a function of r_k) over $(0,\infty)$, it is easy to see that, for every $0 < h < \frac{1}{2}$, there exists a unique solution r_k, and A_k is obtained from (B.11). With these values, we can write

$$(\mathfrak{v}_k^{(2)})_x(x_i) = 2\left[\cos(2r_k\pi x_i) + \frac{1}{A_k}\sin(2r_k\pi x_i) - 1\right], \quad 0 \le i \le N.$$

The boundary condition $(\mathfrak{v}_k^{(2)})_x(x_N) = 0$ at $x_N = 1$ now follows from (B.11).

The discrete second-order and fourth-order derivatives are given by using (10.31) and (10.58),

- $\delta_x^2\mathfrak{v}_k^{(2)}(x_i) = \dfrac{4\sin^2(r_k\pi h)}{h^2}\Big[\cos(2r_k\pi x_i) - A_k\sin(2r_k\pi x_i)\Big]$, $1 \le i \le N-1$,
- $\delta_x^4\mathfrak{v}_k^{(2)}(x_i)$

$$= \frac{12}{h^4}\left[4\sin^2(r_k\pi h) - \frac{\sin^2(2r_k\pi h)}{1-\frac{2}{3}\sin^2(r_k\pi h)}\right]\{A_k\sin(2r_k\pi x_i) - \cos(2r_k\pi x_i)\}$$
$$= \frac{12}{h^4}\frac{4\sin^4(r_k\pi h)}{3-2\sin^2(r_k\pi h)}\{A_k\sin(2r_k\pi x_i) - \cos(2r_k\pi x_i)\}, \ 1 \le i \le N-1.$$

It follows that $\mathfrak{v}_k^{(2)}$ satisfies the eigenfunction equation (B.8) with associated eigenvalue

$$\tilde{\mu}_k^{(2)} = -\frac{12}{h^2}\frac{\sin^2(r_k\pi h)}{3-2\sin^2(r_k\pi h)}. \tag{B.16}$$

We assert that these functions are linearly independent. The basic observation here is that the eigenvalues $\left\{\mu_k^{(1)}, \mu_k^{(2)}\right\}$ are pairwise distinct. Thus, let $\mathfrak{v} \neq 0$ be a solution to (B.8) and let $\mathfrak{w} \neq 0$ be a solution to

$$\nu\delta_x^2\mathfrak{w} = \delta_x^4\mathfrak{w}, \quad \nu \in \mathbb{R},$$

where both $\mathfrak{v}$, $\mathfrak{w}$, and their Hermitian derivatives have zero boundary values. By the self-adjointness of the operator δ_x^4 (see Lemma 10.9 in Section 10.5) we have

$$(\delta_x^4\mathfrak{v}, \mathfrak{w})_h = (\mathfrak{v}, \delta_x^4\mathfrak{w})_h,$$

so that

$$\mu(\delta_x^2 \mathfrak{v}, \mathfrak{w})_h = \nu(\mathfrak{v}, \delta_x^2 \mathfrak{w})_h.$$

In view of the self-adjointness and the coercivity of the operator δ_x^2 [see Equation (9.32)] we know that $(\delta_x^2 \mathfrak{v}, \mathfrak{w})_h$ is a scalar product on $l_{h,0}^2$. Since $\mu \neq \nu$ implies $(\mathfrak{v}, \delta_x^2 \mathfrak{w})_h = 0$ we conclude that the whole set is linearly independent. ∎

Remark B.3. If we fix k and let $h \to 0$ it is easy to see that $\tilde{\mu}_k^{(1)} \to \mu_k^{(1)}$ and $\tilde{\mu}_k^{(2)} \to \mu_k^{(2)}$.

Indeed, in this case we also have $r_k \to q_k$ and $A_k \to \frac{1}{q_k \pi}$ so that the two sides of Equation (B.11) converge to the corresponding sides in Equation (B.7).

Chapter 13

Fully Discrete Approximation of the Navier–Stokes Equations

In Chapter 12 we introduced the Navier–Stokes equations in the streamfunction formulation and the associated semi-discrete second-order compact approximation in space (12.8). In this chapter we consider two topics. The first is a systematic derivation of fourth-order (rather than second-order) discrete approximations for all spatial derivatives. The second topic is the temporal discretization of the semi-discrete scheme, thus producing a fully discrete match to the equations.

13.1 Fourth-order approximation in space

In this section we enhance the spatial discretization in (12.8) by providing fourth-order approximations to all spatial derivatives in (12.1).

We rewrite the second-order semi-discrete scheme (12.8) adding a nonvanishing driving force f^*.

$$(13.1)\qquad \begin{cases} \dfrac{d}{dt}\Delta_h \psi_{i,j}(t) + C_h(\psi(t))_{i,j} - \nu\Delta_h^2\psi_{i,j}(t) = f^*_{i,j}(t), \\ \qquad\qquad 1 \le i,j \le N-1, \\ \psi_{i,j}(t) = 0, \quad \{i = 0, N, \quad 0 \le j \le N\} \\ \qquad\qquad or \quad \{j = 0, N, \quad 0 \le i \le N\}, \\ \psi_{x,i,j}(t) = 0, \quad i = 0, N, \quad 0 \le j \le N, \\ \psi_{y,i,j}(t) = 0, \quad j = 0, N \quad 0 \le i \le N. \end{cases}$$

In this system the discrete operators are the five-point Laplacian $\Delta_h = \delta_x^2 + \delta_y^2$ and the nine-point biharmonic operator Δ_h^2 (11.11). The discrete convective term $C_h(\psi)$ is given by [see (12.9)]

$$(13.2)\qquad C_h(\psi) = -\psi_y\Delta_h\psi_x + \psi_x\Delta_h\psi_y.$$

In this section we intend to replace the three operators Δ_h, Δ_h^2 and C_h with fourth-order approximations to the respective operators in (12.1).

A fourth-order approximation to the Laplacian operator, denoted by $\tilde{\Delta}_h$, can be derived from Equations (11.123)(c1) and (11.123)(c2),

$$\tilde{\Delta}_h \psi = 2(a_{2,0} + a_{0,2}) = 2\delta_x^2 \psi - \delta_x \psi_x + 2\delta_y^2 \psi - \delta_y \psi_y. \tag{13.3}$$

The fourth-order approximation to the biharmonic operator, denoted by $\tilde{\Delta}_h^2$, is the one already introduced in (11.130),

$$\tilde{\Delta}_h^2 \psi = \delta_x^4 \psi + \delta_y^4 \psi + 2\,\delta_x^2 \delta_y^2 \psi - \frac{h^2}{6}(\delta_x^2 \delta_y^4 \psi + \delta_y^2 \delta_x^4 \psi). \tag{13.4}$$

We now turn to the approximation of the convective term. In Subsection 13.1.1 we consider the "no-slip" boundary condition and in Subsection 13.1.2 we consider general boundary conditions.

13.1.1 *A fourth-order convective term: "no-slip" boundary conditions*

We will first study the truncation error in the approximation of the convective term in (12.1)

$$C(\psi) = (\nabla^\perp \psi) \cdot \nabla(\Delta \psi) = -\psi_y \partial_x \Delta \psi + \psi_x \partial_y \Delta \psi \tag{13.5}$$

by the discrete approximation (13.2). Let ψ be a smooth function and ψ^* its associated grid function (see Section 8.2). In view of (13.2), noting that ψ_x^* is the Hermitian derivative in the x direction of ψ^*, we have

$$\begin{aligned} &C_h(\psi^*) - \big(C(\psi)\big)^* \\ &= -\psi_y^* \Delta_h \psi_x^* + \psi_x^* \Delta_h \psi_y^* + (\partial_y \psi \cdot \partial_x \Delta \psi)^* - (\partial_x \psi \cdot \partial_y \Delta \psi)^* \\ &= (\partial_y \psi)^* ((\Delta \partial_x \psi)^* - \Delta_h \psi_x^*) \\ &\quad - (\partial_x \psi)^* ((\Delta \partial_y \psi)^* - \Delta_h \psi_y^*) + O(h^4). \end{aligned} \tag{13.6}$$

Here we have used the fourth-order accuracy of the Hermitian derivative [see (10.29)].

Next, we inspect the accuracy of the terms involving third-order derivatives. Our treatment is analogous to that accorded to the fourth-order discrete biharmonic operator $\tilde{\Delta}_h^2$ following (11.130). We omit grid indices for clarity.

We therefore need to estimate the terms $(\Delta \partial_x \psi)^* - \Delta_h \psi_x^*$ and $(\Delta \partial_y \psi)^* - \Delta_h \psi_y^*$. Considering the first we write

$$(\Delta \partial_x \psi)^* - \Delta_h \psi_x^* = (\partial_x^2 \partial_x \psi)^* - \delta_x^2 \psi_x^* + (\partial_y^2 \partial_x \psi)^* - \delta_y^2 \psi_x^*. \tag{13.7}$$

It suffices to estimate the first term, namely,

$$(\partial_x^2\partial_x\psi)^* - \delta_x^2\psi_x^* = \underbrace{(\partial_x^2\partial_x\psi)^* - \delta_x^2(\partial_x\psi)^*}_{I_1} + \underbrace{\delta_x^2(\partial_x\psi)^* - \delta_x^2\psi_x^*}_{I_2}. \tag{13.8}$$

For I_1 we have the estimate

$$I_1 \overset{(9.12)}{=} -\frac{h^2}{12}\partial_x^5\psi + O(h^4). \tag{13.9}$$

The estimate for I_2 is more delicate, as it involves the action of δ_x^2 on the difference between the grid values of $\partial_x\psi$ and the Hermitian derivative of ψ^*.

Operating on this difference with the Simpson operator σ_x [see (10.10)] we have

$$\begin{aligned}\sigma_x I_2 &= \sigma_x\delta_x^2((\partial_x\psi)^* - \psi_x^*) = \delta_x^2(\sigma_x(\partial_x\psi)^* - \sigma_x\psi_x^*)\\ &= \delta_x^2(\sigma_x(\partial_x\psi)^* - \delta_x\psi^*)\\ &\overset{(10.7)}{=} \delta_x^2\left[\frac{h^4}{180}(\partial_x^5\psi) + O(h^6)\right]\\ &= \frac{h^4}{180}\left[(\partial_x^7\psi)^* + O(h^2)\right] + \delta_x^2\, O(h^6) = O(h^4).\end{aligned} \tag{13.10}$$

Operating with σ_x^{-1} on the final equation and using the boundedness of σ_x^{-1} [see (10.24)], we obtain that $I_2 = O(h^4)$.

Thus, the h^2 term in the truncation error is due only to I_1. Collecting similar terms in (13.6), we find

$$\begin{aligned}C_h(\psi^*) - \big(C(\psi)\big)^* &= \frac{h^2}{12}\Big(- \partial_y\psi\cdot\partial_x(\partial_x^4\psi + \partial_y^4\psi) + \partial_x\psi\cdot\partial_y(\partial_x^4\psi + \partial_y^4\psi)\Big)^*\\ &\quad + O(h^4).\end{aligned} \tag{13.11}$$

Since the velocity $(u, v) = (-\partial_y\psi, \partial_x\psi)$ is divergence-free, the term in parenthesis on the right-hand side of the above equation can be written in conservative form as follows.

$$\begin{aligned}&-\partial_y\psi\cdot\partial_x(\partial_x^4\psi + \partial_y^4\psi) + \partial_x\psi\cdot\partial_y(\partial_x^4\psi + \partial_y^4\psi)\\ &= \partial_x\big(- \partial_y\psi(\partial_x^4\psi + \partial_y^4\psi)\big) + \partial_y\big(\partial_x\psi(\partial_x^4\psi + \partial_y^4\psi)\big).\end{aligned} \tag{13.12}$$

The partial derivatives on the right-hand side of this equation are approximated as follows.

The continuous differentiation operators ∂_x and ∂_y are approximated by the second-order discrete operators δ_x and δ_y, respectively.

The functions $\partial_x^4 \psi$ and $\partial_y^4 \psi$ are approximated by the fourth-order approximations $\delta_x^4 \psi^*$ and $\delta_y^4 \psi^*$, respectively. In addition $\partial_x \psi$ and $\partial_y \psi$ are approximated by the fourth-order Hermitian derivatives ψ_x^* and ψ_y^*, respectively. Since the lowest-order truncation error is due to the approximation of ∂_x and ∂_y by δ_x and δ_y, we obtain [see (9.12)]

$$\begin{aligned}&\Big(\partial_x(-\partial_y\psi\cdot(\partial_x^4\psi+\partial_y^4\psi))+\partial_y(\partial_x\psi\cdot(\partial_x^4\psi+\partial_y^4\psi))\Big)^*\\ &\quad=\delta_x(-\psi_y^*(\delta_x^4\psi^*+\delta_y^4\psi^*))+\delta_y(\psi_x^*(\delta_x^4\psi^*+\delta_y^4\psi^*))+O(h^2).\end{aligned} \tag{13.13}$$

Therefore, a fourth-order approximation of the convective term $C(\psi)$ can be written [using (13.11)–(13.13)] by

$$\begin{aligned}\tilde{C}_h(\psi^*) &= -\,\psi_y^*\Delta_h\psi_x^*+\psi_x^*\Delta_h\psi_y^*\\ &\quad-\frac{h^2}{12}\left(\delta_x(-\psi_y^*(\delta_x^4\psi^*+\delta_y^4\psi^*))+\delta_y(\psi_x^*(\delta_x^4\psi^*+\delta_y^4\psi^*))\right)\\ &=\big(C(\psi)\big)^*+O(h^4).\end{aligned} \tag{13.14}$$

Therefore, for any grid function $\boldsymbol{\psi}$, we define $\tilde{C}_h(\boldsymbol{\psi})$ as

$$\begin{aligned}\tilde{C}_h(\boldsymbol{\psi}) &= -\,\boldsymbol{\psi}_y\Delta_h\boldsymbol{\psi}_x+\boldsymbol{\psi}_x\Delta_h\boldsymbol{\psi}_y\\ &\quad-\frac{h^2}{12}\left(\delta_x(-\boldsymbol{\psi}_y(\delta_x^4\boldsymbol{\psi}+\delta_y^4\boldsymbol{\psi}))+\delta_y(\boldsymbol{\psi}_x(\delta_x^4\boldsymbol{\psi}+\delta_y^4\boldsymbol{\psi}))\right).\end{aligned} \tag{13.15}$$

This approximation involves high-order finite differences, appearing in the term

$$J(\boldsymbol{\psi}) := \delta_x\big(-\boldsymbol{\psi}_y(\delta_x^4\boldsymbol{\psi}+\delta_y^4\boldsymbol{\psi})\big)+\delta_y\big(\boldsymbol{\psi}_x(\delta_x^4\boldsymbol{\psi}+\delta_y^4\boldsymbol{\psi})\big). \tag{13.16}$$

We show now that in the special case of the "no-slip" boundary condition we can evaluate $J(\boldsymbol{\psi})$ at each interior point, including near-boundary points. Consider for example the term $\delta_x(-\boldsymbol{\psi}_y(\delta_x^4\boldsymbol{\psi}+\delta_y^4\boldsymbol{\psi}))$ at a point next to the left or right sides of the square. This requires knowing $\boldsymbol{\psi}_y\ \delta_x^4\boldsymbol{\psi}$ and $\boldsymbol{\psi}_y\ \delta_y^4\boldsymbol{\psi}$ on the boundary. Along the left and right sides the "no-slip" condition is $\mathfrak{u} = -\boldsymbol{\psi}_y = 0$, so that $-\boldsymbol{\psi}_y(\delta_x^4\boldsymbol{\psi}+\delta_y^4\boldsymbol{\psi})$ vanishes. Thus, $\delta_x(-\boldsymbol{\psi}_y(\delta_x^4\boldsymbol{\psi}+\delta_y^4\boldsymbol{\psi}))$ is computable near the left or the right sides. Next to the top or bottom sides, no problem arises when computing the value of $\delta_x(-\boldsymbol{\psi}_y(\delta_x^4\boldsymbol{\psi}+\delta_y^4\boldsymbol{\psi}))$, since δ_x operates in the x direction and $\delta_x^4\boldsymbol{\psi}, \delta_y^4\boldsymbol{\psi}$ may be evaluated at interior points. Similar considerations hold for $\delta_y(-\boldsymbol{\psi}_x(\delta_x^4\boldsymbol{\psi}+\delta_y^4\boldsymbol{\psi}))$.

Thus, we obtain the following fourth-order semi-discrete Navier–Stokes system

$$
\begin{cases}
\dfrac{d}{dt}\tilde{\Delta}_h \boldsymbol{\psi}_{i,j}(t) + \tilde{C}_h(\boldsymbol{\psi}(t))_{i,j} - \nu \tilde{\Delta}_h^2 \boldsymbol{\psi}_{i,j}(t) = f^*(x_i, y_j, t), \\
\hfill 1 \le i, j \le N-1, \\
\boldsymbol{\psi}_{i,j}(t) = 0, \quad \{i = 0, N, \quad 0 \le j \le N\} \\
\hfill \mathit{or} \quad \{j = 0, N, \quad 0 \le i \le N\}, \\
\boldsymbol{\psi}_{x,i,j}(t) = 0, \quad i = 0, N, \quad 0 \le j \le N, \\
\boldsymbol{\psi}_{y,i,j}(t) = 0, \quad j = 0, N \quad 0 \le i \le N.
\end{cases}
\tag{13.17}
$$

13.1.2 *A fourth-order convective term: General boundary conditions*

The discrete operator $\tilde{C}_h(\boldsymbol{\psi})$ in (13.15) was shown to be a fourth-order approximation of the convective term for the "no-slip" boundary condition. In this subsection, we will construct a fourth-order approximation for general boundary conditions. As in Subsection 13.1.1, a fourth-order approximation of the convective term $C(\psi)$ [see (13.5)] is guaranteed if the terms $\partial_x \Delta \psi$ and $\partial_y \Delta \psi$ are approximated to fourth order. Using the fact that the Hermitian gradient (ψ_x^*, ψ_y^*) approximates $(\partial_x \psi, \partial_y \psi)$ within fourth-order accuracy, we only need to construct a fourth-order approximation to $\partial_x(\Delta \psi)$ and $\partial_y(\Delta \psi)$. We consider for example the term

$$\partial_x \Delta \psi = \partial_x^3 \psi + \partial_y^2 \partial_x \psi. \tag{13.18}$$

First we approximate $\partial_x^3 \psi$ for some fixed $y = y_j$.

As in (10.50) we construct a fourth-order polynomial interpolating ψ^* at x_{i-1}, x_i, x_{i+1} and $(\partial_x \psi)^*$ at x_{i-1}, x_{i+1}. The coefficient a_3 of this polynomial is given by

$$6a_3 = \frac{3}{2h^2}((\partial_x \psi)^*_{i+1,j} + (\partial_x \psi)^*_{i-1,j} - 2\delta_x \psi^*_{i,j}).$$

We can consider $6a_3$ as an approximation to $\partial_x^3 \psi(x_i, y_j)$. However, using a Taylor expansion, it may be readily verified that this results in a second-order approximation to $\partial_x^3 \psi$. Since we are looking for a fourth-order approximation, we will try to obtain it by constructing a fifth-order polynomial $p(x, y_j)$,

$$
\begin{aligned}
p(x, y_j) &= a_0 + a_1(x - x_i) + a_2(x - x_i)^2 \\
&\quad + a_3(x - x_i)^3 + a_4(x - x_i)^4 + a_5(x - x_i)^5,
\end{aligned}
\tag{13.19}
$$

which interpolates both ψ^* and $(\partial_x\psi)^*$ at (x_{i-1}, y_j), (x_i, y_j), (x_{i+1}, y_j). Its third-order derivative is
(13.20)
$$\partial_x^3 p(x_i, y_j) = 6a_3 = \frac{3}{2h^2}\left\{10\delta_x\psi^*_{i,j} - \left[(\partial_x\psi)^*_{i+1,j} + 8(\partial_x\psi)^*_{i,j} + (\partial_x\psi)^*_{i-1,j}\right]\right\}$$
$$= \frac{3}{2h^2}\left(10\delta_x\psi^* - h^2\delta_x^2(\partial_x\psi)^* - 10(\partial_x\psi)^*\right)_{i,j} =: (\tilde{\psi}^*_{xxx})_{i,j}.$$
Using a Taylor expansion, it can be verified that $(\tilde{\psi}^*_{xxx})_{i,j}$, as defined above, is a fourth-order approximation to $(\partial_x^3\psi)(x_i, y_j)$. More precisely
$$(\tilde{\psi}^*_{xxx})_{i,j} = \partial_x^3\psi(x_i, y_j) - \frac{6}{7!}h^4\partial_x^7\psi(x_i, y_j) + O(h^6). \tag{13.21}$$
We now consider the term $\partial_y^2\partial_x\psi$ of (13.18). By a straightforward Taylor expansion, this term may be approximated to fourth-order accuracy by
$$\tilde{\psi}^*_{yyx} = \delta_y^2(\partial_x\psi)^* + \delta_x\delta_y^2\psi^* - \delta_x\delta_y(\partial_y\psi)^*. \tag{13.22}$$
Therefore, combining (13.20) and (13.22), $\partial_x\Delta\psi$ is approximated within fourth-order accuracy by
$$\begin{aligned}\widetilde{\partial_x\Delta_h}\psi^* &= \tilde{\psi}^*_{xxx} + \tilde{\psi}^*_{yyx} \\ &= \frac{3}{2}\left(10\frac{\delta_x\psi^* - (\partial_x\psi)^*}{h^2} - \delta_x^2(\partial_x\psi)^*\right) \\ &\quad + \delta_y^2(\partial_x\psi)^* + \delta_x\delta_y^2\psi^* - \delta_x\delta_y(\partial_y\psi)^*.\end{aligned} \tag{13.23}$$
Approximating $\partial_y\Delta\psi$ in a similar fashion, we obtain the following fourth-order approximation to the convective term
$$\begin{aligned}&\tilde{C}'_h(\psi^*, (\partial_x\psi)^*, (\partial_y\psi)^*) \\ &= -\psi^*_y\Big(\Delta_h(\partial_x\psi)^* + \frac{5}{2}(6\frac{\delta_x\psi^* - (\partial_x\psi)^*}{h^2} - \delta_x^2(\partial_x\psi)^*) \\ &\quad + \delta_x\delta_y^2\psi^* - \delta_x\delta_y(\partial_y\psi)^*\Big) \\ &\quad + \psi^*_x\Big(\Delta_h(\partial_y\psi)^* + \frac{5}{2}(6\frac{\delta_y\psi^* - (\partial_y\psi)^*}{h^2} - \delta_y^2(\partial_y\psi)^*) \\ &\quad + \delta_y\delta_x^2\psi^* - \delta_y\delta_x(\partial_x\psi)^*\Big) \\ &= \left(C(\psi)\right)^* + O(h^4).\end{aligned} \tag{13.24}$$
This equation is valid since $(\partial_x\psi)^*$ and $(\partial_y\psi)^*$ are the *exact values* of the first-order derivatives of ψ. Note that by the definition of the Hermitian derivative (10.11), we have
$$6\frac{\delta_x\psi^* - \psi^*_x}{h^2} - \delta_x^2\psi^*_x = 0, \quad 6\frac{\delta_y\psi^* - \psi^*_y}{h^2} - \delta_y^2\psi^*_y = 0.$$

Hence, if we replace $(\partial_x\psi, \partial_y\psi)$ in (13.24) by the Hermitian gradient (ψ_x^*, ψ_y^*), we obtain

(13.25)
$$\begin{aligned}\tilde{C}_h'(\psi^*, \psi_x^*, \psi_y^*) &= -\psi_y^*\left(\delta_x^2\psi_x^* + [\delta_y^2\psi_x^* + \delta_x\delta_y^2\psi^* - \delta_x\delta_y\psi_y^*]\right)\\ &\quad + \psi_x^*\left(\delta_y^2\psi_y^* + [\delta_x^2\psi_y^* + \delta_y\delta_x^2\psi^* - \delta_y\delta_x\psi_x^*]\right)\\ &= C(\psi^*) + O(h^2).\end{aligned}$$

Our objective is therefore to replace in (13.24) the exact derivatives $((\partial_x\psi)^*, (\partial_y\psi)^*)$ with discrete approximations $(\tilde{\psi}_x^*, \tilde{\psi}_y^*)$, while keeping the fourth-order accuracy. To this end we will use the Padé approximation [36, Equation (90)], having the following form

(13.26)
$$\frac{1}{3}(\tilde{\psi}_x^*)_{i+1,j} + (\tilde{\psi}_x^*)_{i,j} + \frac{1}{3}(\tilde{\psi}_x^*)_{i-1,j} = \frac{14}{9}\frac{\psi_{i+1,j}^* - \psi_{i-1,j}^*}{2h} + \frac{1}{9}\frac{\psi_{i+2,j}^* - \psi_{i-2,j}^*}{4h}.$$

The local truncation error $\tilde{\psi}_x^*$ in (13.26) is

(13.27)
$$(\tilde{\psi}_x^*)_{i,j} = (\partial_x\psi)_{i,j}^* + h^6\frac{1}{2100}(\partial_x^7\psi)_{i,j}^* + O(h^8).$$

Replacing $\partial_x\psi$ in (13.20) by (13.26), we obtain a modification of the approximation $\tilde{\psi}_{xxx}^*$, for which we retain the same notation,

(13.28)
$$(\tilde{\psi}_{xxx}^*)_{i,j} = (\partial_x^3\psi)_{i,j}^* + \frac{h^4}{120}(\partial_x^7\psi)_{i,j}^* + O(h^6).$$

At near-boundary points a one-sided approximation for $\partial_x\psi$ [36, Equation (93)] is applied. For the near-boundary point $i = 1$ we have

(13.29)
$$\frac{1}{10}(\tilde{\psi}_x^*)_{0,j} + \frac{6}{10}(\tilde{\psi}_x^*)_{1,j} + \frac{3}{10}(\tilde{\psi}_x^*)_{2,j} = \frac{-10\psi_{0,j}^* - 9\psi_{1,j}^* + 18\psi_{2,j}^* + \psi_{3,j}^*}{30h};$$

similarly for $i = N-1$. The term $\partial_y\psi$ is approximated in the same fashion. To summarize, a fourth-order approximation of the convective term for general boundary conditions is defined by

(13.30)
$$\begin{aligned}\tilde{C}_h'(\boldsymbol{\psi}) = &-\boldsymbol{\psi}_y\Big(\Delta_h\tilde{\boldsymbol{\psi}}_x + \frac{5}{2}\Big(6\frac{\delta_x\boldsymbol{\psi} - \tilde{\boldsymbol{\psi}}_x}{h^2} - \delta_x^2\tilde{\boldsymbol{\psi}}_x\Big) + \delta_x\delta_y^2\boldsymbol{\psi} - \delta_x\delta_y\tilde{\boldsymbol{\psi}}_y\Big)\\ &+ \boldsymbol{\psi}_x\Big(\Delta_h\tilde{\boldsymbol{\psi}}_y + \frac{5}{2}\Big(6\frac{\delta_y\boldsymbol{\psi} - \tilde{\boldsymbol{\psi}}_y}{h^2} - \delta_y^2\tilde{\boldsymbol{\psi}}_y\Big) + \delta_y\delta_x^2\boldsymbol{\psi} - \delta_y\delta_x\tilde{\boldsymbol{\psi}}_x\Big),\end{aligned}$$

where $\boldsymbol{\psi}_x$ and $\boldsymbol{\psi}_y$ are the Hermitian derivatives defined in (11.9). The approximate derivative $\tilde{\boldsymbol{\psi}}_x$ is defined in (13.26) and (13.29), and analogously for $\tilde{\boldsymbol{\psi}}_y$.

13.2 A time-stepping discrete scheme

13.2.1 *Implicit-explicit scheme for the Navier–Stokes system*

In this subsection we present a time-stepping methodology designed as a full discretization of the semi-discrete systems (13.1) and (13.17) in time. We give here a detailed scheme for (13.1); a similar treatment can be accorded to the system (13.17).

Choosing an explicit scheme for (13.1) or (13.17) would force a time step Δt to be proportional to h^2. In order to relax this constraint, we select a scheme that is implicit for the viscous term and explicit for the convective term. A natural time-stepping scheme for (13.1) consists of a combination of an implicit *Crank–Nicolson* scheme for the viscous term and an explicit *modified Euler* scheme for the convective term [65, 119],

$$
(13.31)\quad
\begin{cases}
\left(\Delta_h - \nu\frac{\Delta t}{4}\Delta_h^2\right)\boldsymbol{\psi}_{i,j}^{n+1/2} = \Delta_h\boldsymbol{\psi}_{i,j}^n - \frac{\Delta t}{2}C_h(\boldsymbol{\psi}_{i,j}^n)\\
\qquad +\nu\frac{\Delta t}{4}\Delta_h^2\boldsymbol{\psi}_{i,j}^n + \frac{\Delta t}{2}(f^*)_{i,j}^{n+1/4},\\
\left(\Delta_h - \nu\frac{\Delta t}{2}\Delta_h^2\right)\boldsymbol{\psi}_{i,j}^{n+1} = \Delta_h\boldsymbol{\psi}_{i,j}^n - \frac{\Delta t}{2}C_h(\boldsymbol{\psi}_{i,j}^{n+1/2})\\
\qquad +\nu\frac{\Delta t}{2}\Delta_h^2\boldsymbol{\psi}_{i,j}^n + \Delta t(f^*)_{i,j}^{n+1/2},
\end{cases}
$$

where the discrete convective term [see(13.2)] is

$$
(13.32)\qquad C_h(\boldsymbol{\psi}^n)_{i,j} = -\boldsymbol{\psi}_{y,i,j}^n\Delta_h\boldsymbol{\psi}_{x,i,j}^n + \boldsymbol{\psi}_{x,i,j}^n\Delta_h\boldsymbol{\psi}_{y,i,j}^n;
$$

similarly for $C_h(\boldsymbol{\psi}^{n+1/2})_{i,j}$. This scheme belongs to the family of Implicit-Explicit (IMEX) schemes (see [6, 7] and the Notes to this chapter). Using the spatial discrete approximate operators described in Section 12.2, we obtain a second-order scheme in time and space. Namely, if ψ is the exact solution of

$$
(13.33)\quad
\begin{cases}
\partial_t(\Delta\psi(x,y,t)) + C(\psi(x,y,t)) - \nu\Delta^2(\psi(x,y,t)) = f(x,y,t),\\
\qquad (x,y)\in\Omega,\ \ t>0\\
\psi = \frac{\partial\psi}{\partial n} = 0,\ (x,y)\in\partial\Omega,\ \ t>0\\
\psi(x,y,0) = \psi_0(x,y),\ \ (x,y)\in\Omega,
\end{cases}
$$

then its associated grid function $\psi_{i,j}^*(t_n = n\Delta t)$ satisfies (13.31) up to an error $O((\Delta t)^2 + h^2)$. Note that we apply the fully discrete scheme (13.31) at all interior points. At boundary points the given boundary conditions are imposed. These conditions determine $\boldsymbol{\psi}^n$, $\boldsymbol{\psi}_x^n$ and $\boldsymbol{\psi}_y^n$ on the boundary.

13.2.2 *Stability analysis*

In this subsection we consider the stability of the second-order scheme (13.31). We do so in terms of the linear equation

$$\Delta\psi_t = \bar{C}(\psi) + \nu\Delta^2\psi, \tag{13.34}$$

where $\bar{C}(\psi)$ is *a linear* convection term $\bar{C}(\psi) = a\Delta(\partial_x\psi) + b\Delta(\partial_y\psi)$ with a and b real constants. Note that for simplicity we take $\bar{C}(\psi)$ to be the analog of -$C(\psi)$ in (13.5). We consider the following approximation $\bar{C}_h(\boldsymbol{\psi})$ to $\bar{C}(\psi)$

$$\bar{C}_h(\boldsymbol{\psi}) = a\Delta_h\boldsymbol{\psi}_x + b\Delta_h\boldsymbol{\psi}_y, \tag{13.35}$$

where $\boldsymbol{\psi}_x$ and $\boldsymbol{\psi}_y$ are the Hermitian derivatives of the grid function $\boldsymbol{\psi}$. The analog of the semi-discrete system (12.8) is

(13.36)

$$\begin{cases} \dfrac{d}{dt}\Delta_h\psi_{i,j}(t) = \bar{C}_h(\boldsymbol{\psi})_{i,j}(t) + \nu\Delta_h^2\psi_{i,j}(t), \quad 1 \le i,j \le N-1, \\ \psi_{i,j}(t) = 0, \quad \{i = 0, N, \quad 0 \le j \le N\} \quad or \quad \{j = 0, N, \quad 0 \le i \le N\}, \\ \psi_{x,i,j}(t) = 0, \quad i = 0, N, \quad 0 \le j \le N, \\ \psi_{y,i,j}(t) = 0, \quad j = 0, N \quad 0 \le i \le N. \end{cases}$$

The time discretization corresponding to (13.31) is

$$\begin{cases} \dfrac{\Delta_h\boldsymbol{\psi}^{n+1/2} - \Delta_h\boldsymbol{\psi}^n}{\Delta t/2} = \bar{C}_h(\boldsymbol{\psi}^n) + \dfrac{\nu}{2}(\Delta_h^2\boldsymbol{\psi}^n + \Delta_h^2\boldsymbol{\psi}^{n+1/2}), \\ \dfrac{\Delta_h\boldsymbol{\psi}^{n+1} - \Delta_h\boldsymbol{\psi}^n}{\Delta t} = \bar{C}_h(\boldsymbol{\psi}^{n+1/2}) + \dfrac{\nu}{2}(\Delta_h^2\boldsymbol{\psi}^{n+1} + \Delta_h^2\boldsymbol{\psi}^n). \end{cases} \tag{13.37}$$

The von Neumann stability analysis [156, 169] of (13.37) requires the computation of its amplification factors. Some details are given in the Notes to this chapter and here we content ourselves by summarizing some basic results [14, 15]. This analysis is performed using a periodic setting on a uniform grid of mesh size h. Let the dimensionless coefficient μ be defined by

$$\mu = \frac{\nu\Delta t}{h^2}. \tag{13.38}$$

Proposition 13.1. *The difference scheme (13.37) is stable in the von Neumann sense under the sufficient condition*

$$(a^2 + b^2)\frac{\Delta t^2}{h^2} \le \min\left(\frac{4}{9}\mu, \frac{4}{27} + \frac{2}{27}\mu\right). \tag{13.39}$$

Remark 13.2. From (13.39) we infer that a sufficient condition for the stability of the difference scheme (13.37) is

$$(a^2 + b^2)\Delta t \leq \frac{2}{27}\nu. \tag{13.40}$$

Remark 13.3. Observe that in the non-convective case, $a = b = 0$, the scheme is unconditionally stable, as may be expected. Thus, the presence of lower-order convective terms makes it necessary to limit the time step by the size of the viscosity coefficient, a rather surprising situation.

13.3 Numerical results

We display here results for several analytical solutions of the Navier–Stokes system in streamfunction formulation

$$\Delta\psi_t + \nabla^{\perp}\psi \cdot \nabla\Delta\psi - \nu\Delta^2\psi = f. \tag{13.41}$$

The numerical calculations were carried out by implementing the second-order or fourth-order compact schemes in space and with second- or higher-order time-stepping schemes, as described in the preceding sections. The calculated errors are displayed in three different forms:

- Discrete L^2_h error [see (8.23)]

$$|\mathfrak{e}|_h = |\psi^* - \boldsymbol{\psi}|_h. \tag{13.42}$$

- Discrete relative L^2_h error

$$|\mathfrak{e}_r|_h = |\mathfrak{e}|_h / |\psi^*|_h. \tag{13.43}$$

- Maximum error for the first-order derivatives of ψ [see (8.9)]

$$|\mathfrak{e}_v|_\infty = |(\partial_x\psi)^* - \boldsymbol{\psi}_x|_\infty = |v^* - \mathfrak{v}|_\infty, \tag{13.44}$$

$$|\mathfrak{e}_u|_\infty = |(\partial_y\psi)^* - \boldsymbol{\psi}_y|_\infty = |u^* - \mathfrak{u}|_\infty. \tag{13.45}$$

The calculations are carried out on three increasingly refined meshes. The rate of convergence of an error e_h, measured between calculations with mesh sizes $h_2 < h_1$, is by definition $\left(\log \frac{e_{h_1}}{e_{h_2}}\right) / \left(\log \frac{h_1}{h_2}\right)$.

13.3.1 *Test case 1: Green vortex*

As a first example we have chosen the following exact solution

$$\psi(x, y, t) = -0.5e^{-2\nu t} \sin x \sin y, \quad (x, y) \in [0, \pi]^2 \tag{13.46}$$

of the Navier–Stokes equations [39]. In this case $\Delta\psi = -2\psi$ so that the convective term of (13.41) vanishes. We have picked $\nu = 1$ (Re = 1).

In Table 13.1, we display results using the second-order scheme in space and time (13.31). In order to get optimal accuracy in space, the time step was selected as $\Delta t = O(h^2)$. In this periodic problem we actually obtain fourth-order accuracy in space instead of the expected second-order accuracy.

Table 13.1 Compact second-order scheme (13.31) for $\psi = -0.5e^{-2t} \sin x \sin y$ on $[0, \pi] \times [0, \pi]$ with $f = 0$. The time step is $\Delta t = O(h^2)$.

Mesh		17×17	Rate	33×33	Rate	65×65
$t = 1$	$\lvert e \rvert_h$	2.27(-5)	4.00	1.41(-6)	3.99	8.87(-8)
	$\lvert e_r \rvert_h$	2.13(-4)	3.97	1.35(-5)	4.01	8.34(-7)
	$\lvert e_u \rvert_\infty$	1.79(-5)	3.98	1.13(-6)	4.00	7.08(-8)
$t = 2$	$\lvert e \rvert_h$	3.19(-6)	4.00	2.00(-7)	4.00	1.25(-8)
	$\lvert e_r \rvert_h$	2.22(-4)	4.00	1.39(-5)	4.00	8.69(-7)
	$\lvert e_u \rvert_\infty$	2.52(-6)	3.98	1.60(-7)	4.00	9.99(-9)
$t = 3$	$\lvert e \rvert_h$	4.33(-7)	4.00	2.71(-8)	3.99	1.70(-9)
	$\lvert e_r \rvert_h$	2.22(-4)	4.00	1.39(-5)	4.00	8.71(-7)
	$\lvert e_u \rvert_\infty$	3.43(-7)	3.99	2.16(-8)	4.00	1.35(-9)
$t = 4$	$\lvert e \rvert_h$	5.86(-8)	4.00	3.67(-9)	4.00	2.29(-10)
	$\lvert e_r \rvert_h$	2.22(-4)	4.00	1.39(-5)	4.00	8.71(-7)
	$\lvert e_u \rvert_\infty$	4.62(-8)	3.98	2.93(-9)	4.00	1.83(-10)

13.3.2 *Test case 2*

Next we chose a non-periodic problem where the forcing function in the Navier–Stokes equation is chosen as $f(x, y, t) = -16\nu e^{-\nu t}(x^2 + y^2 + 4)$, so that the exact solution to (13.41) is

$$\psi(x, y, t) = e^{-\nu t}(x^2 + y^2)^2, \quad (x, y) \in [0, 1]^2. \tag{13.47}$$

The corresponding velocity components are

$$\begin{cases} u = -\partial_y \psi = -4y(y^2 + x^2)e^{-\nu t}, \\ v = \partial_x \psi = 4x(x^2 + y^2)e^{-\nu t}. \end{cases} \tag{13.48}$$

In Table 13.2 we display results using the second-order scheme in space and time. The results are shown for different time levels and for different meshes. We observe a second-order convergence rate, as expected.

Table 13.2 Compact second-order scheme (13.31) for $\psi = e^{-t}(x^2+y^2)^2$ with $f = -16e^{-t}(x^2+y^2+4)$ on $[0,1]\times[0,1]$. The time step is $\Delta t = O(h^2)$.

Mesh		17×17	Rate	33×33	Rate	65×65
$t=0.25$	$\|e\|_h$	6.202(−5)	1.99	1.564(−5)	2.00	3.903(−6)
	$\|e_r\|_h$	8.176(−5)		1.895(−5)		4.535(−6)
	$\|e_u\|_\infty$	2.070(−4)	1.99	5.224(−5)	2.00	1.304(−5)
$t=0.5$	$\|e\|_h$	7.632(−5)	2.00	1.908(−5)	2.00	4.762(−6)
	$\|e_r\|_h$	1.030(−4)		2.368(−5)		5.671(−6)
	$\|e_u\|_\infty$	2.572(−4)	2.00	6.431(−5)	2.00	1.605(−5)
$t=0.75$	$\|e\|_h$	7.896(−5)	2.01	1.964(−5)	2.00	4.904(−6)
	$\|e_r\|_h$	1.091(−4)		2.498(−5)		5.984(−6)
	$\|e_u\|_\infty$	2.667(−4)	2.00	6.664(−5)	2.01	1.657(−5)
$t=1$	$\|e\|_h$	7.818(−5)	2.01	1.945(−5)	2.00	4.856(−6)
	$\|e_r\|_h$	1.110(−4)		2.535(−5)		6.072(−6)
	$\|e_u\|_\infty$	2.643(−4)	2.01	6.576(−5)	2.00	1.642(−5)

Table 13.3 Compact fourth-order scheme for the Navier–Stokes system with exact solution $\psi = (1-x^2)^3(1-y^2)^3e^{-t}$ on $[0,1]\times[0,1]$. The convective term is (13.30) with second-order time-stepping scheme, $\Delta t = O(h^2)$.

Mesh		17×17	Rate	33×33	Rate	65×65
$t=0.25$	$\|e\|_h$	1.5792(-7)	3.95	1.0232(-7)	4.03	6.5236(-9)
	$\|e_r\|_h$	6.5463(-6)		4.0380(-7)		2.5141(-8)
	$\|e_u\|_\infty$	4.7907(-5)	3.97	3.0612(-6)	3.97	1.9347(-7)
$t=0.5$	$\|e\|_h$	1.2315(-7)	3.95	7.9827(-8)	3.97	5.0902(-9)
	$\|e_r\|_h$	6.5554(-6)		4.0450(-7)		2.5188(-8)
	$\|e_u\|_\infty$	3.7309(-5)	3.97	2.3840(-6)	3.98	1.5067(-7)
$t=0.75$	$\|e\|_h$	9.5991(-7)	3.95	6.2235(-8)	3.97	3.9688(-9)
	$\|e_r\|_h$	1.0965(-4)		4.0993(-7)		2.5217(-8)
	$\|e_u\|_\infty$	2.9055(-5)	3.97	1.8566(-6)	3.98	1.1173(-7)
$t=1$	$\|e\|_h$	7.5671(-7)	3.95	4.8500(-8)	3.97	3.0930(-9)
	$\|e_r\|_h$	1.0967(-4)		4.0520(-7)		2.5235(-8)
	$\|e_u\|_\infty$	2.9132(-5)	3.97	1.4459(-6)	3.98	9.1385(-7)

13.3.3 *Test case 3*

In Table 13.3 we give numerical results for the exact solution $\psi(x,y,t) = (1-x^2)^3(1-y^2)^3e^{-t}$ on the square $[0,1]\times[0,1]$, using the fourth-order approximation (13.30) to the convective term. Observe that fourth-order accuracy is achieved.

13.3.4 *Test case 4*

In Table 13.4 we display numerical results for the fourth-order spatial scheme applied to the exact solution $\psi(x,y,t) = (1-x^2)^3(1-y^2)^3e^{-t}$ on the square $[0,1]\times[0,1]$. The time-stepping scheme that we used is as in [14], which is almost third-order accurate in time. This temporal scheme was suggested in [167] (see the Notes to this chapter). In order to obtain fourth-order accuracy in time, we chose Δt to be $O(h^{4/3})$. Indeed, the table shows that this high-order accuracy is achieved although, in contrast to test case 1, the boundary conditions are non-homogeneous.

Table 13.4 High order compact scheme in space and time for the Navier–Stokes equations with exact solution: $\psi(x,y,t) = (1-x^2)^3(1-y^2)^3e^{-t}$ on $[0,1]\times[0,1]$. The convective term is (13.30) with high-order time-stepping scheme, $\Delta t = O(h^{4/3})$.

Mesh		17×17	Rate	33×33	Rate	65×65
$t=0.25$	$\lvert e\rvert_h$	1.5022(-6)	3.92	9.9168(-8)	3.87	6.7763(-9)
	$\lvert e_r\rvert_h$	6.2153(-6)		3.9197(-7)		2.6112(-8)
	$\lvert e_u\rvert_\infty$	4.8052(-5)	3.97	3.0614(-6)	3.98	1.9378(-7)
$t=0.5$	$\lvert e\rvert_h$	1.4466(-6)	3.95	9.3439(-8)	3.92	6.1550(-9)
	$\lvert e_r\rvert_h$	7.7001(-6)		4.7348(-7)		3.0451(-8)
	$\lvert e_u\rvert_\infty$	3.7321(-5)	3.97	2.3877(-6)	3.98	1.5096(-7)
$t=0.75$	$\lvert e\rvert_h$	1.1674(-6)	3.96	7.5132(-8)	3.94	4.8817(-9)
	$\lvert e_r\rvert_h$	7.9635(-6)		4.8884(-7)		3.1027(-8)
	$\lvert e_u\rvert_\infty$	2.9106(-5)	3.97	1.8592(-6)	3.98	1.1175(-7)
$t=1$	$\lvert e\rvert_h$	9.0495(-7)	3.96	5.8434(-8)	3.95	3.7702(-9)
	$\lvert e_r\rvert_h$	7.9423(-6)		4.8819(-7)		3.0765(-8)
	$\lvert e_u\rvert_\infty$	2.2612(-5)	3.97	1.4477(-6)	3.98	9.1540(-7)

13.4 Notes for Chapter 13

- Refer to the Introduction to Part II for a discussion of numerical approximations for Navier–Stokes equations.
- Fourth-order finite difference schemes for Navier–Stokes problems have recently attracted a lot of attention [64, 95, 99, 103, 106, 126, 130].
- Fully discrete *compact* difference schemes (not necessarily fourth-order) for the Navier–Stokes equations in various formulations have recently been suggested [55, 56] (vorticity-steamfunction formulation), [30] (velocity–pressure formulation), [90, 91] (streamfunction–velocity formulation), [29] (projection methods). See also the references therein.
- Observe that our scheme uses a fully centered approximation to the convective term without any stabilization procedure. This is in con-

trast to approximations motivated by hyperbolic treatments of the convective term, using upwinding or slope limiters [31, 119].

- The temporal scheme (13.31) is a so-called Implicit-Explicit (IMEX) scheme. A theory for such schemes can be found in [100, 113]. In [15] a linear von Neumann stability analysis is derived for (13.31). The time-step restriction is as in (13.39). Let us briefly show the principle of the proof for a simple model. Consider the one-dimensional time-dependent equation

$$\partial_t \partial_x^2 \psi - \nu \partial_x^4 \psi = a \partial_x^3 \psi, \tag{13.49}$$

which is a simplified linear one-dimensional version of (12.1) with constant velocity a and diffusion coefficient $\nu > 0$. A spatial approximation of (13.49), which is in the spirit of (12.8), is

$$\frac{d}{dt}\delta_x^2 \psi - \nu \delta_x^4 \psi = a \delta_x^2 \psi_x. \tag{13.50}$$

The simplest temporal IMEX scheme for (13.50) is given by

$$\frac{\delta_x^2 \psi^{n+1} - \delta_x^2 \psi^n}{\Delta t} - \nu \delta_x^4 \psi^{n+1} = a \delta_x^2 \psi_x^n, \tag{13.51}$$

which can be rewritten as

$$\left(\delta_x^2 - \nu \Delta t \cdot \delta_x^4\right) \psi^{n+1} = \delta_x^2 \psi^n + a \Delta t \cdot \delta_x^2 \psi_x^n. \tag{13.52}$$

The stability of (13.52) can be established by a von Neumann stability analysis [156, 169] as follows. We compute the amplification factor of (13.52) on a mode $\exp(i\alpha x)$. Defining the phase angle $\theta = \alpha h \in [0, 2\pi)$ where h is the grid size, every discrete operator (applied to ψ) is expressed as a "symbol" multiplying the Fourier transform $\hat{\psi}(\theta)$. The symbol corresponding to the Hermitian derivative ψ_x is

$$\widehat{\psi_x}(\theta) = H_x \hat{\psi}(\theta) = i \frac{3 \sin \theta}{h(2 + \cos \theta)} \hat{\psi}(\theta). \tag{13.53}$$

The symbol for $\delta_x \psi_x$ is

$$\widehat{\delta_x \psi_x}(\theta) = K_x \hat{\psi}(\theta) = -\frac{3 \sin^2 \theta}{h^2(2 + \cos \theta)} \hat{\psi}(\theta). \tag{13.54}$$

Similarly, the symbol for δ_x^2 is

$$\widehat{\delta_x^2 \psi}(\theta) = 2 \frac{\cos \theta - 1}{h^2} \hat{\psi}(\theta). \tag{13.55}$$

Therefore, the symbol corresponding to $-h^2 \delta_x^2$, denoted by $A_1(\theta)$, is

$$A_1(\theta) = 2(1 - \cos \theta). \tag{13.56}$$

Next, the symbol of the three-point biharmonic operator δ_x^4 is

$$J_x(\theta) = \frac{12}{h^4}\frac{(1-\cos\theta)^2}{2+\cos\theta}, \tag{13.57}$$

so that the symbol of $h^2\nu\delta_x^4$ is

$$B_1(\theta) = h^2\nu J_x = 12\nu h^{-2}\frac{(1-\cos\theta)^2}{2+\cos\theta}. \tag{13.58}$$

The symbol $C_1(\theta)$ for the convective operator $ih^2a\delta_x^2\psi_x$ is

$$C_1(\theta) = \frac{6a}{h}\frac{(1-\cos\theta)\sin\theta}{2+\cos\theta}. \tag{13.59}$$

Using the above notation, the amplification factor $g(\theta)$ of (13.52) is

$$g(\theta) = \frac{A_1(\theta) + i\Delta t C_1(\theta)}{A_1(\theta) + \Delta t B_1(\theta)}. \tag{13.60}$$

The (strong) von Neumann stability condition is [169]

$$\sup_{\theta\in[0,2\pi)} |g(\theta)| \le 1. \tag{13.61}$$

Therefore for all $\theta \in [0, 2\pi)$ the following condition must hold

$$A_1^2(\theta) + \Delta t^2 C_1^2(\theta) \le A_1^2(\theta) + 2\Delta t A_1(\theta)B_1(\theta) + \Delta t^2 B_1^2(\theta). \tag{13.62}$$

A sufficient condition for (13.61) is

$$\Delta t^2 C_1^2(\theta) \le 2\Delta t A_1(\theta) B_1(\theta), \quad 0 \le \theta < 2\pi \tag{13.63}$$

Invoking (13.56), (13.58) and (13.59), this is equivalent to

$$a^2\Delta t \max_{\theta\in[0,2\pi)} \frac{\sin^2\theta}{2+\cos\theta} \le \frac{4}{3}\nu. \tag{13.64}$$

Using the fact that

$$\max_{\theta\in[0,2\pi)} \frac{\sin^2\theta}{2+\cos\theta} = \max_{x\in[-1,1]} \frac{1-x^2}{2+x} = 4 - 2\sqrt{3}, \tag{13.65}$$

we finally deduce that a sufficient condition for (13.61) is

$$a^2\Delta t \le \frac{1}{3(1-\frac{\sqrt{3}}{2})}\nu. \tag{13.66}$$

This condition is of the form (13.39). Refer to [14, 15] for full details in two dimensions.

- The implicit-explicit (IMEX) scheme (13.31) is second-order accurate in time. Another IMEX scheme which is "almost" third-order accurate was given in [14] following [167].

- Refer to [1] for a stability analysis for non-periodic boundary conditions for several one-dimensional models.
- The implementation of partially implicit schemes such as (13.31) requires solving linear systems associated with discrete biharmonic problems. Due to ill-conditioning, such linear systems are notoriously difficult to solve [4]. Nonetheless it turns out that the resolution of the linear problems in (13.31) can be handled by a fast solver (of the FFT family), at least in rectangular regions. In order to simplify the explanation let us consider again the model problem (13.49). Due to the matrix form of the finite difference operators δ_x^2 in (9.51) and δ_x^4 in (10.118), then Equation (13.52) can be expressed as the linear system for Ψ^{n+1},

(13.67)
$$\Bigg(\underbrace{\left[\frac{T^2}{h^2}+\frac{6\nu\Delta t}{h^2}P^{-1}T^2\right]}_{B}+\underbrace{\left[\frac{36\nu\Delta t}{h^4}\Big(V_1V_1^T+V_2V_2^T\Big)\right]}_{RR^T}\Bigg)\Psi^{n+1}=G^n.$$

Here G^n is the vector corresponding to the right-hand side of (13.52) and the matrix R is

$$R=\frac{6\sqrt{\nu\Delta t}}{h^2}[V_1,V_2]. \tag{13.68}$$

Now consider the Sherman–Morrison formula [82, Chapter 2].

Sherman–Morrison Formula

Suppose that $A, B \in \mathbb{M}_N(\mathbb{R})$ are two invertible matrices, such that

$$A = B + RS^T, \tag{13.69}$$

with $R, S \in \mathbb{M}_{N,n}(\mathbb{R})$, $n \le N$. Then, the inverse of A may be written as

$$A^{-1} = B^{-1} - B^{-1}R(I + S^TB^{-1}R)^{-1}S^TB^{-1}, \tag{13.70}$$

provided that $I + S^TB^{-1}R \in \mathbb{M}_n(\mathbb{R})$ is invertible.

Observe that if $n \ll N$ in (13.69) then the matrix A is a low-rank perturbation of the matrix B. Hence, if B can be easily inverted, then (13.70) provides an efficient way to compute A^{-1}. It is this property that we invoke to express Ψ^{n+1} in (13.67) in terms of G^n as

$$\Psi^{n+1} = B^{-1}G^n - B^{-1}R(I + R^TB^{-1}R)^{-1}RB^{-1}G^n. \tag{13.71}$$

Note that

(i) $B^{-1}G^n$ is easily obtained using the fact the eigenvectors of B in (13.67) are actually the vectors Z^k in (9.55). Here an FFT algorithm is used to perform the forward and backward decompositions of the solution decomposed in the spectral basis $\{Z^k\}_{1 \le k \le N-1}$.

(ii) The linear system $(I + R^T B^{-1} R)Y = RB^{-1}G^n$ can be easily solved, because it is "small" compared to the size $N - 1$ of the matrix B.

- The fast algorithm described above refers to the one-dimensional model equation (13.49). In [13] the same methodology was implemented for the fourth-order (two-dimensional) numerical solution of (13.31).
- A numerical linear stability analysis of Navier–Stokes flows has been performed [5, 141, 154].

Chapter 14

Numerical Simulations of the Driven Cavity Problem

In this chapter we consider the two-dimensional driven cavity problem using the numerical schemes developed in the preceding chapters. This is a classic benchmark problem for direct numerical simulations of the incompressible Navier–Stokes system.

The driven cavity problem consists of a flow of a viscous incompressible fluid in the square $[0,1] \times [0,1]$. On the top boundary $y = 1$ the fluid is driven to the right with horizontal velocity $(u, v) = (1, 0)$. On the other sides of the square the "no-slip" condition is imposed, namely, $(u, v) = (0, 0)$.

This problem is well documented in the literature. Refer to [57, 92] for historical remarks and to [165] for the physical significance of such flows. Reference results may be found, for example, in [23, 76, 138]. The work of Ghia, Ghia and Shin [76] gave accurate numerical results for the steady-state solution associated with a wide array of Reynolds numbers. It is a well-known reference for the stationary problem since the streamfunction contours as well as the values of the velocity components are very well detailed. It relies on the vorticity-streamfunction formulation of the time-independent Navier–Stokes system in two dimensions.

The Reynolds number Re is defined as $\mathrm{Re} = UL/\nu$, where U is a typical velocity of the fluid, L is a typical length of the square and ν is the kinematic viscosity. We pick $U = L = 1$ so that $\mathrm{Re} = 1/\nu$. It is observed in physical and numerical experiments that as the Reynolds number increases the time-dependent flow develops instabilities and that in some cases it does not approach a steady state [154]. In fact it is believed that beyond a certain critical Reynolds number the time-dependent flow undergoes a Hopf bifurcation and develops a time-periodic behavior. The critical Reynolds number is believed to be $\mathrm{Re} \simeq 8000$ [8, 57, 145]. The numerical results presented here show that at Re = 10000, the transition to a time-periodic solution has occurred.

In Section 14.1 we present numerical results using the second-order semi-discrete scheme (12.8) and its fully discrete analog (13.31).

In Section 14.2 we display numerical results obtained with the high-order scheme (13.17), where the convective term was approximated by (13.30). The double-driven cavity problem is addressed in Section 14.3.

14.1 Second-order scheme for the driven cavity problem

In this section we describe numerical results obtained by implementing the (fully discrete) second-order scheme in space and time as in Equation (13.31). We display results for Reynolds numbers ranging from Re = 400 to Re = 10000.

14.1.1 *Driven cavity with* Re = 400 *and* 1000

We first present numerical simulations for Re = 400 and Re = 1000, which are considered to be low Reynolds numbers. The computed values for the streamfunction ψ at Re = 400 are given in Table 14.1, using different grids and time levels. In particular the maximal value of ψ and the location $(\bar{x}, \bar{y})$ where it is attained are shown. The minimal value of ψ is also given. The meshes consist of 65×65, 81×81, and 97×97 points and the time levels are $t = 10, 20, 40, 60$. We observe that the highest value of the streamfunction at $t = 60$ is 0.1136 on the finest grid. The maximum of ψ occurs at $(\bar{x}, \bar{y}) = (0.5521, 0.6042)$ and the minimum of ψ is -6.498×10^{-4}. In addition, observe that the maximal value of ψ stabilizes at $t = 40$ and that the location $(\bar{x}, \bar{y})$ of the primary vortex remains constant beyond $t = 20$. This value may be compared to the steady-state values calculated by Ghia, Ghia and Shin [76]; their maximal value of ψ is 0.1139 and it occurs at (0.5547, 0.6055); their minimal value of the streamfunction is -6.424×10^{-4}.

In Figure 14.3(a) we plot streamfunction contours at $t = 60$ obtained with the 97×97 mesh. In Figure 14.4(a) our calculated graphs of the velocity components $u(0.5, y)$ and $v(x, 0.5)$ are plotted as solid curves at $t = 60$. They are compared with the values obtained in [76] (marked by an "o"). Note the excellent match between the results. Table 14.2 parallels Table 14.1 for Re = 1000. The displayed time levels are $t = 20, 40, 60, 80$,

Table 14.1 Compact second-order scheme for the driven cavity problem, Re = 400. Ghia, Ghia and Shin [76]: $\max \psi = 0.1139$ at $(0.5547, 0.6055)$, $\min \psi = -6.424(-4)$.

Time	Quantity	65×65	81×81	97×97
10	$\max \psi$	0.1053	0.1057	0.1059
	$(\bar{x}, \bar{y})$	(0.5781, 0.6250)	(0.5750, 0.6250)	(0.5833, 0.6354)
	$\min \psi$	−4.786(−4)	−4.758(−4)	−4.749(−4)
20	$\max \psi$	0.1124	0.1128	0.1130
	$(\bar{x}, \bar{y})$	(0.5625, 0.6094)	(0.5625, 0.6125)	(0.5521, 0.6042)
	$\min \psi$	−6.333(−4)	−6.371(−4)	−6.361(−4)
40	$\max \psi$	0.1131	0.1134	0.1136
	$(\bar{x}, \bar{y})$	(0.5625, 0.6094)	(0.5500, 0.6000)	(0.5521, 0.6042)
	$\min \psi$	−6.513(−4)	−6.5148(−4)	−6.498(−4)
60	$\max \psi$	0.1131	0.01134	0.1136
	$(\bar{x}, \bar{y})$	(0.5625, 0.6094)	(0.5500, 0.6000)	(0.5521, 0.6042)
	$\min \psi$	−6.514(−4)	−6.5155(−4)	−6.498(−4)

Table 14.2 Streamfunction formulation: compact scheme for the driven cavity problem, Re = 1000. Ghia, Ghia and Shin [76]: $\max \psi = 0.1179$ at $(0.5313, 0.5625)$, $\min \psi = -0.0017$.

Time	Quantity	65×65	81×81	97×97
20	$\max \psi$	0.1129	0.1139	0.1143
	$(\bar{x}, \bar{y})$	(0.5469, 0.5781)	(0.5375, 0.5750)	(0.5417, 0.5729)
	$\min \psi$	−0.0015	−0.0015	−0.0015
40	$\max \psi$	0.1160	0.1169	0.1175
	$(\bar{x}, \bar{y})$	(0.5312, 0.5625)	(0.5250, 0.5625)	(0.5312, 0.5625)
	$\min \psi$	−0.0017	−0.0017	−0.0017
60	$\max \psi$	0.1160	0.1171	0.1177
	$(\bar{x}, \bar{y})$	(0.5312, 0.5625)	(0.5250, 0.5625)	(0.5312, 0.5625)
	$\min \psi$	−0.0017	−0.0017	−0.0017
80	$\max \psi$	0.1160	0.1172	0.1178
	$(\bar{x}, \bar{y})$	(0.5312, 0.5625)	(0.5250, 0.5625)	(0.5312, 0.5625)
	$\min \psi$	−0.0017	−0.0017	−0.0017

and the grids remain as before. We still observe that for each of the grids the flow quantities converge to a steady state as time progresses. For the latest time on the finest grid our calculated maximal value of ψ is 0.1178, located at $(\bar{x}, \bar{y}) = (0.5312, 0.5625)$. Once again it should be compared to a calculated maximal value of 0.1179, located at $(0.5313, 0.5625)$ in [76]. Our calculated minimum value of the streamfunction is -0.0017, exactly as in [76].

Figures 14.3(b) and 14.4(b) parallel Figures 14.3(a) and 14.4(a). In Figure 14.3(b) we display streamfunction contours at $t = 80$, with a 97×97 mesh. In Figure 14.4(b) the velocity components $u(0.5, y)$ and $v(x, 0.5)$ are plotted as solid curves at $t = 80$. Again the match is excellent with the corresponding values in [76] (marked by "o" in this figure).

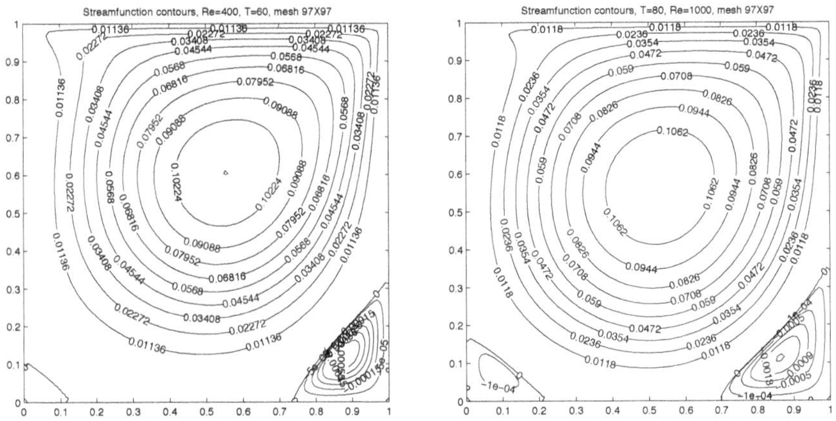

(a) Re = 400 (b) Re = 1000

Fig. 14.3 Driven cavity for Re = 400, 1000: calculated streamfunction contours.

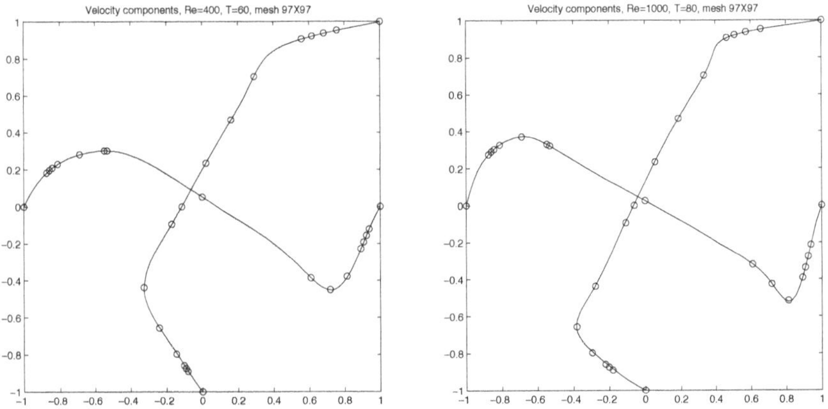

(a) Re = 400 (b) Re = 1000

Fig. 14.4 Driven cavity for Re = 400, 1000: calculated velocity components (solid curves). Ghia, Ghia and Shin results [76] are marked by "o."

14.1.2 *Driven cavity with* Re = 3200 *and* Re = 5000

The values Re = 3200 and Re = 5000 are considered medium Reynolds numbers.

Results for Re = 3200 on a 81×81 mesh and a 97×97 mesh are given in Table 14.5, using once again the second-order scheme (13.31). At the latest time ($t = 360$) on the finest grid the maximal value of ψ as calculated by our scheme is 0.1174, compared to the value 0.1204 in [76]. This maximal value is attained at $(\bar{x}, \bar{y}) = (0.5208, 0.5417)$, compared to $(0.5165, 0.5469)$ in [76]. The minimal calculated value of the streamfunction is -0.0027, compared to -0.0031 in [76].

The calculated streamfunction contours at $t = 360$ are plotted in Figure 14.7(a) using the 97×97 mesh. The corresponding velocity components $u(0.5, y)$ and $v(x, 0.5)$ are plotted in Figure 14.8(a) as solid curves (again at $t = 360$). They may be compared with the values in [76] marked by "o." Again we observe an excellent match.

Table 14.5 Streamfunction formulation: compact scheme for the driven cavity problem, Re = 3200. Ghia, Ghia and Shin results: $\max\psi = 0.1204$ at $(0.5165, 0.5469)$, $\min\psi = -0.0031$.

Time	Quantity	81×81	97×97
40	$\max\psi$	0.1157	0.1145
	$(\bar{x}, \bar{y})$	(0.5125, 0.5500)	(0.5104, 0.5417)
	$\min\psi$	−0.0024	−0.0025
80	$\max\psi$	0.1152	0.1154
	$(\bar{x}, \bar{y})$	(0.5125, 0.5375)	(0.5208, 0.5417)
	$\min\psi$	−0.0026	−0.0027
160	$\max\psi$	0.1155	0.1169
	$(\bar{x}, \bar{y})$	(0.5125, 0.5375)	(0.5208, 0.5417)
	$\min\psi$	−0.0026	−0.0027
200	$\max\psi$	0.1155	0.1172
	$(\bar{x}, \bar{y})$	(0.5125, 0.5375)	(0.5208, 0.5417)
	$\min\psi$	−0.0027	−0.0027
240	$\max\psi$	0.1156	0.1173
	$(\bar{x}, \bar{y})$	(0.5125, 0.5375)	(0.5208, 0.5417)
	$\min\psi$	−0.0027	−0.0027
360	$\max\psi$	0.1156	0.1174
	$(\bar{x}, \bar{y})$	(0.5125, 0.5375)	(0.5208, 0.5417)
	$\min\psi$	−0.0027	−0.0027

Table 14.6 Streamfunction formulation: compact scheme for the driven cavity problem, Re $= 5000$. Ghia, Ghia and Shin results: $\max \psi = 0.11897$ at $(0.5117, 0.5352)$, $\min \psi = -0.0031$.

Time	Quantity	81×81	97×97
40	$\max \psi$	0.0936	0.0983
	$(\bar{x}, \bar{y})$	(0.4875, 0.6125)	(0.5114, 0.6146)
	$\min \psi$	−0.0029	−0.0030
80	$\max \psi$	0.1007	0.1010
	$(\bar{x}, \bar{y})$	(0.5000, 0.5125)	(0.5312, 0.5312)
	$\min \psi$	−0.0027	−0.0029
120	$\max \psi$	0.1060	0.1068
	$(\bar{x}, \bar{y})$	(0.5125, 0.5375)	(0.5104, 0.5417)
	$\min \psi$	−0.0028	−0.0028
160	$\max \psi$	0.1095	0.1105
	$(\bar{x}, \bar{y})$	(0.5125, 0.5375)	(0.5104, 0.5312)
	$\min \psi$	−0.0028	−0.0028
200	$\max \psi$	0.1117	0.1127
	$(\bar{x}, \bar{y})$	(0.5125, 0.5375)	(0.5104, 0.5312)
	$\min \psi$	−0.0028	−0.0029
240	$\max \psi$	0.1131	0.1141
	$(\bar{x}, \bar{y})$	(0.5125, 0.5375)	(0.5104, 0.5417)
	$\min \psi$	−0.0028	−0.0029
280	$\max \psi$	0.1139	0.1150
	$(\bar{x}, \bar{y})$	(0.5125, 0.5375)	(0.5104, 0.5417)
	$\min \psi$	−0.0028	−0.0029
400	$\max \psi$	0.1149	0.1160
	$(\bar{x}, \bar{y})$	(0.5125, 0.5375)	(0.5104, 0.5417)
	$\min \psi$	−0.0028	−0.0029

Finally, in Table 14.6 we display results for Re $= 5000$. At the latest time $t = 400$ on the finest grid the maximal value of ψ is 0.1160. This should be compared to the steady state value of 0.11897 in [76]. The calculated location of the maximal value is $(\bar{x}, \bar{y}) = (0.5104, 0.5417)$, compared to $(0.5117, 0.5352)$ in [76]. The calculated minimum value of the streamfunction is -0.0029, compared to the value of -0.0031 in [76]. For this case the calculated streamfunction contours at $t = 400$ are plotted in Figure 14.7(b) using the 97×97 mesh.

The corresponding velocity components $u(0.5, y)$ and $v(x, 0.5)$ are plotted in Figure 14.8(b) as solid curves (again at $t = 400$). They may be compared with the values in [76] marked by "o." Note that there is a discrepancy in the values of the streamfunction while the velocity values are a better match.

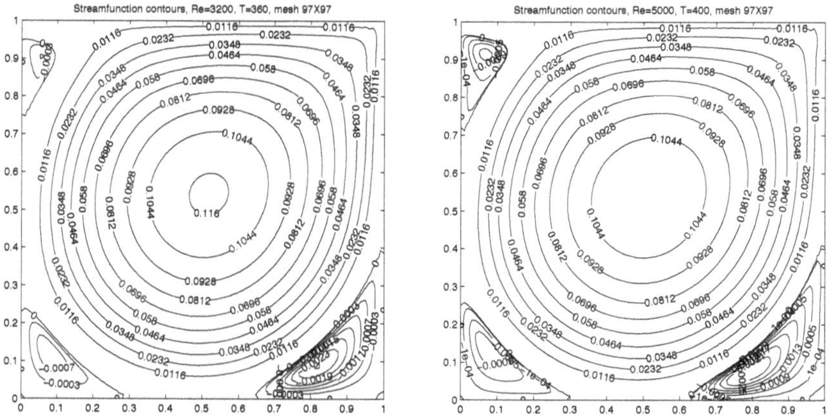

(a) Re = 3200 (b) Re = 5000

Fig. 14.7 Driven cavity for Re = 3200, 5000: calculated streamfunction contours.

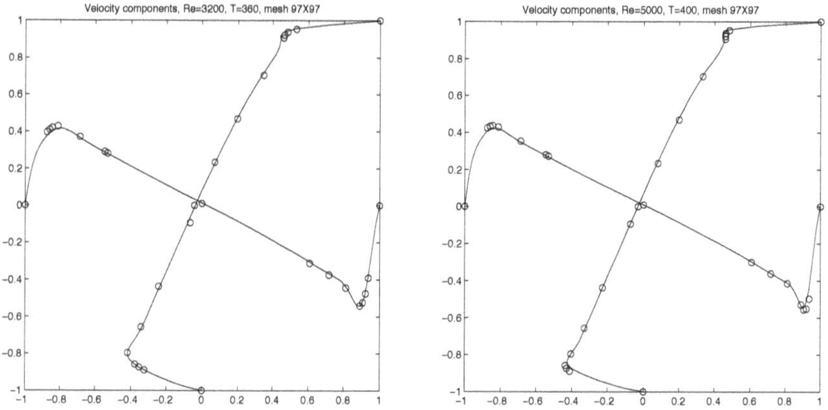

(a) Re = 3200 (b) Re = 5000

Fig. 14.8 Driven cavity for Re = 3200, 5000: calculated velocity components. Ghia, Ghia and Shin [76] results are marked by "o."

14.1.3 *Driven cavity with* Re = 7500 *and* Re = 10000

The values Re = 7500 and Re = 10000 are considered as high Reynolds numbers. As mentioned in the beginning of this chapter, steady-state solutions are not necessarily achieved as limits of the time-dependent flow. In order to carry out the time dependent calculation, always using the scheme

(13.31), we adopted the following strategy: the initial condition was taken to be the calculated flow pattern at $t = 400$ for $\mathrm{Re} = 5000$.

Starting from this initial value we advance to $t = 560$ for $\mathrm{Re} = 7500$ using the 97×97 mesh. The maximal calculated value of ψ is now 0.1175, which may be compared to 0.11998 obtained in [76], obtained for the steady-state system. The location of the calculated maximal value is $(\bar{x}, \bar{y}) = (0.5104, 0.5312)$ compared to $(0.5117, 0.5322)$ in [76]. The minimal calculated value of the streamfunction is -0.003, whereas the value -0.0033 was given in [76]. The calculated streamfunction contours at $t = 560$ are plotted in Figure 14.9(a) using the 97×97 mesh. The corresponding velocity components $u(0.5, y)$ and $v(x, 0.5)$ are plotted in Figure 14.10(a) as solid curves (again at $t = 560$). They may be compared with the values in [76] marked by "o." Again the match is seen as excellent.

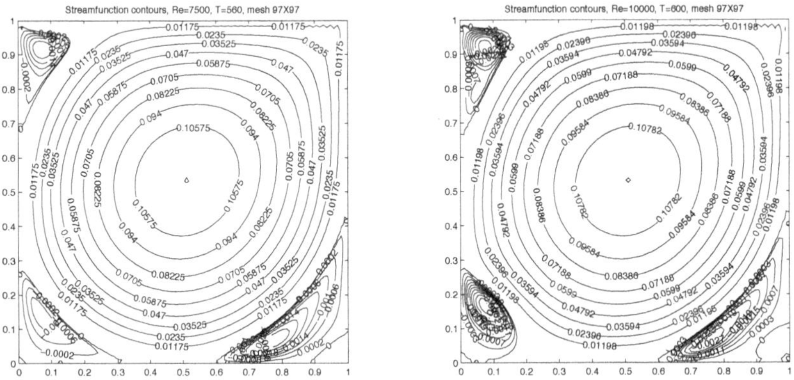

(a) Re = 7500 (b) Re = 10000

Fig. 14.9 Driven cavity for Re = 7500, 10000: calculated streamfunction contours.

Starting again for the calculated flow with $\mathrm{Re} = 5000$ at $t = 400$, we advance to $t = 500$ using $\mathrm{Re} = 10000$ and a 97×97 grid. The maximal calculated value of ψ is 0.1190. This should be compared to 0.1197 in [76]. The location of the calculated maximal value is $(\bar{x}, \bar{y}) = (0.5104, 0.5312)$, compared to $(0.5117, 0.5333)$ in [76]. The minimal value of the streamfunction is -0.0033, whereas the value -0.0034 was found in [76]. Figure 14.9(b) displays the streamfunction contours. The corresponding velocity components $u(0.5, y)$ and $v(x, 0.5)$ are plotted in Figure 14.10(b) as solid curves (at $t = 500$). They may be compared with values obtained in [76] (marked by "o"). Notice the very good match of the velocity values.

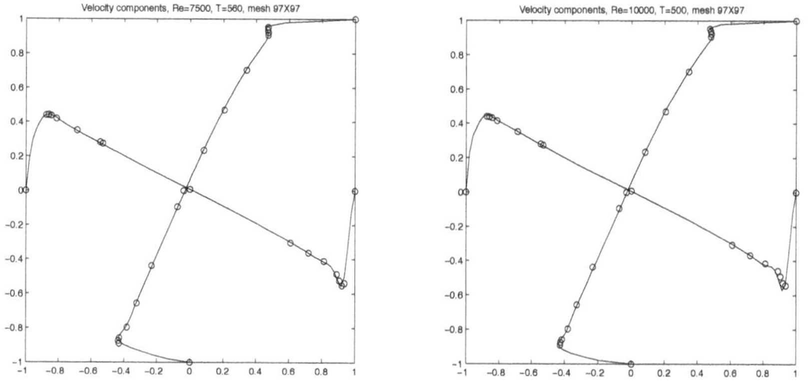

(a) Re = 7500 (b) Re = 10000

Fig. 14.10 Driven cavity for Re = 7500, 10000: calculated velocity components. Ghia, Ghia and Shin [76] results are marked by "o."

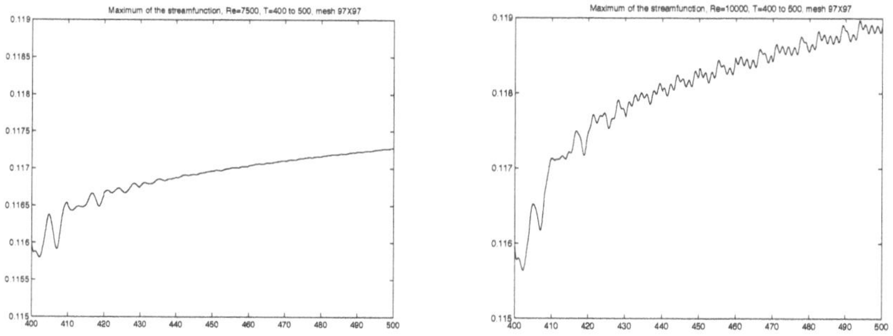

(a) Re = 7500 (b) Re = 10000

Fig. 14.11 Driven cavity for Re = 7500, 10000: calculated maximal value of the streamfunction, t=400 to 500.

Figure 14.11(a) indicates that the maximal value of the streamfunction grows from $t = 400$ to $t = 500$ for Re = 7500 and approaches a horizontal asymptote. However, Figure 14.11(b) shows that it is not the case for Re = 10000. In fact, a steady state has not been reached. A similar phenomenon has been observed [119, 145, 166] for Re = 8500 or above. It seems that in this case the time dependent solution does not approach a steady state. Instead it bifurcates into a time-periodic solution. It is commonly interpreted as a Hopf bifurcation of the steady-state solution as the parameter Re crosses a critical value around 8500. A rigorous analysis of this bifurcation remains an open issue.

Table 14.12 Streamfunction formulation: fourth-order compact scheme for the driven cavity problem, Re = 1000. Ghia, Ghia and Shin [76] results: $\max\psi = 0.1179$ at $(0.5313, 0.5625)$, $\min\psi = -0.0017$.

Time	Quantity	33 × 33	65 × 65	129 × 129
20	$\max\psi$	0.1107526	0.1142693	0.1150295
	$(\bar{x}, \bar{y})$	(0.531250, 0.562500)	(0.546875, 0.578125)	(0.539062, 0.570312)
	$\min\psi$	−0.1466347(−2)	−0.1464712(−2)	−0.1493240(−2)
40	$\max\psi$	0.1146979	0.1175422	0.1182664
	$(\bar{x}, \bar{y})$	(0.531250, 0.562500)	(0.531250, 0.562500)	(0.531250, 0.562500)
	$\min\psi$	−0.1675910(−2)	−0.1696769(−2)	−0.1713428(−2)
60	$\max\psi$	0.1152327	0.1179083	0.1186733
	$(\bar{x}, \bar{y})$	(0.531250, 0.562500)	(0.531250, 0.562500)	(0.531250, 0.562500)
	$\min\psi$	−0.1685384(−2)	−0.1705089(−2)	−0.1722033(−2)
80	$\max\psi$	0.1153669	0.1180018	0.1187779
	$(\bar{x}, \bar{y})$	(0.531250, 0.562500)	(0.531250, 0.562500)	(0.531250, 0.562500)
	$\min\psi$	−0.1686994(−2)	−0.1706326(−2)	−0.1723401(−2)
100	$\max\psi$	0.1154009	0.1180258	0.1188048
	$(\bar{x}, \bar{y})$	(0.531250, 0.562500)	(0.531250, 0.562500)	(0.531250, 0.562500)
	$\min\psi$	−0.1687387(−2)	−0.1706626(−2)	−0.1723737(−2)
120	$\max\psi$	0.1154095	0.1180319	0.1188117
	$(\bar{x}, \bar{y})$	(0.531250, 0.562500)	(0.531250, 0.562500)	(0.531250, 0.562500)
	$\min\psi$	−0.1687486(−2)	−0.1706703(−2)	−0.1723823(−2)

Table 14.13 Streamfunction formulation: fourth-order compact scheme for the driven cavity problem, Re = 5000. Ghia, Ghia and Shin [76] results: $\max\psi = 0.11897$ at $(0.5117, 0.5352)$, $\min\psi = -0.0031$.

Time	Quantity	65 × 65	129 × 129	257 × 257
40	$\max\psi$	0.9900489(−1)	0.1210794	0.1222390
	$(\bar{x}, \bar{y})$	(0.5000000.468750)	(0.5156250.539062)	(0.5156250.535156)
	$\min\psi$	−0.2087525(−2)	−0.3036084(−2)	−0.3067078(−2)
80	$\max\psi$	0.1033521	0.1211404	0.1221542
	$(\bar{x}, \bar{y})$	(0.5156250.531250)	(0.5156250.539062)	(0.5156250.535156)
	$\min\psi$	−0.2903847(−2)	−0.3036400(−2)	−0.3066649(−2)
120	$\max\psi$	0.1094742	0.1211780	0.1221018
	$(\bar{x}, \bar{y})$	(0.5156250.531250)	(0.5156250.539062)	(0.5156250.535156)
	$\min\psi$	−0.2986051(−2)	−0.3036596(−2)	−0.3066370(−2)
160	$\max\psi$	0.1133213	0.1212012	0.1220695
	$(\bar{x}, \bar{y})$	(0.5156250.531250)	(0.5156250.539062)	(0.5156250.535156)
	$\min\psi$	−0.3005411(−2)	−0.3036717(−2)	−0.3066180(−2)
200	$\max\psi$	0.1157160	0.1212156	0.1220495
	$(\bar{x}, \bar{y})$	(0.5156250.531250)	(0.5156250.539062)	(0.5156250.535156)
	$\min\psi$	−0.3019430(−2)	−0.3036792(−2)	−0.3066055(−2)
240	$\max\psi$	0.1171963	0.1212244	0.1220372
	$(\bar{x}, \bar{y})$	(0.5156250.531250)	(0.5156250.539062)	(0.5156250.535156)
	$\min\psi$	−0.3027428(−2)	−0.3036838(−2)	−0.3065978(−2)
280	$\max\psi$	0.1181093	0.1212299	0.12202960
	$(\bar{x}, \bar{y})$	(0.5156250.531250)	(0.5156250.539062)	(0.5156250.535156)
	$\min\psi$	−0.3032461(−2)	−0.3036867(−2)	−0.3065930(−2)
400	$\max\psi$	0.1192325	0.1212366	0.12202020
	$(\bar{x}, \bar{y})$	(0.5156250.531250)	(0.5156250.539062)	(0.5156250.535156)
	$\min\psi$	−0.3038565(−2)	−0.3036902(−2)	−0.3065873(−2)

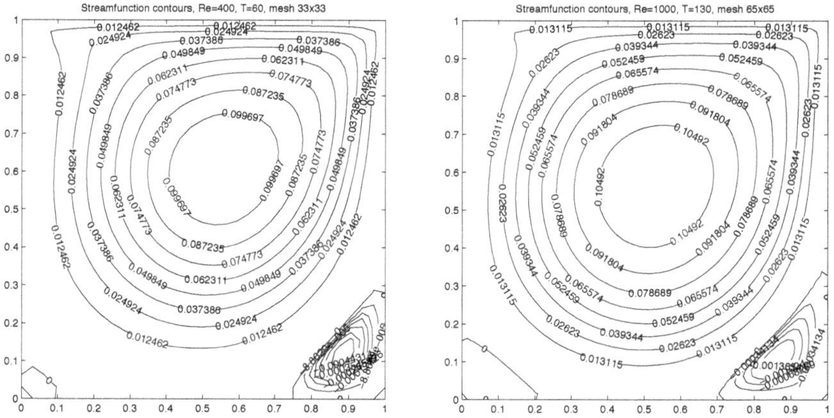

(a) Re = 400 (b) Re = 1000

Fig. 14.14 Driven cavity for Re = 400, 1000: calculated streamfunction contours using the fourth-order scheme.

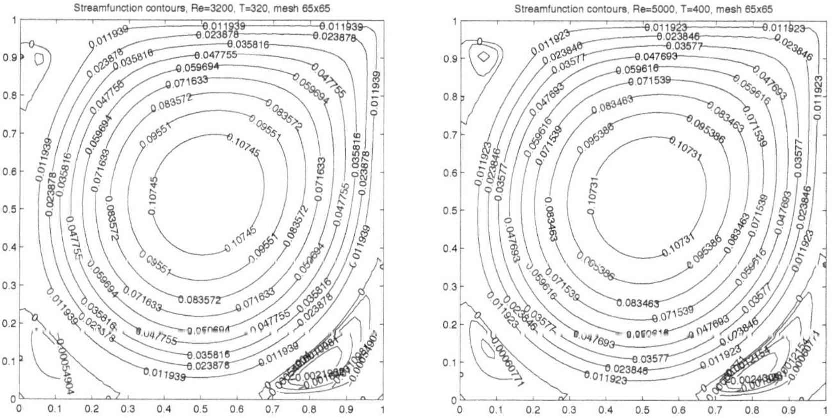

(a) Re = 3200 (b) Re = 5000

Fig. 14.15 Driven cavity for Re = 3200, 5000: calculated streamfunction contours using the fourth-order scheme.

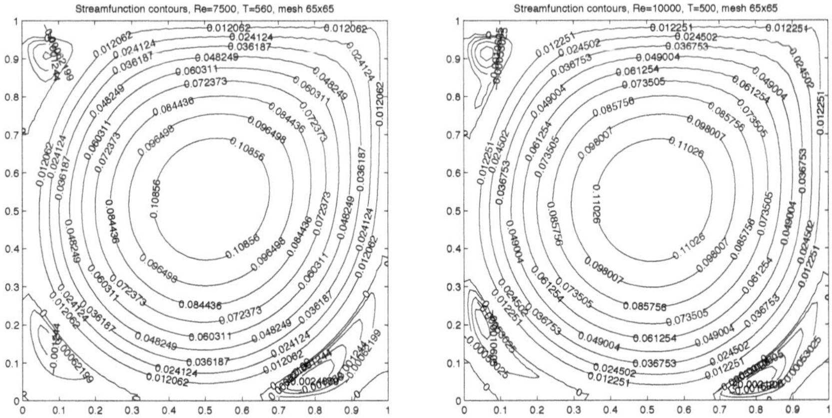

(a) Re = 7500 (b) Re = 10000

Fig. 14.16 Driven cavity for Re = 7500, 10000: calculated streamfunction contours using the fourth-order scheme.

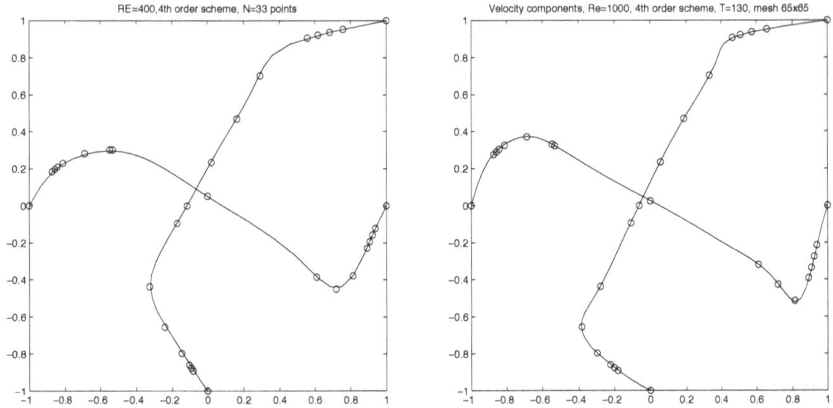

(a) Re = 400 (b) Re = 1000

Fig. 14.17 Driven cavity for Re = 400, 1000: velocity components using the fourth-order scheme. (a) Re = 400, 33×33 grid (solid curve), Ghia, Ghia and Shin with 129×129 ("o"). (b) Re = 1000, 65×65 grid (solid curve), Ghia, Ghia and Shin with 129×129 ("o").

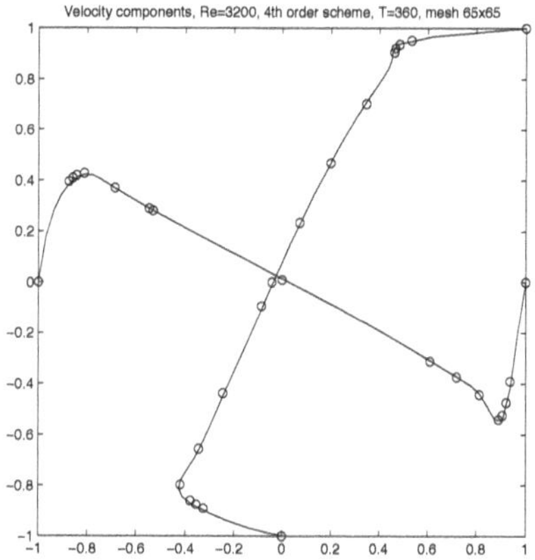

(a) Re = 3200

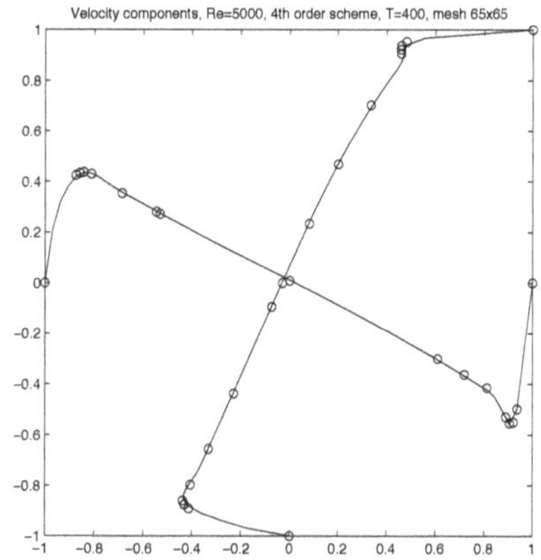

(b) Re = 5000

Fig. 14.18 Driven cavity for Re = 3200, 5000: calculated velocity components using the fourth-order scheme. (a): Re = 3200, 65×65 grid (solid curve), Ghia, Ghia and Shin with 129×129 ("o"). (b): Re = 5000, 65×65 grid (solid curve), Ghia, Ghia and Shin with 257×257 ("o").

14.2 Fourth-order scheme for the driven cavity problem

In the preceding section we discussed the driven cavity problem using a second-order scheme. In this section we display results obtained with the fourth-order scheme (13.17), where the convective term was approximated by (13.30). The discretization in time is carried out as in [14] (see also Notes for Chapter 13). Subsection 14.2.1 deals with Reynolds numbers up to Re = 10000. As has been noted, these are the cases where the time-dependent solution approaches a steady state. Thus we focus here on the improvement of the accuracy, relative to the second-order scheme. In Subsection 14.2.2 we take a closer look at the asymptotic behavior of the solution for Re = 10000.

14.2.1 *The convergence to steady state, from low to high Reynolds numbers*

In Table 14.12 we display flow quantities for Re = 1000. These results should be compared with the ones in Table 14.2 (obtained with the second-order scheme). Similarly, in Table 14.13 we display flow quantities for Re = 5000, which should be compared with the ones in Table 14.6. We see that for the same grid size, the current results are closer to the steady state in [76].

In Figure 14.14 we display streamfunction contours for Re $= 400, 1000$ using the 33×33 and 65×65 grids, respectively. These contours should be compared with the ones obtained with the second-order scheme using the 97×97 grid (see Figure 14.3). In Figure 14.15 streamfunction contours for Re $= 3200$ and Re $= 5000$ are shown. Similarly, Figure 14.16 displays the contours for Re $= 7500$ and Re $= 10000$. The fourth-order scheme clearly captures the levels of the streamfunction, even on very coarse grids.

We will now focus on the shape of the horizontal and vertical values of the velocity components $u(0.5, y)$ and $v(x, 0.5)$. Unlike the figures of the functions plotted in Section 14.1, the results are given in the sideways Tables 14.25 to 14.28. They refer to sufficiently large times where the solution can be considered to be numerically close to a steady state. Tables 14.25 and 14.26 (at the end of this chapter) compare the horizontal and vertical velocities at $x = 0.5$ and $y = 0.5$ obtained with the fourth-order scheme for Re $= 1000$ on 33×33, 65×65, and 129×129 grids. The results compare very favorably with the ones obtained with finer grids in [76] and [23]. In Tables 14.27 and 14.28 (at the end of this chapter), we display similar results for Re $= 5000$ on 65×65, 129×129, and 257×257 grids. The results are very good, even for the coarse grid 65×65.

In Figure 14.17 we display results for the velocity components $u(0.5, y)$ and $v(x, 0.5)$ with Re $= 400, 1000$ using the grids 33×33 and 65×65, respectively. They are compared with the values obtained in [76] using the grid 129×129. In Figure 14.18 we display results for $u(0.5, y)$ and $v(x, 0.5)$ with Re $= 3200, 5000$, using the grid 65×65. These are compared with [76] using the grids 129×129 and 257×257, respectively. In Figure 14.19, we display results for $u(0.5, y)$ and $v(x, 0.5)$ for Re $= 7500, 10000$, using the grid 65×65, compared with [76], which uses a 257×257 mesh.

14.2.2 *Behavior of the scheme for* Re $= 10000$

In this subsection, we consider the time evolution of the maximal value of the streamfunction ψ for Re $= 10000$. This evolution is recorded in Figure 14.11(b) where the second-order scheme was used, which we now compare with the result of the fourth-order scheme displayed in Figure 14.20(a). In the former case a 97×97 grid was used, while in the latter a coarser grid 65×65 is used. In both cases the time evolution is plotted over the interval $400 < t < 500$. In the present case (more accurate) we extended the calculation much further in time so as to be able to capture the asymptotic behavior, which is displayed in Figure 14.20(b). The time scale here is much finer, spanning the interval $2712 < t < 2722$. The emerging plot is quite remarkable, demonstrating clearly the time-periodic character of the solution for large time.

In [145] the bifurcation of driven cavity solutions was studied. It was observed that periodic solutions appear at Reynolds number Re $= 8500$. A numerical study of the first bifurcation in terms of the Reynolds number

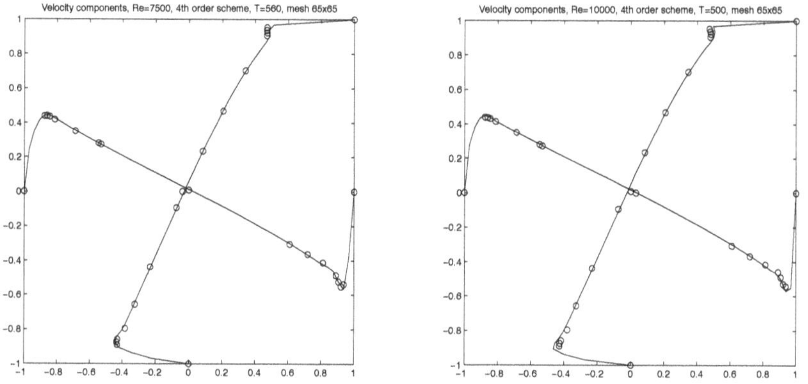

(a) Re $= 7500$ (b) Re $= 10000$

Fig. 14.19 Driven cavity for Re $= 7500, 10000$: calculated velocity components using the fourth-order scheme. (a) Re $= 7500$, 65×65 grid (solid curve), Ghia, Ghia and Shin with 257×257 ("o"). (b) Re $= 10000$, 65×65 grid (solid curve), Ghia, Ghia and Shin with 257×257 ("o").

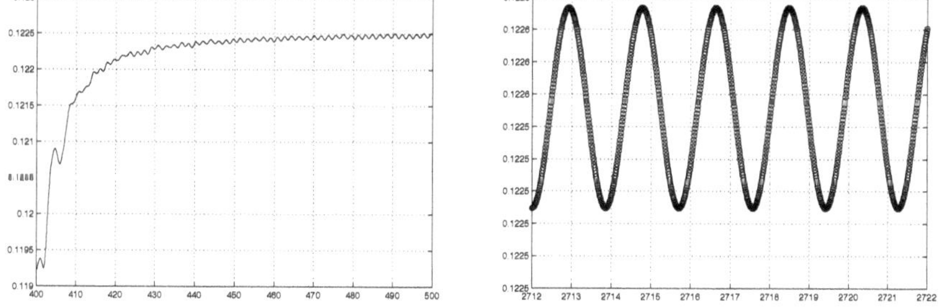

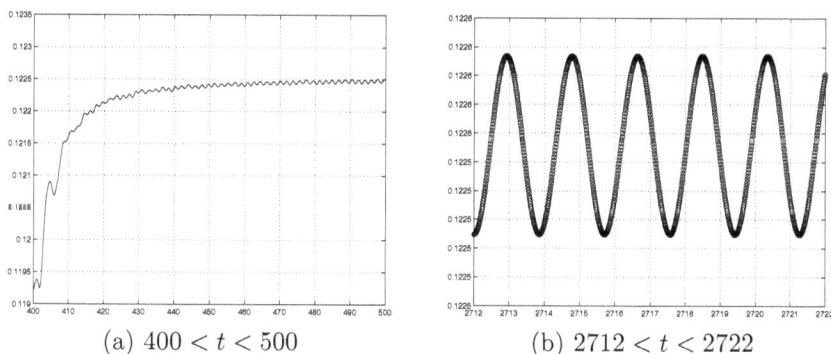

(a) $400 < t < 500$ (b) $2712 < t < 2722$

Fig. 14.20 Driven cavity for Re $= 10000$: maximal calculated value of the streamfunction with the fourth-order scheme, 65×65 grid, $\Delta t = 1/90$. (a) $400 < t < 500$, (b) $2712 < t < 2722$.

[8] indicates that it occurs at a Reynolds number which is close to 8018. This result sharpens the critical value Re $=$ 8000 [67]. The latter uses a finite element discretization of the Navier–Stokes system and an eigenvalue analysis of the linearized system. As in most of the calculations for the driven cavity problem, the primitive variable formulation is used.

14.3 Double-driven cavity problem

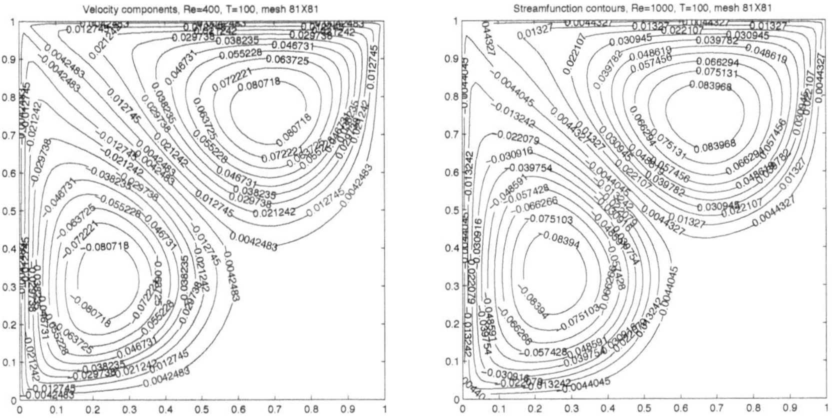

(a) Re $= 400$ (b) Re $= 1000$

Fig. 14.21 Double-driven cavity for Re $= 400, 1000$: calculated streamfunction contours.

A similar problem that we considered has the same geometry Ω, but the fluid is driven also in the negative y direction on the left side of Ω. Thus, $(u, v) = (1, 0)$ for $y = 1$ and $(u, v) = (0, -1)$ for $x = 0$. We set Re $= 400, 1000, 3200, 5000$ and we used the second-order scheme. Figures 14.21 and 14.22 show the calculated streamfunction contours at $t = 100$ for the various Reynolds numbers. Clearly, the exact solution is symmetric with respect to the diagonal $x + y = 1$. At Re $= 3200$ the computed solution is clearly seen to be asymmetric when using the 81×81 or 97×97 grids. Numerical experiments [145] corroborate this observation but place the critical Reynolds number (for symmetry breaking) somewhere between 4000 and 5000. In Figure 14.23 the maximum of the streamfunction from $t = 0$ to $t = 200$ is displayed for Re $= 3200$ and Re $= 5000$. The same quantity is displayed in Figure 14.24 spanning the time interval $200 < t < 400$ and using a finer scale. It can be seen that there is no steady state for either case.

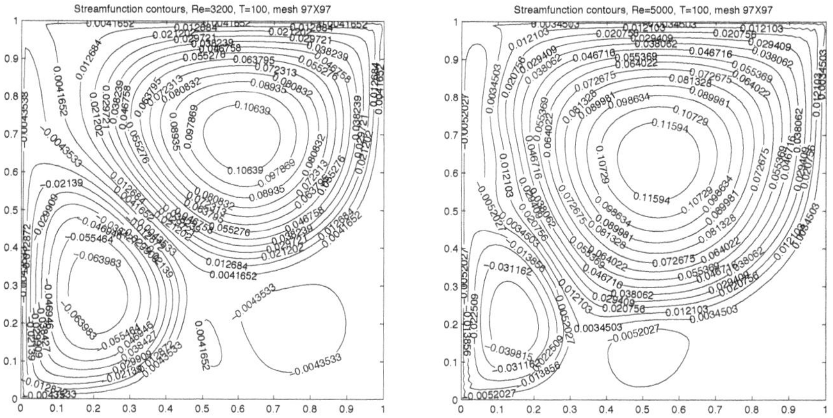

(a) Re = 3200 (b) Re = 5000

Fig. 14.22 Double-driven cavity for Re = 3200, 5000: calculated streamfunction contours.

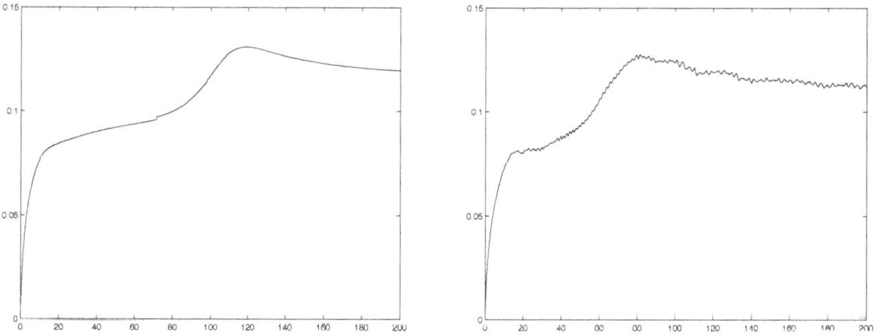

(a) Re = 3200 (b) Re = 5000

Fig. 14.23 Double-driven cavity for Re = 3200, 5000: maximal calculated value of the streamfunction, $0 < t < 200$. (a) Re = 3200, (b) Re = 5000.

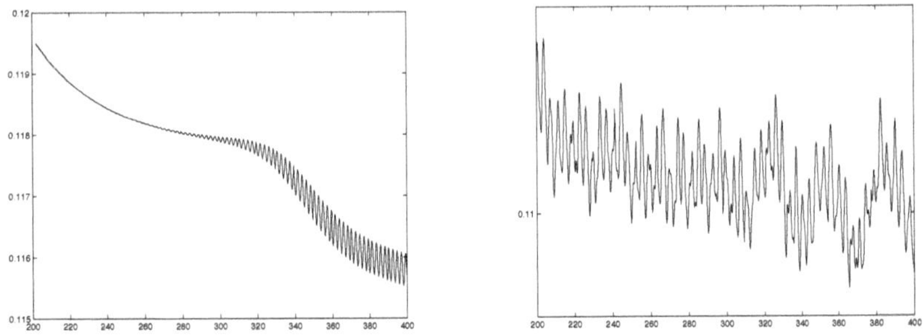

(a) Re = 3200 (b) Re = 5000

Fig. 14.24 Double-driven cavity for Re = 3200, 5000: maximal calculated value of the streamfunction, $200 < t < 400$. (a) Re = 3200, (b) Re = 5000.

14.4 Notes for Chapter 14

- The paper [76] has served as a common reference for steady-state solutions of the driven cavity problem in a square. Updated results have been reported [23] for Re = 1000 using a spectrally accurate method. These results were recently confirmed up to five digits [138] using a parallel splitting algorithm and a very fine grid 5000×5000. Observe that the results in Tables 14.25 and 14.26 using the 129×129 grid and our fourth-order scheme conform with [23] and [138] up to three digits.
- See early references on the problem [81, 161] and recent studies of driven cavity flows [22, 31, 158].
- Numerical studies of bifurcation of driven cavity solutions have been analyzed in [2, 67, 83, 92, 104, 148, 159, 166]. A numerical study of the first bifurcation in terms of the Reynolds number can be found in [8].
- Three-dimensional driven cavity results are given in [3, 24, 88, 114, 138].
- For further applications in fluid dynamics problems, including lid-driven cavity flows, using the pure streamfunction formulation of Chapter 13 (and modifications thereof), refer to [105, 146, 173, 174].

Table 14.25 Horizontal velocity through the vertical centerline of the cavity at Re = 1000. Fourth-order scheme using 33×33, 65×65, and 129×129 grids, compared to Ghia, Ghia, Shin [76] and Botella and Peyret [23] results.

y	u, 129×129, [76]	u, 160 modes [23]	$u = -\psi_y$, 33×33	$u = -\psi_y$, 65×65	$u = -\psi_y$, 129×129
0.0000	0.00000	0.00000	0.00000	0.00000	0.00000
0.0547	−0.18109	−0.18129	−0.16005	−0.17776	−0.18089
0.0625	−0.20196	−0.20233	−0.17951	−0.19848	−0.20190
0.0703	−0.22220	−0.22290	−0.19878	−0.21877	−0.22243
0.1015	−0.29730	−0.30046	−0.27305	−0.2954	−0.29978
0.1719	−0.38289	−0.38857	−0.37318	−0.38517	−0.38814
0.2813	−0.27805	−0.28037	−0.28104	−0.27960	−0.28028
0.4531	−0.10648	−0.10820	−0.10999	−0.10800	−0.10814
0.5000	−0.06080	−0.06206	−0.06473	−0.06213	−0.06203
0.6172	0.05702	0.05702	0.05238	0.05625	0.05695
0.7344	0.18719	0.18867	0.18215	0.18712	0.18849
0.8516	0.33304	0.33722	0.32804	0.33442	0.33679
0.9531	0.46604	0.47233	0.46464	0.46840	0.47171
0.9609	0.51117	0.51693	0.50744	0.51311	0.51662
0.9688	0.57492	0.58084	0.57000	0.57636	0.57972
0.9766	0.65928	0.66442	0.64024	0.66100	0.66323
1.0000	1.00000	1.00000	1.00000	1.00000	1.00000

Table 14.26 Vertical velocity for the horizontal centerline of the cavity at Re = 1000. Fourth-order scheme using 33 × 33, 65 × 65 and 129 × 129 grids, compared to Ghia, Ghia and Shin [76] and Botella and Peyret [23] results.

x	v, 129 × 129 [76]	v, 160 modes [23]	$v = \psi_x$, 33 × 33	$v = \psi_x$, 65 × 65	$v = \psi_x$, 129 × 129
0.00000	0.00000	0.00000	0.00000	0.00000	0.00000
0.06250	0.27485	0.28071	0.26716	0.27732	0.28022
0.07030	0.29012	0.29627	0.28231	0.29277	0.29577
0.07810	0.30353	0.30991	0.29548	0.30629	0.30940
0.09375	0.32627	0.33304	0.31742	0.32920	0.33246
0.15625	0.37095	0.37602	0.36268	0.37345	0.37646
0.22655	0.33075	0.33399	0.32620	0.33203	0.33379
0.23440	0.32235	0.32536	0.31809	0.32348	0.32512
0.50000	0.02526	0.02580	0.02530	0.02577	0.02582
0.80470	-0.31966	−0.32021	-0.31463	−0.31851	−0.31996
0.85940	-0.42665	−0.42645	-0.42347	−0.42507	−0.42622
0.90625	-0.51550	−0.52644	-0.50576	−0.52167	−0.52586
0.94530	-0.39188	−0.41038	-0.38042	−0.40387	−0.40968
0.95310	-0.33714	−0.35532	-0.32702	−0.34941	−0.35469
0.96095	-0.27669	−0.29369	-0.26808	−0.28816	−0.29274
0.96875	-0.21388	−0.22792	-0.20728	−0.22426	−0.22791
1.00000	0.00000	0.00000	0.00000	0.00000	0.00000

Table 14.27 Horizontal velocity for the vertical centerline of the cavity at Re = 5000. Fourth-order scheme using 65×65, 129×129 and 257×257 grids, compared to Ghia, Ghia and Shin [76].

y	u, 257×257 [76]	$u = -\psi_y$, 65×65	$u = -\psi_y$, 129×129	$u = -\psi_y$, 257×257
0.0000	0.00000	0.00000	0.00000	0.00000
0.0547	−0.41165	−0.39629	−0.41150	−0.41580
0.0625	−0.42901	−0.41873	−0.43186	−0.43584
0.0703	−0.43643	−0.43131	−0.44183	−0.44534
0.1015	−0.40435	−0.41148	−0.41454	−0.41664
0.1719	−0.33050	−0.33244	−0.33592	−0.33799
0.2813	−0.22855	−0.23263	−0.23456	−0.23608
0.4531	−0.07404	−0.07579	−0.07573	−0.07619
0.5000	−0.03039	−0.03256	−0.03199	−0.03212
0.6172	0.08183	0.07874	0.08051	0.08123
0.7344	0.20087	0.20008	0.20293	0.20451
0.8516	0.33556	0.34173	0.34559	0.34802
0.9531	0.46036	0.47171	0.47429	0.47772
0.9609	0.45992	0.46983	0.47368	0.47710
0.9688	0.46120	0.47049	0.47404	0.47741
0.9766	0.48223	0.48815	0.49245	0.49588
1.0000	1.00000	1.00000	1.00000	1.00000

Table 14.28 Vertical velocity through the horizontal centerline of the cavity at Re = 5000. Fourth-order scheme using 65×65, 129×129 and 257×257 grids, compared to Ghia, Ghia and Shin [76].

x	v, 257×257 [76]	$v = \psi_x$, 65×65	$v = \psi_x$, 129×129	$v = \psi_x$, 257×257
0.00000	0.00000	0.00000	0.00000	0.00000
0.06250	0.42447	0.42028	0.42983	0.43348
0.07030	0.43329	0.43072	0.43945	0.44294
0.07810	0.43648	0.43552	0.44334	0.44663
0.09375	0.42951	0.43144	0.43742	0.44026
0.15625	0.35368	0.35872	0.36276	0.36501
0.22655	0.28066	0.28417	0.28735	0.28921
0.23440	0.27280	0.27603	0.27910	0.28092
0.50000	0.00945	0.01188	0.01170	0.01174
0.80470	-0.30018	−0.30431	−0.30817	−0.31019
0.85940	-0.36214	−0.36948	−0.37379	−0.37614
0.90625	-0.41442	−0.42263	−0.42708	−0.42971
0.94530	-0.52876	−0.53590	−0.53809	−0.54040
0.95310	-0.55408	−0.55832	−0.56577	−0.56941
0.96095	-0.55069	−0.55229	−0.56500	−0.57029
0.96875	-0.49774	−0.49889	−0.51364	−0.51985
1.00000	0.00000	0.00000	0.00000	0.00000

Bibliography

[1] Abarbanel, S., Ditkowski, A. and Gustafsson, B. (2000). On error bound of finite difference approximations for partial differential equations, *J. Sci. Comput.* **15**, pp. 79–116.

[2] Abouhamza, A. and Pierre, R. (2003). A neutral stability curve for incompressible flows in rectangular driven cavity, *Math. Comput. Modelling* **38**, pp. 141–157.

[3] Albensoeder, S. and Kuhlman, H. (2005). Accurate three-dimensional lid-driven cavity flow, *J. Comput. Phys.* **206**, pp. 536–558.

[4] Altas, I., Dym, J., Gupta, M. M. and Manohar, R. P. (1998). Multigrid solution of automatically generated high-order discretizations for the biharmonic equation, *SIAM J. Sci. Comput.* **19**, pp. 1575–1585.

[5] Armbruster, D., Nikolaenko, B., Smaoui, N. and Chossat, P. (1996). Symmetries and dynamics for 2-D Navier–Stokes flow, *Physica D* **95**, pp. 81–93.

[6] Ascher, U., Ruuth, S. and Spiteri, R. (1997). Implicit-explicit Runge–Kutta methods for time-dependent partial differential equations, *Appl. Numer. Math.* **25**, pp. 151–167.

[7] Ascher, U., Ruuth, S. and Wetton, T. (1995). Implicit-explicit methods for time-dependent partial differential equations, *SIAM J. Numer. Anal.* **32**, pp. 797–823.

[8] Auteri, F., Parolini, N. and Quartapelle, L. (2005). Numerical investigation on the stability of the singular driven cavity flow, *J. Comput. Phys.* **202**, pp. 488–506.

[9] Ben-Artzi, M. (1994). Global solutions of two-dimensional Navier–Stokes and Euler equations, *Arch. Rat. Mech. Anal.* **128**, pp. 329–358.

[10] Ben-Artzi, M. (2003). Planar Navier–Stokes equations, vorticity approach, in S. J. Fridlander and D. Serre (eds.), *Handbook of Mathematical Fluid Dynamics*, Vol. II, chap. 5 (Elsevier), pp. 143–168.

[11] Ben-Artzi, M., Chorev, I., Croisille, J.-P. and Fishelov, D. (2009a). A compact difference scheme for the biharmonic equation in planar irregular domains, *SIAM J. Numer. Anal.* **47**, pp. 3087–3108.

[12] Ben-Artzi, M., Croisille, J.-P. and Fishelov, D. (2006). Convergence of a compact scheme for the pure streamfunction formulation of the unsteady Navier–Stokes system, *SIAM J. Numer. Anal.* **44**, pp. 1997–2024.

[13] Ben-Artzi, M., Croisille, J.-P. and Fishelov, D. (2008). A fast direct solver for the biharmonic problem in a rectangular grid, *SIAM J. Sci. Comput.* **31**, pp. 303–333.
[14] Ben-Artzi, M., Croisille, J.-P. and Fishelov, D. (2009b). A high order compact scheme for the pure-streamfunction formulation of the Navier–Stokes Equations, *J. Sci. Comput.* **42**, pp. 216–250.
[15] Ben-Artzi, M., Croisille, J.-P., Fishelov, D. and Trachtenberg, S. (2005). A pure-compact scheme for the streamfunction formulation of Navier–Stokes equations, *J. Comput. Phys.* **205**, pp. 640–664.
[16] Ben-Artzi, M., Fishelov, D. and Trachtenberg, S. (2001). Vorticity dynamics and numerical resolution of Navier–Stokes equations, *Math. Model. and Numer. Anal.* **35**, pp. 313–330.
[17] Ben-Artzi, M. and Koch, H. (1999). Decay of mass for a semilinear parabolic equation, *Comm. PDE* **24**, pp. 869–881.
[18] Ben-Artzi, M., Souplet, P. and Weissler, F. B. (2002). The local theory for viscous Hamilton–Jacobi equations in Lebesgue spaces, *J. Math. Pures Appl.* **81**, pp. 343–378.
[19] Biagioni, H. and Gramchev, T. (1996). Navier–Stokes equation with singular initial data and forcing term, *Matematica Contemporanea* **10**, pp. 1–20.
[20] Bialecki, B. (2003). A fast solver for the orthogonal spline collocation solution of the biharmonic Dirichlet problem on rectangles, *J. Comput. Phys.* **191**, pp. 601–621.
[21] Bjørstad, P. (1983). Fast numerical solution of the biharmonic Dirichlet problem on rectangles, *SIAM J. Numer. Anal.* **20**, pp. 59–71.
[22] Botella, O. (1997). On the solution of the Navier–Stokes equations using Chebyshev projection schemes with third-order accuracy in time, *Comput. Fluids* **26**, pp. 107–116.
[23] Botella, O. and Peyret, R. (1998). Benchmark spectral results on the lid-driven cavity flow, *Comput. Fluids* **27**, pp. 421–433.
[24] Bouffanais, R., Deville, M., Fischer, P., Leriche, E. and Weill, D. (2006). Large-eddy simulation of the lid-driven cubic cavity flow by the spectral element method, *J. Scient. Comput.* **27**, pp. 151–162.
[25] Boyce, W. E. and DiPrima, R. C. (2003). *Elementary Differential Equations and Boundary Value Problems*, 7th edn. (Wiley).
[26] Brezis, H. (1994). Remarks on the preceding paper by M. Ben-Artzi: "Global solutions of two-dimensional Navier–Stokes and Euler equations", *Arch. Rat. Mech. Anal.* **128**, pp. 359–360.
[27] Brezis, H. and Cazenave, T. (1996). A nonlinear heat equation with singular initial data, *J. Anal. Math* **68**, pp. 277–304.
[28] Brown, B., Davies, E., Jimack, P. K. and Mihajlovic, M. (2000). On the accurate finite element solution of a class of fourth order eigenvalue problems, *Proc. R. Soc. Lond. A* **456**, pp. 1505–1521.
[29] Brown, D., Cortez, R. and Minion, M. (2001). Accurate projection methods for the incompressible Navier–Stokes equations, *J. Comput. Phys.* **168**, pp. 464–499.

[30] Brüger, A., Gustafsson, B., Lötstedt, P. and Nilsson, J. (2005). High order accurate solution of the incompressible Navier–Stokes equations, *J. Comput. Phys.* **203**, pp. 49–71.

[31] Bruneau, C.-H. and Saad, M. (2006). The 2D lid-driven cavity revisited, *Computers and fluids* **35**, pp. 326–348.

[32] Bubnovitch, V., Rosas, C. and Moraga, N. (2002). A stream function implicit difference scheme for 2D incompressible flows of Newtonian fluids, *Int. J. Numer. Meth. Eng.* **53**, pp. 2163–2184.

[33] Canuto, C., Hussaini, M., Quarteroni, A. and Zang, T. (2007). *Spectral Methods Evolution to Complex Geometries and Applications to Fluid Dynamics, Series in Scientific Computation* (Springer).

[34] Carlen, E. and Loss, M. (1993). Sharp constant in Nash's inequality, *Duke Math. J.* **71**, pp. 213–215.

[35] Carlen, E. and Loss, M. (1995). Optimal smoothing and decay estimates for viscously damped conservation laws, with applications to the 2-D Navier–Stokes equation, *Duke Math. J.* **81**, pp. 135–157.

[36] Carpenter, M. H., Gottlieb, D. and Abarbanel, S. (1993). The stability of numerical boundary treatments for compact high-order finite difference schemes, *J. Comput. Phys.* **108**, pp. 272–295.

[37] Carpio, A. (1994). Asymptotic behavior for the vorticity equations in dimensions two and three, *Comm. PDE* **19**, pp. 827–872.

[38] Cayco, M. and Nicolaides, R. (1986). Finite element technique for optimal pressure recovery from stream function formulation of viscous flows, *Math. Comp.* **46**, pp. 371–377.

[39] Chorin, A. J. (1968). Numerical solution of the Navier–Stokes equations, *Math. Comp.* **22**, pp. 745–762.

[40] Chorin, A. J. (1973). Numerical study of slightly viscous flow, *J. Fluid Mech.* **57**, pp. 785–796.

[41] Chorin, A. J. (1978). Vortex sheet approximation of boundary layers, *J. Comput. Phys.* **27**, pp. 428–442.

[42] Chorin, A. J. (1980). Vortex models and boundary layer instability, *SIAM J. Sci. Stat. Comput.* **1**, pp. 1–21.

[43] Chorin, A. J. (1994). *Vorticity and Turbulence, Applied Mathematical Sciences*, Vol. 103 (Springer-Verlag).

[44] Chorin, A. J. and Marsden, J. E. (1990). *A Mathematical Introduction to Fluid Mechanics*, 2nd edn. (Springer-Verlag).

[45] Christara, C. and Ng, K. (2002). Fast Fourier transform solvers and preconditioners for quadratic spline collocation, *BIT Numer. Math* **42**, pp. 702–739.

[46] Collatz, L. (1960). *The Numerical Treatment of Differential Equations*, 3rd edn. (Springer-Verlag).

[47] Constantin, P. and Foias, C. (1988). *Navier–Stokes Equations* (The University of Chicago Press).

[48] Conte, S. D. and de Boor, C. (1980). *Elementary Numerical Analysis: An Algorithmic Approach*, 3rd edn. (McGraw-Hill).

[49] Cottet, G.-H. (1986). Equations de Navier–Stokes dans le plan avec tourbillon initial mesure, *Comptes Rendus Acad. Sci. Paris Sér. I Math.* **303**, pp. 105–108.

[50] Cottet, G.-H. and Koumoutsakos, P. (2000). *Vortex Methods. Theory and Practice* (Cambridge University Press).

[51] Dafermos, C. M. (2000). *Hyperbolic Conservation Laws in Continuum Physics* (Springer).

[52] Dean, E. J., Glowinski, R. and Pironneau, O. (1991). Iterative solution of the stream function-vorticity formulation of the Stokes problem, application to the numerical simulation of incompressible viscous flow, *Comput. Meth. Appl. Mech. Eng.* **87**, pp. 117–155.

[53] Deville, M., Fischer, P. and Mund, O. (2002). *High-Order Methods For Incompressible Flows* (Cambridge University Press).

[54] Doering, D. and Gibbon, J. (1995). *Applied Analysis of the Navier–Stokes Equations* (Cambridge University Press).

[55] E, W. and Liu, J.-G. (1996a). Essentially compact schemes for unsteady viscous incompressible flows, *J. Comput. Phys.* **126**, pp. 122–138.

[56] E, W. and Liu, J.-G. (1996b). Vorticity boundary condition and related issues for finite difference scheme, *J. Comput. Phys.* **124**, pp. 368–382.

[57] Erturk, E. (2009). Discussions on driven cavity flow, *Int. J. Numer. Meth. Fluids* **60**, pp. 275–294.

[58] Evans, L. C. (1998). *Partial Differential Equations* (AMS).

[59] Eymard, R., Gallouët, T., Herbin, R. and Linke, A. (2012). Finite volume schemes for the biharmonic problem on general meshes, *Math. Comp.* **81**, pp. 2019–2048.

[60] Eymard, R., Herbin, R. and Rhoudaf, M. (2011). Approximation of the biharmonic problem using P^1 finite elements, *J. of Num. Math.* **19**, pp. 1–26.

[61] Fabes, E. and Stroock, D. (1986). A new proof of Moser's parabolic Harnack inequality using the old ideas of Nash, *Arch. Rat. Mech. Anal.* **96**, pp. 327–338.

[62] Finden, W. F. (2008). An error term and uniqueness for Hermite–Birkhoff interpolation involving only function values and/or first derivative values, *J. Comput. and Appl. Math.* **212**, pp. 1–15.

[63] Fishelov, D., Ben-Artzi, M. and Croisille, J.-P. (2003). A compact scheme for the streamfunction formulation of Navier–Stokes equations, in V. Kumar, M. Gavrilova, C. Tan and P. L'Ecuyer (eds.), *Computational Science and its Applications, ICCSA 2003*, no. 2667 in Lecture Notes in Computer Science (Springer), pp. 809–817.

[64] Fishelov, D., Ben-Artzi, M. and Croisille, J.-P. (2010). Recent developments in the pure streamfunction formulation of the Navier–Stokes system, *J. Sci. Comput.* **45**, pp. 238–258.

[65] Fishelov, D., Ben-Artzi, M. and Croisille, J.-P. (2011). Highly accurate discretization of the Navier–Stokes equations in streamfunction formulation, in J. S. Hesthaven and E. M. Ronquist (eds.), *Spectral and High Order Methods for Partial Differential Equations*, 9th ICOSAHOM (Springer), pp. 189–197.

[66] Fishelov, D., Ben-Artzi, M. and Croisille, J.-P. (2012). Recent advances in the study of a fourth-order compact scheme for the one-dimensional biharmonic equation, *J. Sci. Comput.* **53**, 1, pp. 55–79.

[67] Fortin, A., Jardak, M., Gervais, J.-J. and Pierre, R. (1997). Localization of Hopf bifurcations in fluid flow problems, *Int. J. Numer. Meth. Fluids* **24**, pp. 1185–1210.

[68] Friedlander, S., Pavlović, N. and Shvydkoy, R. (2006). Nonlinear instability for the Navier–Stokes equations, *Comm. Math. Phys.* **264**, pp. 335–347.

[69] Friedman, A. (1964). *Partial Differential Equations of Parabolic Type* (Prentice-Hall).

[70] Gallagher, I. and Gallay, T. (2005). Uniqueness for the two-dimensional Navier–Stokes equation with a measure as initial vorticity, *Math. Ann.* **332**, pp. 287–327.

[71] Gallagher, I., Gallay, T. and Lions, P.-L. (2005). On the uniqueness of the solution of the two-dimensional Navier–Stokes equation with a Dirac mass as initial vorticity, *Math. Nachr.* **278**, pp. 1665–1672.

[72] Gallagher, I. and Planchon, F. (2002). On global infinite energy solutions to the Navier–Stokes equations in two dimensions, *Arch. Rat. Mech. Anal.* **161**, pp. 307–337.

[73] Gallay, T. and Wayne, C. E. (2002). Invariant manifolds and the long-time asymptotics of the Navier–Stokes and vorticity equations on $\mathbf{R}^2$, *Arch. Rat. Mech. Anal.* **163**, pp. 209–258.

[74] Gallay, T. and Wayne, C. E. (2005). Global stability of vortex solutions of the two-dimensional Navier–Stokes equation, *Comm. Math. Phys.* **255**, pp. 97–129.

[75] Germain, P. (2006). Equations de Navier–Stokes en deux dimensions: existence et comportement asymptotique de solutions d'énergie infinie, *Bull. Sci. Math.* **130**, pp. 123–151.

[76] Ghia, U., Ghia, K. N. and Shin, C. T. (1982). High-Re solutions for incompressible flow using the Navier–Stokes equations and a multigrid method, *J. Comput. Phys.* **48**, pp. 387–411.

[77] Giga, M.-H., Giga, Y. and Saal, J. (2010). *Nonlinear Partial Differential Equations: Asymptotic Behavior of Solutions and Self-Similar Solutions* (Birkhäuser Verlag).

[78] Giga, Y. and Kambe, T. (1988). Large time behavior of the vorticity of two-dimensional viscous flow and its application to vortex formation, *Comm. Math. Phys.* **117**, pp. 549–568.

[79] Giga, Y., Miyakawa, T. and Osada, H. (1988). Two-dimensional Navier–Stokes flow with measures as initial vorticity, *Arch. Rat. Mech. Anal.* **104**, pp. 223–250.

[80] Girault, V. and Raviart, P. (1986). *Finite Element methods for Navier–Stokes Equation: Theory and Algorithms*, no. 5 in *Springer Series in Computational Mathematics* (Springer).

[81] Goda, K. (1979). A multistep technique with implicit difference schemes for calculating two- or three-dimensional cavity flows, *J. Comput. Phys.* **30**, pp. 76–95.

[82] Golub, G. and Van Loan, C. (1996). *Matrix Computations*, 3rd edn. (John Hopkins University Press).
[83] Goodrich, J., Gustafson, K. and Halasi, K. (1990). Hopf bifurcation in the driven cavity, *J. Comput. Phys.* **90**, pp. 219–261.
[84] Goodrich, J. and Soh, W. Y. (1989). Time-dependent viscous incompressible Navier–Stokes equations: The finite difference Galerkin formulation and streamfunction algorithms, *J. Comput. Phys.* **84**, pp. 207–241.
[85] Gottlieb, D. and Orszag, S. (1977). *Numerical Analysis of Spectral Methods: Theory and Applications* (SIAM).
[86] Gresho, P. M. (1991). Incompressible Fluid Dynamics: Some Fundamental Formulation Issues, *Annu. Rev. Fluid Mech.* **23**, pp. 413–453.
[87] Gresho, P. M. and Sani, R. L. (2000). *Incompressible Flow and the Finite Element Method: Vol. 2, Isothermal Laminar Flow* (Wiley).
[88] Guermond, J.-L., Migeon, C., Pineau, G. and Quartapelle, L. (2002). Start-up flows in a three-dimensional rectangular driven cavity of aspect ratio 1:1:2 at Re=1000, *J. Fluid Mech.* **450**, pp. 169–199.
[89] Gunzburger, M. D. (1989). *Finite Element Methods for Viscous Incompressible Flows: A Guide to Theory, Practice and Algorithms* (Academic Press).
[90] Gupta, M. (1991). High accuracy solutions of incompressible Navier–Stokes equations, *J. Comput. Phys.* **93**, pp. 343–359.
[91] Gupta, M. M. and Kalita, J. C. (2005). A new paradigm for solving Navier–Stokes equations: streamfunction-velocity formulation, *J. Comput. Phys.* **207**, pp. 52–68.
[92] Gustafson, K. (1995). Theory and computation of periodic solutions of autonomous partial differential equation boundary value problems, with application to the driven cavity problem, *Math. Comput. Modelling* **22**, pp. 57–75.
[93] Gustafsson, B. (2008). *High Order Difference Methods for Time Dependent PDE* (Springer-Verlag).
[94] Hackbusch, W. (1992). *Elliptic Differential Equations: Theory and Numerical Treatment, Springer Series in Computational Mathematics*, Vol. 15 (Springer-Verlag).
[95] Henshaw, W., Kreiss, H. and Reyna, L. (1994). A fourth-order accurate difference approximation for the incompressible Navier–Stokes equations, *Computers and Fluids* **23**, pp. 575–593.
[96] Hesthaven, J., Gottlieb, S. and Gottlieb, D. (2007). *Spectral Methods for Time-Dependent Problems, Cambridge Monographs on Applied and Computational Mathematics*, Vol. 21 (Cambridge University Press).
[97] Hesthaven, J. S. and Waburton, T. (2008). *Nodal Discontinuous Galerkin Methods: Algorithms, Analysis and Applications*, no. 54 in Texts in Applied Mathematics (Springer-Verlag).
[98] Hörmander, L. (1997). *Lectures on Nonlinear Hyperbolic Differential Equations*, 4th edn. (Springer-Verlag).
[99] Hou, T. and Wetton, B. (2009). Stable fourth order stream-function methods for incompressible flows with boundaries, *J. Comput. Math.* **27**, pp. 441–458.

[100] Hundsdorfer, W. and Verwer, J. (2010). *Numerical Solution of Time-Dependent Advection-Diffusion-Reaction Equations, Springer Series in Computational Mathematics*, Vol. 33, 2nd edn. (Springer).

[101] Iserles, A. (1996). *A First Course in the Numerical Analysis of Differential Equations* (Cambridge University Press).

[102] John, F. (1982). *Partial Differential Equations, Applied Mathematical Sciences*, Vol. 1, 4th edn. (Springer).

[103] Johnston, H. and Krasny, R. (2002). Fourth-order finite difference simulation of a differentially heated cavity, *Int. J. Numer. Meth. Fluids* **40**, pp. 1031–1037.

[104] Kalita, J. and Gupta, M. (2010). A streamfunction-velocity approach for 2D transient incompressible viscous flows, *Int. J. Numer. Meth. Fluids* **62**, pp. 237–266.

[105] Kalita, J. and Sen, S. (2010). Biharmonic computation of the flow past an impulsively started circular cylinder at Re=200, in *Lecture Notes in Engineering and Computer Science*, 2185, (1) (Springer), pp. 1805–1810.

[106] Kampanis, N. A. and Ekaterinaris, J. (2006). A staggered grid, high-order accurate method for the incompressible Navier–Stokes equations, *J. Comput. Phys.* **215**, pp. 589–613.

[107] Kato, T. (1986). Remarks on the Euler and Navier–Stokes Equations in $\mathbf{R}^2$, in *Proc. Symp. Pure. Math.*, Vol. 45 (part 2) (AMS), pp. 1–7.

[108] Kato, T. (1994). The Navier–Stokes equation for an incompressible fluid in $\mathbf{R}^2$ with a measure as the initial vorticity, *Diff. and Integral Equations* **7**, pp. 949–966.

[109] Kato, T. and Fujita, H. (1962). On the nonstationary Navier–Stokes system, *Rend. Sem. Math. Univ. Padova* **32**, pp. 243–260.

[110] Kato, T. and Ponce, G. (1986). Well-posedness of the Euler and Navier–Stokes equations in the Lebesgue spaces $L_s^p(\mathbf{R}^2)$, *Revista Mat. Iberoamericana* **2**, pp. 73–88.

[111] Kato, T. and Ponce, G. (1987). On nonstationary flows of viscous and ideal fluids in $L_s^p(\mathbf{R}^2)$, *Duke Math. J.* **55**, pp. 487–499.

[112] Kato, T. and Ponce, G. (1988). Commutator estimates and the Euler and Navier–Stokes equations, *Comm. Pure Appl. Math.* **41**, pp. 891–907.

[113] Kennedy, C. A. and Carpenter, M. H. (2003). Additive Runge–Kutta schemes for convection-diffusion-reaction equations, *Appl. Numer. Math.* **44**, pp. 139–181.

[114] Kim, J. and Moin, P. (1985). Application of a fractional-step method to incompressible Navier–Stokes equations, *J. Comput. Phys.* **59**, pp. 308–323.

[115] Kobayashi, M. and Pereira, J. (2005). A computational streamfunction method for the two-dimensional incompressible flows, *Int. J. Numer. Meth. Eng.* **62**, pp. 1950–1981.

[116] Koch, H. and Tataru, D. (2001). Well-posedness for the Navier–Stokes equations, *Adv. in Math.* **157**, pp. 22–35.

[117] Kosma, Z. (2000). A computing laminar incompressible flows over a backward-facing step using Newton iterations, *Mech. Research Comm.* **27**, pp. 235–240.

[118] Kuntzmann, J. (1959). *Méthodes numériques. Interpolation - Dérivées* (Dunod).

[119] Kupferman, R. (2001). A central-difference scheme for a pure streamfunction formulation of incompressible viscous flow, *SIAM J. Sci. Comput.* **23**, pp. 1–18.

[120] Kwak, D. and Kiris, C. (2009). CFD for incompressible flows at NASA Ames, *Comput. Fluids* **38**, pp. 504–510.

[121] Ladyzhenskaya, O. A. (1963). *The Mathematical Theory of Viscous Incompressible Flow* (Gordon and Breach).

[122] Lai, M.-J. and Wenston, P. (2000). Bivariate spline method for numerical solution of steady-state Navier–Stokes equations over polygons in stream function formulation, *Numer. Meth. Part. Diff. Eqs* **16**, pp. 147–183.

[123] Lamb, H. (1993). *Hydrodynamics*, 6th edn. (Cambridge University Press).

[124] Landau, L. D. and Lifshitz, E. M. (1959). *Fluid Mechanics* (Pergamon Press).

[125] Leray, J. (1933). Etudes de diverses équations intégrales non linéaires et de quelques problèmes que pose l'hydrodynamique, *J. Math. Pures Appl.* **12**, pp. 1–82.

[126] Li, M. and Tang, T. (2001). A compact fourth-order finite difference scheme for unsteady viscous incompressible flows, *J. Sci. Comput.* **16**, pp. 29–45.

[127] Lieb, E. (1983). Sharp constants in the Hardy-Littlewood-Sobolev and related inequalities, *Annals Math.* **118**, pp. 349–374.

[128] Lions, J. L. and Prodi, G. (1959). Un théoréme d'existence et d'unicité dans les équations de Navier–Stokes en dimension 2, *Comptes Rendus Acad. Sci. Paris Sér. I Math.* **248**, pp. 3519–3521.

[129] Liu, J.-G. and Shu, C.-W. (2000). A high-order discontinuous Galerkin method for 2D incompressible flows, *J. Comput. Phys.* **160**, pp. 577–596.

[130] Liu, J.-G., Wang, C. and Johnston, H. (2003). A fourth-order scheme for incompressible Boussinesq equations, *J. Sci. Comput.* **18**, pp. 253–285.

[131] Lomtev, I. and Karniadakis, G. (1999). A discontinuous Galerkin method for the Navier–Stokes equations, *Int. J. Numer. Meth. Fluids* **29**, pp. 587–603.

[132] Maekawa, Y. (2005). On spatial decay estimates for derivatives of vorticities of the two-dimensional Navier–Stokes flow, *Proceedings Equadiff* **11**, pp. 223–228.

[133] Majda, A. J. and Bertozzi, A. L. (2002). *Vorticity and Incompressible Flow* (Cambridge University Press).

[134] Marchioro, C. and Pulvirenti, M. (1982). Hydrodynamics in two dimensions and vortex theory, *Comm. Math. Phys.* **84**, pp. 483–503.

[135] McGrath, F. (1968). Nonstationary plane flow of viscous and ideal fluids, *Arch. Rat. Mech. Anal.* **27**, pp. 328–348.

[136] McKee, S., Tome, M., Ferreira, V., Cuminato, J., Castelo, A., Sousa, F. and Mangiavacchi, N. (2008). The MAC method, *Computers and Fluids* **37**, pp. 907–930.

[137] Meleshko, V. (2003). Selected topics in the history of the two-dimensional biharmonic problem, *Appl. Mech. Review* **56**, pp. 33–85.

[138] Minev, P. and Guermond, J.-L. (2012). Start-up flow in a three-dimensional lid-driven cavity by means of a massively parallel direction splitting algorithm, *Int. J. Numer. Meth. Fluids* **68**, 7, pp. 856–871.
[139] Miyakawa, T. and Schonbek, M. (2001). On optimal decay rates for weak solutions to the Navier–Stokes equations in $\mathbf{R}^n$, *Math. Bohem.* **126**, pp. 443–455.
[140] Miyakawa, T. and Yamada, M. (1992). Planar Navier–Stokes flows in a bounded domain with measures as initial data, *Hiroshima Math. J.* **22**, pp. 401–420.
[141] Non, E., Pierre, R. and Gervais, J.-J. (2006). Linear stability of the three-dimensional lid-driven cavity, *Phys. Fluids* **18**, pp. 084103–1–084103–10.
[142] Orszag, S. A. and Israeli, M. (1974). Numerical simulation of viscous incompressible flows, *Ann. Rev. Fluid Mech.* **6**, pp. 281–318.
[143] Osada, H. (1987). Diffusion processes with generators of generalized divergence form, *J. Math. Kyoto Univ.* **27**, pp. 597–619.
[144] Owen, M. P. (1996). *Topics in the spectral theory of fourth order elliptic differential operators*, Ph.D. thesis, Kings College London.
[145] Pan, T. W. and Glowinski, R. (2000). A projection/wave-like equation method for the numerical simulation of incompressible viscous fluid flow modeled by the Navier–Stokes equations, *Comput. Fluid Dynamics* **9**, pp. 28–42.
[146] Pandit, S. K. (2008). On the use of compact streamfunction-velocity formulation of steady Navier–Stokes equations on geometries beyond rectangular, *J. Scient. Comput.* **36**, pp. 219–242.
[147] Patera, A. T. (1984). A spectral element method for fluid dynamics: Laminar flow in a channel expansion, *J. Comput. Phys.* **54**, pp. 468–488.
[148] Peng, Y.-F., Shiau, Y.-H. and Hwang, R. (2003). Transition in a 2-D–lid-driven cavity flow, *Computers and Fluids* **32**, pp. 337–352.
[149] Pérez-Guerrero, J. S. and Cotta, R. M. (1992). Integral transform solution for the lid-driven cavity flow problem in streamfunction-only formulation, *Int. J. Numer. Meth. Fluids* **15**, pp. 399–409.
[150] Pinchover, Y. and Rubinstein, J. (2005). *Introduction to Partial Differential Equations* (Cambridge University Press).
[151] Ponce, G. (1986). On two dimensional incompressible fluids, *Comm. PDE* **11**, pp. 483–511.
[152] Quartapelle, L. (1993). *Numerical Solution of the Incompressible Navier–Stokes Equations* (Birkhäuser Verlag).
[153] Quartapelle, L. and Valz-Gris, F. (1981). Projection conditions on the vorticity in viscous incompressible flows, *Int. J. Numer. Meth. Fluids* **1**, pp. 129–144.
[154] Ramanan, N. and Homsy, G. M. (1994). Linear stability of lid-driven cavity flow, *Phys. Fluids* **6**, pp. 2690–2701.
[155] Reed, M. and Simon, B. (1978). *Methods of Modern Mathematical Physics* (Academic Press).
[156] Richtmyer, R. and Morton, K. (1967). *Difference Methods for Initial-Value Problems*, 2nd edn. (John Wiley & Sons).
[157] Saffman, P. G. (1992). *Vortex Dynamics* (Cambridge University Press).

[158] Sahin, M. and Owens, R. G. (2003a). A novel fully implicit finite volume method applied to the lid-driven cavity problem - Part I: High Reynolds number flow calculations, *Int. J. Numer. Meth. Fluids* **42**, pp. 57–77.

[159] Sahin, M. and Owens, R. G. (2003b). A novel fully implicit finite volume method applied to the lid-driven cavity problem - Part II: Linear stability analysis, *Int. J. Numer. Meth. Fluids* **42**, pp. 79–88.

[160] Schonbek, M. E. (1985). L^2 decay for weak solutions of the Navier–Stokes equations, *Arch. Rat. Mech. Anal.* **88**, pp. 209–222.

[161] Schreiber, R. and Keller:, H. B. (1983). Driven cavity flows by efficient numerical techniques, *J. Comput. Phys.* **49**, pp. 310–333.

[162] Serre, D. (1996). Stabilité L^1 pour les lois de conservation visqueuses, *Comptes Rendus Acad. Sci. Paris Sér. I Math.* **323**, pp. 359–363.

[163] Serrin, J. (1962). On the interior regularity of weak solutions of the Navier–Stokes equations, *Arch. Rat. Mech. Anal.* **9**, pp. 187–195.

[164] Shahbazi, K., Fischer, P. F. and Ethier, C. (2007). A high-order discontinuous Galerkin method for the unsteady incompressible Navier–Stokes equations, *J. Comput. Phys.* **222**, pp. 391–407.

[165] Shankar, P. N. and Deshpande, M. D. (2000). Fluid mechanics in the driven cavity, *Annu. Rev. Fluid Mech.* **32**, pp. 93–136.

[166] Shen, J. (1991). Hopf bifurcation of the unsteady regularized driven cavity, *J. Comput. Phys.* **95**, pp. 228–245.

[167] Spalart, P., Moser, R. and Rogers, M. (1991). Spectral methods for the Navier–Stokes equations with one infinite and two periodic directions, *J. Comput. Phys.* **96**, pp. 297–324.

[168] Stephenson, J. W. (1984). Single cell discretizations of order two and four for biharmonic problems, *J. Comput. Phys.* **55**, pp. 65–80.

[169] Strikwerda, J. (2004). *Finite Difference Schemes and Partial Differential Equations*, 2nd edn. (SIAM).

[170] Temam, R. (1969). Sur l'approximation de la solution des équations de Navier–Stokes par la méthode des pas fractionnaires II, *Arch. Rat. Mech. Anal.* **33**, pp. 377–385.

[171] Temam, R. (2001). *Navier–Stokes Equations* (AMS Edition).

[172] Tezduyar, T. E., Liou, J., Ganjoo, D. K. and Behr, M. (1990). Solution techniques for the vorticity-streamfunction formulation of the two-dimensional unsteady incompressible flows, *Int. J. Numer. Meth. in Fluids* **11**, pp. 515–539.

[173] Tian, Z. F. and Yu, P. X. (2011). An efficient compact difference scheme for solving the streamfunction formulation of the incompressible Navier–Stokes equations, *J. Comp. Phys.* **230**, pp. 6404–6419.

[174] Tsai, C. H., Young, D. L. and Hsiang, C. C. (2011). The localized quadrature method for two-dimensional streamfunction formulation of Navier–Stokes equations, *Eng. Anal. with Bound. Elem.* **35**, pp. 1190–1203.

[175] Van Loan, C. (1992). *Computational Frameworks for the Fast Fourier Transform*, *Frontiers in Applied Mathematics*, Vol. 10 (SIAM).

[176] Vishik, M. and Friedlander, S. (2003). Nonlinear instability in two-dimensional ideal fluids: The case of a dominant eigenvalue, *Comm. Math. Phys.* **243**, pp. 261–273.

[177] Wayne, C. E. (2011). Vortices and two-dimensional fluid motion, *Notices of the AMS* **58**, 1, pp. 10–19.
[178] Weinberger, H. F. (1995). *A First Course on Partial Differential Equations with Complex Variables and Transform Methods* (Dover).
[179] Yudovich, V. I. (1989). *The Linearization Method in Hydrodynamical Stability Theory*, American Mathematical Society Translations (AMS).

Index

www.ingramcontent.com/pod-product-compliance
Lightning Source LLC
Chambersburg PA
CBHW050541190326
41458CB00007B/1871
9781848162754